Biomass-based Biocomposites

Editors:

Vijay Kumar Thakur

Amar Singh Singha

SMITHERS RAPRA

A Smithers Group Company

Shawbury, Shrewsbury, Shropshire, SY4 4NR, United Kingdom
Telephone: +44 (0)1939 250383 Fax: +44 (0)1939 251118
http://www.polymer-books.com

First Published in 2013 by

Smithers Rapra Technology Ltd

Shawbury, Shrewsbury, Shropshire, SY4 4NR, UK

A catalogue record for this book is available from the British Library.

ISBN: 978-1-84735-980-3 (hardback)
978-1-84735-981-0 (softback)
978-1-84735-982-7 (ebook)

Typeset by Argil Services

Preface

During the last few years, demand for green products obtained from renewable and sustainable biomass-based resources has increased unceasingly. Plants, including grasses and water plants e.g., arrowhead, cattail and duckweed, together with algae and plant by-products such as straw and agricultural wastes, are some of the most promising and most abundant biobased sources of biopolymers on earth and are potentially suitable for use as components in biocomposites. Biomass-based materials, such as wood and cotton, have been used for thousands of years. However, in recent years, particular attention has been paid to new waste sources such as lignocellulosic waste by-products, which are available in abundant quantities throughout the world. The supply of wood is diminishing and these lignocellulosic wastes can be used to replace it in many applications. Indeed, the utilisation of biomass resources is one of the main concerns in every country. Biomass-based resources which have the potential to be renewable resources will not compete directly with food crops for either livestock or man because they include, for example, wood, sawdust, wood waste, herbaceous plants, agricultural waste, waste paper and sludge from pulp and paper mills. Besides, biomass which is abundantly available and renewable, is suitable to be used as a source of low-cost sustainable and ethical renewable raw materials. The important feature of biomass-based materials is that these can be designed and tailored to meet different requirements, depending upon the applications. Biomass can be used as a renewable replacement for fuels and petrochemicals. Renewability, low cost, eco-friendliness, eas of processing, non-abrasiveness and appropriate mechanical and physicochemical properties are among the most important advantages of using biomass-based materials for the development of green biocomposites. The primary aim of the present book is to give a wide-ranging overview of the basic understanding and potential applications of biomass-based biocomposites in different fields including the biomedical one.

This book is the culmination of the untiring efforts of researchers from different parts of the world with extensive experience in the field of biomass-based materials across different disciplines. The book gives an overview of different kinds of biomass-based biocomposites for different applications. This book is unique in the sense that it deals exclusively with biomass-based biocomposites which are procured from the biopolymers found in nature. In addition, it covers novel topics related to the synthesis, properties, characterisation and applications of different biomass-based biocomposites.

As the editors of this book on biomass-based biocomposites, we have really enjoyed working with the individual authors and appreciate their diligence and patience. We would also like to thank all the publishers who generously gave their permission to reprint material in the present book. We hope that this book will contribute significantly to the basic knowledge of students and researchers all around the globe working in the field of biomass based biocomposites.

We would like to thank Helene Chavaroche (Commissioning Editor) along with the publisher, Smithers Rapra, for invaluable help in the organisation of the editing process.

Dr. Vijay Kumar Thakur, Ph.D., P.D.F., M.R.S.C

Washington State University, USA

Dr. Amar Singh Singha, Ph.D.

National Institute of Technology, India

2013

Contributors

Nasim Amiralian

Australian Institute for Bioengineering and Nanotechnology, University of Queensland, Old Cooper Road, Brisbane 4072, QLD, Australia

Pratheep Kumar Annamalai

Adolphe Merkle Institute and Fribourg Center for Nanomaterials, University of Fribourg, Rte de l'Ancienne Papeterie, Marly CH-1723, Switzerland

and

Australian Institute for Bioengineering and Nanotechnology, University of Queensland, Old Cooper Road, Brisbane 4072, QLD, Australia

Pramendra Kumar Bajpai

Department of Mechanical and Industrial Engineering, Indian Institute of Technology Roorkee, 247667, India

Sri Bandyopadhyay

School of Materials Science and Engineering University of New South Wales, Sydney, 2052, Australia

Ana Baptista

CENIMAT/I3N and Departamento de Ciência dos Materiais, Faculdade de Ciências e Tecnologia da Universidade Nova de Lisboa (FCT/UNL), Portugal

Camila C. Barros

Volta Redonda Center University - UniFOA, Volta Redonda/RJ, Brazil

Mahesh V. Biyani

Adolphe Merkle Institute and Fribourg Center for Nanomaterials, University of Fribourg, Rte de l'Ancienne Papeterie, Marly CH-1723, Switzerland

Johann Bödecker

Miniwiz Sustainable Energy Development Ltd., Taipei, Taiwan 106, Republic of China

Ruxanda Bodîrlău

Institute of Macromolecular Chemistry 'Petru Poni', Aleea Grigore Ghica Voda, nr. 41A 700487 Iasi, Romania

João Paulo Borges

CENIMAT/I3N and Departamento de Ciência dos Materiais, Faculdade de Ciências e Tecnologia da Universidade Nova de Lisboa (FCT/UNL), Portugal

Raluca Nicoleta Darie

Institute of Macromolecular Chemistry 'Petru Poni', Aleea Grigore Ghica Voda, nr. 41A 700487 Iasi, Romania

Kishore Debnath

Department of Mechanical and Industrial Engineering, Indian Institute of Technology Roorkee, 247 667, India

Vikas Dhawan

Department of Mechanical and Industrial Engineering, Indian Institute of Technology Roorkee, 247667, India

Alain Dufresne

The International School of Paper, Print Media and Biomaterials (Pagora), Grenoble Institute of Technology, Saint Martin d'Hères CS 10065 - 38402, France

Akshay Dvivedi

Department of Mechanical and Industrial Engineering, Indian Institute of Technology Roorkee, 247 667, India

Isabel Ferreira

CENIMAT/I3N and Departamento de Ciência dos Materiais, Faculdade de Ciências e Tecnologia da Universidade Nova de Lisboa (FCT/UNL), Portugal

Svetlana Genieva

Assen Zlatarov University, 8010 Burgas, Bulgaria

Mehdi Jorfi

Adolphe Merkle Institute and Fribourg Center for Nanomaterials, University of Fribourg, Rte de l'Ancienne Papeterie, Marly CH-1723, Switzerland

Daniella R. Mulinari

Technology College - UERJ, Resende- RJ, Brazil and Volta Redonda Center University - UniFOA, Volta Redonda/RJ, Brazil

Elias Nassiopoulos

Centre for Automotive Technology, Cranfield University, Bedfordshire, MK43 0AL, UK

James Njuguna

Institute for Innovation, Design & Sustainability, School of Engineering, Robert Gordon University, Aberdeen, AB10 7GJ, UK

Mariana Pereda

The International School of Paper, Print Media and Biomaterials (Pagora), Grenoble Institute of Technology, Saint Martin d'Hères CS 10065 - 38402, France

Fei-ling Pua

Department of Mechanical and Manufacturing Engineering, Faculty of Engineering, Universiti Putra Malaysia, 43400 UPM Serdang, Selangor, Malaysia

Arup Kumar Rana

School of Materials Science and Engineering University of New South Wales, Sydney, 2052, Australia

Ana Maria Resmeriţă

Institute of Macromolecular Chemistry 'Petru Poni', Aleea Grigore Ghica Voda, nr. 41A 700487 Iasi, Romania

George J.M. Rocha

Brazilian Laboratory for Science and Technology in Bioethanol - CTBE, Campinas/SP, Brazil

Salit Mohd Sapuan

Department of Mechanical and Manufacturing Engineering, Faculty of Engineering, Universiti Putra Malaysia, 43400 UPM Serdang, Selangor, Malaysia

Amar Singh Singha

Department of Chemistry, National Institute of Technology Hamirpur, 177005 HP, India

Inderdeep Singh

Department of Mechanical and Industrial Engineering, Indian Institute of Technology Roorkee, 247667, India

Iuliana Spiridon

Institute of Macromolecular Chemistry 'Petru Poni', Aleea Grigore Ghica Voda, nr. 41A 700487 Iasi, Romania

Carnen-Alice Teacă

Institute of Macromolecular Chemistry 'Petru Poni', Aleea Grigore Ghica Voda, nr. 41A 700487 Iasi, Romania

Manju Kumari Thakur

Division of Chemistry, Govt. Degree College Sarkaghat, Himachal Pradesh University Shimla, 171005, India

Vijay Kumar Thakur

School of Mechanical and Materials Engineering, PO BOX 642920, Washington State University, Pullman, WA 99164-2920, USA

Sevdalina Turmanova

Assen Zlatarov University, 8010 Burgas, Bulgaria

Lyubomir Vlaev

Assen Zlatarov University, 8010 Burgas, Bulgaria

Shu-Kai Yeh

Department of Chemical Engineering and Biotechnology, National Taipei University of Technology, Taipei, Taiwan 106, Republic of China

Christopher C. Yen

Miniwiz Sustainable Energy Development Ltd., Taipei, Taiwan 106, Republic of China

Contents

1 Biomass-based Biocomposites: A Perspective on the Future 1
1.1 Introduction 1
1.2 Fibre Reinforcements 2
1.2.1 Lignocellulosic Materials 2
1.3 Applications of Biomass-based Composites 4
1.4 Summary 5

2 Development and Applications of Biocomposites from Renewable Resources 11
2.1 International Year of Natural Fibres 11
2.2 Types of Natural Fibre 11
2.3 Jute Fibre 13
2.3.1 Background 13
2.3.2 Nature of the Fibre 13
2.3.3 Advantages of Jute and Other Lignocellulosic Fibres 14
2.3.4 Disadvantages of Jute and Other Lignocellulosic Fibres ... 14
2.3.5 Morphology and Structure of Jute Fibre 15
2.3.6 Composition of Jute Fibre 15
2.4 Glass Fibre 16
2.4.1 Background 16
2.4.2 Comparison with Jute Fibre 17
2.5 Chemical Modification of Lignocellulosic Fibres 17
2.5.1 Acetals 18
2.5.2 Esters 18

2.5.3 Chemical Modification to Improve Dimensional Stability 19

2.5.4 Modification of Fibre-matrix Interaction 19

2.6 Natural Fibre Composites 20

2.6.1 Background 20

2.6.2 Thermoplastic *versus* Thermoset Polymer Matrix............ 21

2.6.3 Natural Fibre Composites as Wood Substitutes................ 22

2.6.4 Variables Influencing the Properties of Polymer Composites 23

2.6.4.1 Aspect Ratio of the Fibre or Filler 23

2.6.4.2 Polymer-fibre Interface and Interphase 24

2.6.4.3 Alignment or Distribution of the Reinforcement.. 24

2.6.4.4 Loading and Processing Techniques 24

2.6.4.5 Wetting, Adhesion and Dispersion 25

2.7 Short Natural Fibre-reinforced Thermoplastic Composites 25

2.7.1 Background 25

2.7.2 Preparation of Fibres 26

2.7.3 Compounding of Fibres 27

2.7.4 Moulding.. 28

2.7.5 Properties of Jute Fibre-reinforced Polypropylene Composites 28

2.7.5.1 Fibres and Other Reagents Used 28

2.7.5.2 Specific Gravity 29

2.7.5.3 Water Absorption... 29

2.7.5.4 Tensile Behaviour 29

2.7.5.5 Flexural Behaviour.. 30

2.7.5.6 Impact Behaviour... 31

2.7.5.7 Tensile Energy Absorption 31

2.7.5.8 Dynamic Mechanical Study............................ 32

2.7.5.9 Correlation of Mechanical Properties with other Factors 34

2.7.5.10 Conclusions 35
2.8 Long Natural Fibre-reinforced Thermosetting Composite Boards and Moulded Items 36
2.8.1 Choice of Thermosetting Resin 36
2.8.2 Compounding of the Resin Solution 36
2.8.3 Impregnation of Fibre 37
2.8.4 Moulding 38
2.8.5 Characteristic Properties 38

3 Cellulose-based Composite Systems for Biomedical Applications 47
3.1 Cellulose-based Composites 47
3.1.1 Biocomposites 47
3.1.2 Cellulose 49
3.1.2.1 Sources of Cellulose as a Raw Material 49
3.1.2.2 Structure 50
3.1.2.3 Cellulose Derivatives 51
3.2 Applications of Cellulose-based Composites 51
3.2.1 Medical Applications 51
3.3 Conclusions 57

4 Primary and Secondary Processing of Biocomposites 61
4.1 Introduction 61
4.2 Primary Processing of Biocomposites: Challenges 63
4.3 Processing Techniques for Biocomposites 64
4.3.1 Hand Lay-up Method 64
4.3.2 Compression Moulding 65
4.3.3 Extrusion and Injection Moulding Processes 68
4.3.4 Resin Transfer Moulding 69
4.3.5 Prepregging of Biocomposites 71
4.3.6 Pultrusion 72
4.3.7 Sheet Moulding Compounds 73

4.4 Applications of Biocomposites 73
4.5 Secondary Processing of Biocomposites 75
4.5.1 The Need for Secondary Processing 75
4.5.2 Drilling of Biocomposites 76
4.5.2.1 Input parameters 77
4.5.2.1.1 Cutting Speed and Feed Rate 77
4.5.2.1.2 Selection of Drill Bit 77
4.5.2.2 Output Parameters 78
4.5.2.2.1 Drilling Forces 78
4.5.2.2.2 Drilling-induced Damage 79
4.6 Concluding Remarks 80

5 Composites of Thermosetting Polymers Reinforced with Natural Fibres .. 85
5.1 Introduction 85
5.2 Natural Fibres 85
5.3 Composites Containing Short Natural Fibres 87
5.4 Thermosetting Polymers Reinforced with Natural Fibres 88
5.4.1 Coconut Fibre-reinforced Polyester Resin 88
5.4.2 Sugarcane Bagasse Fibre-reinforced Polyester Resin 94
5.4.3 Banana Pseudostem Fibre-reinforced Polyester Resin 97

6 The Potential Use of Biomass as Reinforcement in Biocomposites 105
6.1 Introduction 105
6.2 Experimental 106
6.2.1 Materials and Methods 106
6.2.2 Fabrication of Biomass-based Biocomposites 106
6.2.3 Mechanical Characterisation of Biomass-based Biocomposites 107
6.2.4 Morphological and Thermal Behaviour of Biomass-based Biocomposites 107
6.3 Results and Discucssion 107

6.3.1 Mechanical Properties of *Saccaharum Cilliare* Fibre-reinforced Bicomposites 108

6.3.2 Morphological and Thermal Characterisation of the Biocomposites 110

6.4 Conclusions 111

7 Recent Progress in Polymer Natural Fibre Composites Made from Recycled Plastics 115

7.1 Introduction 115

7.2 Background on Natural Fibres 116

7.3 Fibre Markets 119

7.3.1 Classification by Origin 119

7.3.2 Classification by Biological Features 119

7.4 Introduction to Recycled Plastics and Waste Flow Analysis 120

7.5 Natural Fibre Composites made with Recycled Polystyrene and Acrylonitrile Butadiene Styrene 125

7.6 Natural Fibre Composites made with Recycled Nylon and Polyethylene Terephthalate 135

7.7 Case Studies 135

7.7.1 Case Study 1: Mobile Phone Protective Cases 136

7.7.2 Case Study 2: Wine Carrying Case 137

7.7.3 Case Study 3: Sunglasses 137

7.8 Summary 138

8 Joining of Natural Fibre-reinforced Thermoplastic Composites 145

8.1 Introduction 145

8.2 Purpose of Joining 146

8.3 Adhesive Bonding 147

8.3.1 Types of Adhesively Bonded Joints 147

8.3.2 Types of Adhesives 148

8.3.3 Selection Criteria and Guidelines for Adhesives 149

8.3.4 Advantages of Adhesive Bonding 149

8.3.5 Disadvantages of Adhesive Bonding 150
8.3.6 Guidelines for Surface Preparation 150
8.3.7 Design Guidelines for Adhesive Bonding 151
8.3.8 Basic Theory of Adhesive Bonding 152
8.3.8.1 Mechanical Theory 152
8.3.8.2 Adsorption Theory 153
8.3.8.3 Electrostatic Theory 153
8.3.8.4 Diffusion Theory 153
8.3.8.5 Chemical Bonding Theory 153
8.3.9 Failure Mechanisms in Adhesive Bonding 154
8.4 Mechanical Joining 154
8.4.1 Types of Mechanical Joining 155
8.4.1.1 Riveted Joint 156
8.4.1.2 Bolted Joint 157
8.4.1.3 Screw Joint 157
8.4.2 Advantages of Mechanical Joining 158
8.4.3 Disadvantages of Mechanical Joining 159
8.4.4 Design Guidelines for Mechanical Joining 160
8.5 Microwave Joining 160
8.5.1 Microwave Joining of Natural Fibre-reinforced Thermoplastic Composites 161
8.6 Comparisons of Joint Strength under Tensile Loading 162
8.6.1 Joint Strength of Microwave Bonded Joint 162
8.6.2 Joint Strength of Adhesively Bonded Joint 163
8.7 Concluding Remarks 164

9 Natural Plant Fibre Biocomposites for Structural Vehicle Components .. 167
9.1 Introduction 167
9.1.1 Structure of Plant Fibres 168
9.1.2 Chemical Composition 170

9.1.3 Mechanical Properties ... 172
9.1.4 Fibre Modifications ... 175
9.1.5 Selected Natural Fibres ... 175
9.1.5.1 Flax ... 175
9.1.5.2 Hemp ... 177
9.2 Resins ... 179
9.2.1 Selected Natural Resins ... 179
9.2.1.1 Bio-epoxy ... 179
9.2.1.2 Polylactides ... 180
9.3 Mechanical Performance of Natural Fibre Composites ... 182
9.3.1 Stiffness and Strength ... 182
9.3.2 Impact Performance ... 183
9.3.3 Fatigue Behaviour ... 184
9.4 Current Applications of Natural Fibre-reinforced Composites: Overview ... 185
9.5 Conclusions ... 187

10 Biocomposites based on Cellulose Material (Poplar Seed Floss) and High Density Polyethylene: Accelerated Weathering Behaviour ... 193
10.1 Introduction ... 193
10.2 Experimental ... 196
10.2.1 Materials ... 196
10.2.2 Compounding and Processing of the High Density Polyethylene/Poplar Seed Floss Composites ... 197
10.2.3 Accelerated Weathering Tests ... 197
10.2.4 Methods for Characterisation of the Composites ... 197
10.2.4.1 Mechanical Testing ... 198
10.2.4.2 Fourier-Transform Infrared Spectroscopy ... 198
10.2.4.3 Contact Angle Measurements ... 199
10.2.4.4 Water Uptake Measurements ... 199
10.2.4.5 Scanning Electron Microscopy ... 199

10.2.4.6 Differential Scanning Calorimetry Analysis ... 200

10.2.4.7 Thermogravimetric Analysis ... 200

10.2.4.8 Dynamic Rheological Measurements ... 200

10.3 Investigation of the Properties of High Density Polyethylene/Poplar Seed Floss Composites ... 200

10.3.1 Fourier-Transform Infrared Spectroscopy ... 200

10.3.2 Investigation of Surface Properties ... 205

10.3.2.1 Scanning Electron Microscopy Analysis ... 205

10.3.2.2 Contact Angle Measurements ... 207

10.3.2.3 Water Uptake Measurements ... 208

10.3.3 Investigation of Thermal Properties ... 210

10.3.3.1 Thermogravimetric Analysis ... 210

10.3.3.2 Differential Scanning Calorimetry Analysis ... 212

10.3.4 Investigation of Mechanical Properties ... 214

10.3.5 Investigation of Melt Rheology ... 216

10.4 Conclusions ... 223

11 Effect of Physicochemical Conditions on Biomass-based Biocomposites. 229

11.1 Introduction ... 229

11.2 Preparation of *Hibiscus sabdariffa*-based Biocomposites for Physicochemical Studies ... 230

11.3 Physicochemical Studies ... 231

11.4 Results and Discussion ... 232

11.5 Conclusions ... 236

12 Obtaining and Utilisation of Rice Husk Ash as a Filler of Polymers or Adsorbent for Oil Spill Clean-up ... 239

12.1 Introduction ... 239

12.2 Chemical Composition, Constituent, Structure and Properties of Raw Rice Husk ... 240

12.3 Technologies Available for the Thermal Degradation of Rice Husk ... 240

12.4 Physicochemical Characteristics of the Products of Thermal Degradation of Rice Husk 242

12.5 Utilisation of Rice Husk and the Products of its Thermal Degradation as Fillers for Polymers 244

12.6 Polypropylene Composites Filled with Rice Husk Ash 245

12.7 Utilisation of the Products of Thermal Degradation of Rice Husk as an Adsorbent of Crude Oil or Diesel Fuel 248

12.8 Conclusions 251

13 The Potential of Lignin in Biocomposites 259

13.1 Introduction 259

13.1.1 Biocomposites 261

13.1.2 Types of Lignin 263

13.1.2.1 Kraft Lignin 264

13.1.2.2 Soda Lignin 265

13.1.2.3 Other Lignins 265

13.2 Applications of Lignin in Polymer Composites 266

13.2.1 Lignin/Epoxy 266

13.2.2 Lignin/Thermoplastic Composites 268

13.2.3 Lignin-based Biodegradable Composites 269

13.3 Conclusions 271

14 Biopolymer Nanocomposites Reinforced with Nanocrystalline Cellulose 277

14.1 Introduction 277

14.2 Biopolymers 278

14.3 Cellulose Nanocrystals 279

14.4 Nanocomposites of Polymers Directly Derived from Biomass 282

14.4.1 Cellulose Nanocomposites 282

14.4.1.1 All-cellulose Nanocomposites 283

14.4.2 Starch Nanocomposites 285

14.4.3 Chitosan Nanocomposites 286

14.4.4 Other Polysaccharide Nanocomposites 287

14.4.5 Protein Nanocomposites 287

14.5 Nanocomposites of Polymers Synthesised using Biobased Monomers 288

14.5.1 Polylactic Acid Nanocomposites 288

14.5.2 Naturally Occurring Oil-based Polymer Nanocomposites 289

14.6 Nanocomposites of Microbial/Bioengineered Polymers 291

14.6.1 Polyhydroxyalkanoate Nanocomposites 291

14.6.2 Bacterial Cellulose Nanocomposites 292

14.6.2.1 Applications of Bacterial Cellulose based Nanocomposites 293

14.7 Nanocomposites of Non-degradable Biobased Polymers 295

14.8 Concluding Remarks and Future Perspectives 296

15 Cellulose Nanocrystals and Related Polymer Nanocomposites 305

15.1 Introduction 305

15.2 Hierarchical Structure of Lignocellulosic Materials 306

15.3 Crystalline Structure of Cellulose: Young's Modulus 309

15.4 Cellulose Nanocrystals 312

15.4.1 Extraction Process: Acid Hydrolysis 312

15.4.2 Effect of Hydrolysis Conditions on the Properties of Cellulose Nanocrystals 314

15.4.2.1 Hydrolysis Time 315

15.4.2.2 Acid and Cellulose Concentration 315

15.4.2.3 Acid Type 315

15.5 Sources of Cellulose Nanocrystals: Morphological Characteristics 316

15.6 Nanocrystals from Polysaccharides other than Cellulose 318

15.7 Cellulose Nanocomposite Materials 320

15.7.1 Effect of Polymeric Matrix on Cellulose Nanocrystal Dispersion and Processing 320

15.7.1.1 Water-soluble Polymers 320

15.7.1.2 Non-polar Polymers: Alternative Dispersion Methods 321

15.7.1.2.1 Dispersion of Nanoparticles in an Organic Medium 321

15.7.1.2.2 Coating of Nanoparticle Surface321

15.7.1.2.3 Chemical Surface Modification . 322

15.7.2 Processing 323

15.7.2.1 Melt Extrusion and Impregnation Methods .. 323

15.7.2.2 Electrospinning 325

15.7.2.3 Multilayer Films 326

15.7.3 Nanocomposite Properties 326

15.7.3.1 Mechanical and Thermomechanical Properties 326

15.7.3.1.1 Cellulose Nanocrystals 326

15.7.3.1.2 Cellulose Nanocomposites 328

15.7.3.1.3 Morphology and Dimensions of the Nanoparticles 328

15.7.3.1.4 Effect of Processing Method 329

15.7.3.1.5 Effect of Matrix Microstructure and Matrix-filler Interactions ... 329

15.7.3.2 Thermal Properties 330

15.7.3.2.1 Glass Transition Temperature ... 331

15.7.3.2.2 Melting Temperature 331

15.7.3.2.3 Degree of Crystallinity 332

15.7.3.2.4 Thermal Stability 333

15.7.3.3 Swelling Properties 333

15.7.3.4 Barrier Properties 334

15.8 Conclusions and Future Trends 335

Abbreviations 349

Index 355

1 Biomass-based Biocomposites: A Perspective on the Future

Vijay Kumar Thakur, Manju Kumari Thakur and Amar Singh Singha

1.1 Introduction

During the last few years, biomass-based materials derived from different kinds of natural resources have become a topic of interest intensive because of their easy availability and renewable nature [1-3]. Traditionally-used synthetic polymers such as polyethylene and polypropylene are derived from fossil fuels, a non-renewable resource that produces a heavy carbon footprint. These polymers are non biodegradable and persist in the environment for many years after their disposal [4-7]. Furthermore, their environmental impact has been exacerbated by the extensive use of these synthetic polymers in everyday products such as packaging materials, industrial products, medical devices, coatings and hygiene products [8, 9].

The primary concern which arises from using these materials is what to do with these materials or articles made from the synthetic polymers once they are no longer useful. Research is being focused on finding the solutions for the disposal of these synthetic polymers [10-12]. Of the various solutions presently being suggested and practised, incineration represents one of the easiest. However, this process releases toxic substances into the environment which affect human beings to a considerable extent. Another solution for disposing of the synthetic polymers, especially thermoplastic polymers, is recycling. However, this technique is also not very successful because the properties of the recycled products, especially their mechanical properties, are not the same as those of the parent materials. Furthermore a number of materials which are currently available on the market are completely non–recyclable, such as flower pots, hunting bullets, golf discs and biomedical devices including stents and artificial arteries and these are currently either incinerated or disposed of in landfill sites.

However, the economy of diminishing petroleum resources, rising fuel prices, pressures for reducing our dependence on oil, the dwindling availability of landfill sites and the demands for high performance eco-friendly materials, coupled with a growing environmental awareness, has led to interest in the use of renewable resources for making new materials as substitutes for traditional synthetic polymers [13-17]. The interest in using renewable materials is not only attributed to their renewability and environmental friendliness, but also to their easy processing and the tailorable properties. Polymers from natural biomass by-products are especially attractive as replacements for traditional synthetic polymers as the most important of these natural polymers are carbohydrates, such as starch and cellulose, which are readily available at much lower cost than their synthetic counterparts and are also renewable [18-21].

The renewability of different biomass-based materials depends on the availability and life cycle of the raw materials [25-27]. Furthermore, biomass-based materials can be easily degraded through the enzymatic action of bacteria, fungi, algae and other living organisms, into carbon dioxide, methane, water, and/or various inorganic compounds, and therefore do not persist in the environment [28, 29]. By exploiting the synergistic effect of different properties from biobased polymers, the properties of the resulting biocomposite materials can be enormously improved [30, 31]. These types of biopolymers can be classified into three categories: (a) those extracted directly from biomass; (b) those produced through chemical synthesis from biomass monomers; and (c) those, such as microbial polymers, produced directly by natural or genetically modified organisms [33-35].

1.2 Fibre Reinforcements

Biomass-based lignocellulosic by-products have a huge potential to be used as reinforcements for composites due to their biodegradability, renewability, abundance, cost effectiveness and improved chemical functionalities [22-24] compared to synthetic reinforcements. **Figure 1.1** shows the different types of reinforcement which are commonly used in polymer matrices.

1.2.1 Lignocellulosic Materials

Among various biomass-based materials, lignocellulosic biomass has been suggested as the most promising alternative to traditional polymers as one of the essential components of composites to make these materials 'greener' [36, 37]. Lignocellulosic biomass has the best life cycle assessment, as regards its abundance, low cost and high polysaccharide (cellulose and hemicellulose) content. Intensive worldwide research on lignocellulosic biomass-based materials during the last few years will most likely make them an important component of the 'green' materials of the future [38-41].

Lignocellulosic materials are renewable resources which have the potential to be directly or indirectly used for a number of applications such as in green composites and the production of different kinds of biomolecules and commodity chemicals [42-44]. Lignocellulosic biomass-based materials consist mainly of three major components, i.e., cellulose, hemicelluloses, and lignin [45-47]. Each of these plays a significant role in determining the overall properties of the materials. The cell wall polysaccharides of lignocellulosic biomass have been found to be composed of crystalline cellulose fibrils. These fibrils are surrounded by a matrix of non-crystalline hemicelluloses that bind with pectin to cellulose. Cellulose is the prime constituent of lignocellulosic biomass-based materials. It is a linear polymer chain which is formed by joining anhydroglucose units into glucan chains and the anhydroglucose units are linked together by β-(1,4)-glycosidic bonds [48-51]. As a result of this linkage, cellobiose is established as the repeat unit for cellulose chains.

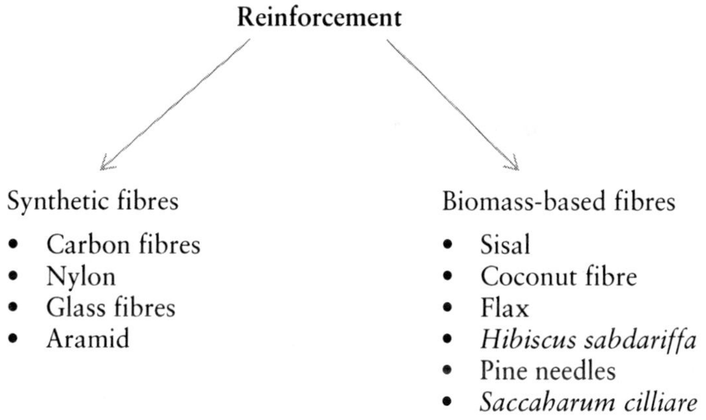

Figure 1.1 Types of reinforcement. Adapted from A. Gandini, *Macromolecules*, 2008, **41**, 24, 9491 [3]; V.K. Thakur, A.S. Singha and M.K. Thakur, *International Journal of Polymer Analysis and Characterization*, 2012, **17**, 1, 48 [15]; P. Tingaut, T. Zimmermann and F. Lopez-Suevos, *Biomacromolecules*, 2010, **11**, 2, 454 [19]; and K. Oksman, A.P. Mathew and M. Sain, *Plastics Rubber and Composites*, 2009, **38**, 9-10, 396 [20]

In contrast to the structure of cellulose, hemicelluloses consist of different monosaccharide units. The polymer chains of hemicelluloses have been found to be short-branched and amorphous. As a result of the amorphous morphology, hemicelluloses are partially soluble in water.

Lignin, the third most important component of the biomass-based materials, is a complex hydrophobic network of phenylpropanoid units. These units are thought to be the result of the oxidative polymerisation of one or more of three types of hydroxycinnamyl alcohol precursors. Lignin plays a significant role in providing the mechanical strength or rigidity to a plant as well as in preventing invasion by pathogens and pests.

Along with these three major components, lignocellulosic biomass also contains extractives, which are organic substances which have low molecular weight and are soluble in neutral solvents. Resins, fats, waxes, salts, minerals, phenolics, fatty acids and alcohols, phytosterols, and other compounds are categorised as extractives.

Lignocellulosic biomass is currently one of the indispensable components of polymer composites. It is frequently used as a reinforcement in polymer matrices (both from renewable and non-renewable resources). At least by replacing portions of petroleum-based synthetic materials by using reinforcement obtained from biomass in the making of composite materials, society's dependence on fossil fuels may be eased [52, 53].

1.3 Applications of Biomass-based Composites

A composite may be defined as the combination of two or more individual materials and it shows selected properties of each constituent. A suitable combination of different constituents results in a superior substance or material from practical point of view [54-57]. Biomass-based composites are used in many different areas such as furniture for the garden and house, parts for cars and technical devices [58-60]. Along with these applications, biomass-based materials are also being used in the preparation of products for medicine and nanotechnology. **Figure 1.2** shows the different potential applications of biomass-based materials.

Depending on the type of application, the biomass-based composite needs to be developed and tested with the focus on the requirements for specific uses. Biomass-based composites prepared using both synthetic and natural polymer matrices receive reinforcement in both strength and stiffness from fillers originating from different renewable resources whilst maintaining the formability and stability of the polymer matrix [61-63]. The main biomass-based natural fibres which are used in the automotive sector are wood, cellulose, flax, jute, sisal, hemp, and kenaf fibres. Depending on the application, the fibre content varies between 20 and 90 wt% [64, 65]. Attempts to develop totally biobased polymer composites to replace traditional synthetic polymers such as thermoplastics have gained some success, but these have not lived up to expectations due to the inability of these biobased polymers to compete economically and to their inferior mechanical properties [66, 67]. On the other hand, biomass-based materials can consistently provide sources of renewable biobased filler that can readily be incorporated into existing polymer systems [68-70]. A number of everyday household consumer products, such as furniture, automotive parts and decking materials, are today being fabricated from biomass-based composites using different processing techniques such as compression moulding, injection moulding and extrusion [71, 72]. In order to improve the fuel consumption and efficiency of vehicles, there has been a continuous effort in the automotive industry to seek high strength raw biomass-based materials that are more eco-friendly. The easy availability, non-abrasiveness, non-toxic nature, light weight and low specific gravity of biomass-based materials as reinforcement makes possible the development of new eco-friendly materials. All these properties of the biomass-based materials thus give the resulting composite materials an added advantage over composites which contain inorganic fillers.

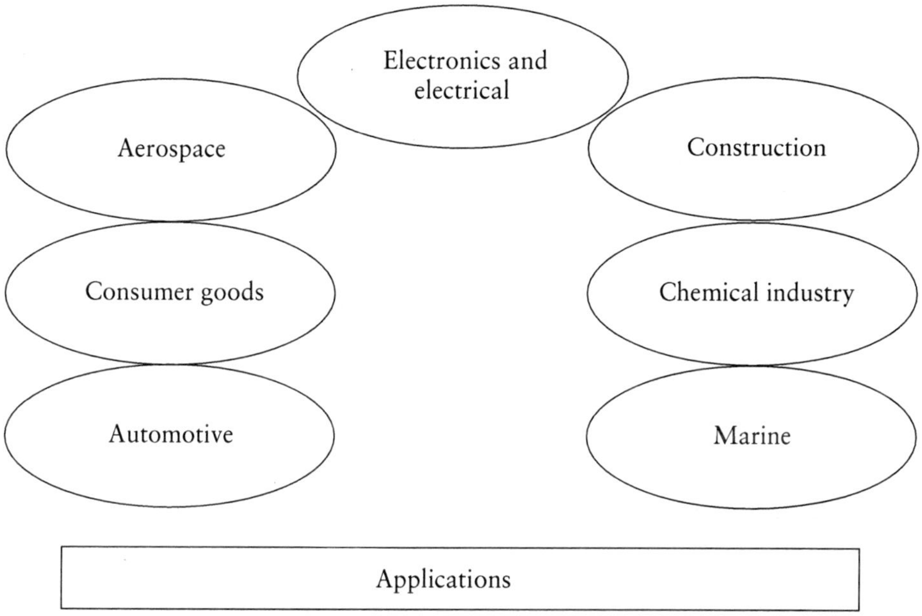

Figure 1.2 Potential applications of biomass-based materials. Adapted from A. Gandini, *Macromolecules*, 2008, **41**, 24, 9491 [3]; V.K. Thakur, A.S. Singha and M.K. Thakur, *International Journal of Polymer Analysis and Characterization*, 2012, **17**, 1, 48 [15]; P. Tingaut, T. Zimmermann and F. Lopez-Suevos, *Biomacromolecules*, 2010, **11**, 2, 454 [19]; and K. Oksman, A.P. Mathew and M. Sain, *Plastics Rubber and Composites*, 2009, **38**, 9-10, 396 [20]

1.4 Summary

To conclude, during the last few years, increasing environmental awareness has prompted the development and use of green materials procured from renewable biomass in a number of consumer products, as part of the effort to promote renewability of resources and to lower the emission of harmful greenhouse gases. Although the research on green materials has become widespread, studies pertinent to materials based upon biomass by-products such as natural cellulosic fibres are quite limited. There is great potential for the composites produced with biomass-based materials to replace existing commercial materials that are less environmentally friendly, and to create new markets in novel applications where traditional synthetic polymer composites failed to penetrate. Research into, and understanding of, the components from which the composites are made, is essential for developing the appropriate composite a specialised application. The recent research on developing new materials from renewable resources, especially from biomass, is being driven by rising environmental concern and demand for alternative resources to petroleum. This has also initiated the use of renewable materials in a number of fields, especially in the automotive, biomedical, materials science and technology fields. However, in order to make the transition from laboratory to industrial production, appropriate new technologies are still required for the effective use of natural biomass. Therefore, this book is primarily focused on understanding the fundamental concepts involved in the fabrication and preparation of composites from different kinds of cellulosic materials such as natural fibres; lignin; rice husk and so on. The book explores the potential of different types of cellulosic fibres and their by-products to prepare green composites and nanocomposites with different characteristic properties that make these biomass-based materials suitable for various technical applications, where biodegradability is important.

Acknowledgement

The authors gratefully acknowledge their respective institutes for the necessary facilities and encouragement for the accomplishment of this project.

References

1. V.K. Thakur, A.S. Singha and M.K. Thakur, *Advances in Polymer Technology*, 2012, DOI: 10.1002/adv.21317.

2. V.K. Thakur, A.S. Singha and M.K. Thakur, *Journal of Polymers and the Environment*, 2012, **20**, 1,164.

3. A. Gandini, *Macromolecules*, 2008, **41**, 24, 9491.

4. P.K. Annamalai and R.P. Singh in *Cellulose Fibers: Bio- and Nano-Polymer Composites*, Eds., K. Susheel, B. S. Kaith and I. Kaur, Springer-Verlag, Berlin and Heidelberg, Germany, 2011, p. 519.

5. C.K. Williams and M.A. Hillmyer, *Polymer Reviews*, 2008, **48**, 1, 1.

6. A. Gandini, *Green Chemistry*, 2011, **13**, 5,1061.

7. V.K. Thakur, A.S. Singha and M.K. Thakur, *Journal of Polymers and the Environment*, 2012, **20**, 2, 412.

8. V.K. Thakur, A.S. Singha and M.K. Thakur, *International Journal of Polymer Analysis and Characterization*, 2012, **17**, 7, 547.

9. L. Yu, K. Dean and L. Li, *Progress in Polymer Science*, 2006, **31**, 6, 576.

10. J.K. Pandey, A.P. Kumar, M. Misra, A.K. Mohanty, L.T. Drzal and R.P. Singh, *Journal of Nanoscience and Nanotechnology*, 2005, **5**, 4, 497.

11. M.A.S.A. Samir, F. Alloin and A. Dufresne, *Biomacromolecules*, 2005, **6**, 2, 612.

12. V.K. Thakur, A.S. Singha and M.K. Thakur, *International Journal of Polymer Analysis and Characterization*, 2012, **17**, 6, 401.

13. V.K. Thakur, A.S. Singha and M.K. Thakur, *International Journal of Polymer Analysis and Characterization*, 2012, **17**, 2, 133.

14. S. Kamel, *Express Polymer Letters*, 2007, **1**, 9, 546.

15. V.K. Thakur, A.S. Singha and M.K. Thakur, *International Journal of Polymer Analysis and Characterization*, 2012, **17**, 1, 48.

16. V.K. Thakur, A.S. Singha and B.N. Mishra, *Journal of Applied Polymer Science*, 2011, **122**, 1, 532.

17. E.S. Takacs and J. Vlachopoulos, *Plastics Engineering*, 2008, **64**, 9, 28.

18. M. Sanchez-Garcia and J. Lagaron, *Cellulose*, 2010, **17**, 5, 987.

19. P. Tingaut, T. Zimmermann and F. Lopez-Suevos, *Biomacromolecules*, 2010, **11**, 2, 454.

20. K. Oksman, A.P. Mathew and M. Sain, *Plastics Rubber and Composites*, 2009, **38**, 9-10, 396.

21. R.A. Holser, *Industrial Crops and Products*, 2008, **27**, 1, 130.

22. A. Tullo, *Chemical and Engineering News*, 2007, **85**, 19, 14.

23. Y.X. Wang, H.F. Tian and L.N. Zhang, *Carbohydrate Polymers*, 2010, **80**, 3, 665.

24. H.D.X. Cao and Ch.M. Li., *Biomacromolecules*, 2007, **8**, 899.

25. V.K. Thakur and A.S. Singha, *International Journal of Polymer Analysis and Characterization*, 2011, **16**, 6, 390.

26. V.K. Thakur and A.S. Singha, *Bulletin of Material Science*, 2010, **33**, 3, 257.

27. V.K. Thakur and A.S. Singha, *Polymer-Plastics Technology and Engineering*, 2010, **49**, 11, 1101.

28. V.K. Thakur and A.S. Singha, *Polymer-Plastics Technology and Engineering*, 2010, **49**, 7, 694

29. S.S. Ray and M. Bousmina, *Progress in Materials Science*, 2005, **50**, 8, 962.

30. O.A. Battista, *Industrial and Engineering Chemistry*, 1950, **42**, 3, 502.

31. H. Ono, Y. Shimaya, K. Sato and T. Hongo, *Polymer Journal*, 2004, **36**, 9, 684.

32. N. Lavoine, I. Desloges, A. Dufresne and J. Bras, *Carbohydrate Polymers*, 2012, **90**, 2, 735.

33. V.K. Thakur and A.S. Singha, *Iranian Polymer Journal*, 2010, **19**, 1, 3.

34. A.S. Singha and V.K. Thakur, *Polymer-Plastics Technology and Engineering*, *2009*, **48**, 7, 736.

35. I. Siro and D. Plackett, *Cellulose*, 2010, **17**, 3, 459.

36. A.S. Singha and V.K. Thakur, *International Journal of Polymer Analysis and Characterization*, 2009, **14**, 4, 301.

37. A. L. Goffin, J. M. Raquez, E. Duquesne, G. Siqueira, Y. Habibi, A. Dufresne and P. Dubois, *Biomacromolecules*, 2011, **12**, 7, 2456.

38. K.J. Edgar, C.M. Buchanan, J.S. Debenham, P.A. Rundquist, B.D. Seiler, M.C. Shelton and D. Tindall, *Progress in Polymer Science*, 2001, **26**, 9, 1605.

39. J. Lu and L.T. Drzal, *Journal of Polymer Science, Part B:Polymer Physics*, 2010, **48**, 2, 153.

40. Y.X. Wang and L.Y. Chen, *Carbohydrate Polymers*, 2011, **83**, 4, 1937.

41. M.N. Angles and A. Dufresne, *Macromolecules*, 2001, **34**, 9, 2921.

42. A.P. Mathew and A. Dufresne, *Biomacromolecules*, 2002, **3**, 3, 609.

43. A.P. Mathew, W. Thielemans and A. Dufresne, *Journal of Applied Polymer Science*, 2008, **109**, 6, 4065.

44. A.S. Singha and V.K. Thakur, *Polymer-Plastics Technology and Engineering*, 2009, **48**, 2, 201.

45. A.S. Singha and V.K. Thakur, *Iranian Polymer Journal*, **17**, 7, 541.

46. A.S. Singha and V.K. Thakur, *International Journal of Polymeric Materials*, 2008, **57**,12, 1059.

47. J.W. Rhim and P.K.W. Ng, *Critical Reviews in Food Science and Nutrition*, 2007, **47**, 4, 411.

48. M.J. John and S. Thomas, *Carbohydrate Polymers*, 2008, **71**, 3, 343.

49. A.S. Singha and V.K. Thakur, *Bulletin of Material Science*, **32**, 1, 49.

50. J. Simon, H.P. Muller, R. Koch and V. Muller, *Polymer Degradation and Stability*, 1998, **59**, 1-3, 107.

51. O. van den Berg, J.R. Capadona and C. Weder, *Biomacromolecules*, 2007, **8**, 4, 1353.

52. A.S. Singha and V.K. Thakur, *Journal of Polymer Materials*, 2009, **26**, 1, 81.

53. A.S. Singha and V.K. Thakur, *Bulletin of Material Science*, 2008, **31**, 5, 991.

54. V. Favier, H. Chanzy and J.Y. Cavaille, *Macromolecules*, 1995, **28**, 18, 6365.

55. V. Favier, G.R. Canova, J.Y. Cavaille, H. Chanzy, A. Dufresne and C. Gauthier, *Polymers for Advanced Technologies*, 1995, **6**, 5, 351.

56. A.L. Goffin, J.M. Raquez, E. Duquesne, G. Siqueira, Y. Habibi, A. Dufresne and P. Dubois, *Polymer*, 2011, **52**, 7, 1532.

57. V.K. Thakur and A.S. Singha in *Nanotechnology in Polymers*, Eds., V.K. Thakur and A.S. Singha, Studium Press, Houston, TX, USA, 2012, p. 519.

58. A.J. de Menezes, G. Siqueira, A.A.S. Curvelo and A. Dufresne, *Polymer*, 2009, **50**, 19, 4552.

59. A.P. Kumar and R.P. Singh, *Bioresource Technology*, 2008, **99**, 18, 8803.

60. A.S. Singha and V.K. Thakur, *E-Journal of Chemistry*, 2009, **6**, 1, 34.

61. A.S. Singha and V.K. Thakur, *E-Journal of Chemistry*, 2009, **6**, 1, 71.

62. Y.S. Lu, L.H. Weng and X.D. Cao, *Carbohydrate Polymers*, 2006, **63**, 2, 198.

63. A. Alemdar and M. Sain, *Composites Science and Technology*, 2008, **68**, 2, 557.

64. H.M.C. Azeredo, K.W.E. Miranda, M.F. Rosa, D.M. Nascimento and M.R. de Moura, *LWT-Food Science and Technology*, 2012, **46**, 1, 294.

65. A.S. Singha and V.K. Thakur, *Polymer and Polymer Composites*, 2009, **17**, 3, 127.

66. A.S. Singha and V.K. Thakur, *Polymer Composites*, 2010, **31**, 3, 459.

67. D.G. Liu, T.H. Zhong, P.R. Chang, K.F. Li and Q.L. Wu, *Bioresource Technology*, 2010, **101**, 7, 2529.

68. A.S. Singha and V.K. Thakur, *International Journal of Polymer Analysis and Characterization*, 2010, **15**, 3, 127.

69. A.S. Singha and V.K. Thakur, *Iranian Polymer Journal*, 2009, **19**, 1, 3.

70. Y.S. Lu, L.H. Weng and X.D. Cao, *Macromolecular Bioscience*, 2005, **5**, 11, 1101.

71. A.S. Singha and V.K. Thakur in *Green Polymer Materials*, Studium Press, Houston, TX, USA, 2011, p.11.

72. D.P.E.M. Teixeira, A.A.S. Curvelo, E. Corradini, M.N. Belgacem and A. Dufresne, *Carbohydrate Polymers*, 2009, **78**, 422.

73. A.S. Singha and V.K. Thakur, *BioResources*, 2009, **4**, 1, 292.

2 Development and Applications of Biocomposites from Renewable Resources

Arup Kumar Rana and Sri Bandyopadhyay

2.1 International Year of Natural Fibres

The International Year for Natural Fibres was celebrated by the United Nations in 2009. The main reason for this was to highlight the beneficial features of natural fibres. Moreover, the livelihood of several millions of people directly or indirectly depend on these fibres. A large section of people, mainly in developing countries, are involved in the cultivation of natural fibres, and in the processing of them either in factories or at home to make industrial or hand crafted products [1]. Since farmers are their own decision makers, they are gradually switching over to more remunerative crops. Therefore, this is the right time to research, develop and implement novel alternative uses of natural fibres beside their traditional uses and at the same time to capitalise on their environmental benefits.

2.2 Types of Natural Fibre

In a broad classification, from the point of view of origin, there are three types of natural fibre. These are vegetable fibres, animal fibres and mineral fibres. Some examples are as follows:

- Vegetable fibres: cotton, jute, flax, hemp, sisal, kenaf, coir, ramie, bagasse, wood fibre.
- Animal fibres: wool, hair (such as cashmere and angora), silk.
- Mineral fibres: wollastonite, asbestos.

Natural fibres of mineral origin are mainly used for industrial insulation purposes, whereas animal fibres are used for domestic insulation, clothing and carpets. The most beneficial fibres from the environmental point of view are those of vegetable origin. Among vegetable fibres, pure cotton is made up of α-cellulose and is very costly. Lignocellulosic fibres are found to be cheap and strong. Lignocellulosics are composed of lignin and cellulose. Lignin is a thermoplastic natural resin and cellulose is a reinforcing fibre. Thus, lignocellulose fibre can be regarded as a self-reinforcing composite fibre. The production, availability and uses of lignocellulosic vegetable fibre are given in **Table 2.1**. The production of natural fibre in the world, including cotton, is around 30 million tonnes, of which lignocellulosics contribute around 5 million tonnes.

Table 2.1 Lignocellulosic vegetable fibres

Name	Main cultivation location	Quantity (tonnes per annum)	Present main use	Origin	By-product
Jute	India, Bangladesh	2,800,000	Bulk packaging, upholstery, geotextiles, composites	Bast fibre	Stalk, leaves
Flax	Western Europe, China, Russia	1,000,000	Linen textiles, paper pulp, insulation, geotextiles, composites	Bast fibre	Seeds, stalk, leaves
Coir	India, Srilanka	500,000	Mattresses, brushes, ropes, geotextiles, insulation, automobile seats	Seed fibre	Shell, pith, wood, leaves
Sisal	Brazil, China, Kenya, Cuba	300,000	Ropes and twine	Leaf fibre	Stem, juice
Hemp	China, Europe, Chile	90,000	Agrotextiles, paper pulp, composites	Bast fibre	Stalk, leaves
Abaca	Philippines, Ecuador	70,000	Paper pulp, composites	Leaf fibre	Leaves, juice

Jute is the second largest, cheapest, and most abundantly available fibre in a form that is ready to use. Jute fibre and its applications in the field of composites are described in detail as a representative member of all the lignocellulosic vegetable fibres. Though there is some difference in properties between different lignocellulosic fibres [2] due to percentage differences in their chemical composition they behave more or less similarly in composites. The properties of different lignocellulosic fibres are given in **Table 2.2.**

Table 2.2 Properties of natural fibre						
Properties	Jute	Flax	Coir	Sisal	Hemp	Abaca
Density (g/cm^3)	1.46	1.4	1.25	1.33	1.48	1.5
Tensile strength ($10E^6 N/m^2$)	400–800	800 –1500	220	600–700	550–900	980
E-modulus (GPa)	10–30	60–80	6	38	70	-
Specific modulus (GPa)	7–21	26–46	5	29	47	-
Elongation at failure (%)	1.8	1.2–1.6	15–25	2–3	1.6	2–4.5
Moisture absorption (%)	12	7	10	11	8	-
- Data not available in the literature						

2.3 Jute Fibre

2.3.1 Background

Commercial jute is an annual plant grown in tropical countries and is obtained from the stem of cultivated varieties of the two species: *Corchorus capsularis* (white jute) and *Corchorus olitorius* (tossa jute). The major jute producing countries are India and Bangladesh. Others are China, Thailand, Nepal, Indonesia, Brazil, Nepal, Burma, Peru and Vietnam. In India, the total area under cultivation is approximately 0.9 million hectares and the yield is 2,000 kg/hectare. Jute ranks next to cotton as the most important natural fibre in the world. Jute, in India, provides a livelihood for around four million farmers and 362,000 workers.

2.3.2 Nature of the Fibre

The major differences between cotton and jute fibres are:

- Jute has 12-14% lignin but cotton contains none.
- Jute fibre units are multiple cells cemented to each other by lignin and hemicelluloses, whereas cotton fibre units are single cells.
- The aspect ratio of jute cells is 98 to 118 and for cotton it is in the order of 1000.

Jute fibre, like other natural fibres, is hygroscopic. In the presence of water the jute filament swells by 23% in diameter, 40% in cross-section and 0.06% in length [3]. Absorption of water vapour changes the dimensions of the filament as well as its mechanical and electrical properties. Very good mechanical properties of the fibre are the most favourable qualities for its load bearing applications.

2.3.3 Advantages of Jute and Other Lignocellulosic Fibres

The major advantages of all the lignocellulosic fibres, including jute, are as follows:

- Annual renewable materials.
- Inexpensive.
- Environmentally friendly.
- Emit a very low level of harmful substances and leave negligible residues when incinerated.
- High level of stability.
- Resistance towards fracture.
- Non-abrasive.
- Low density.
- High strength, low elongation.
- No health hazard.
- Absorb atmospheric carbon dioxide as they grow.

One further aspect, which is of special relevance to applications and processing, is the high degree of porosity usually found in vegetable fibres in general, which is chiefly a result of the empty lumens of the cells. This hollow nature of jute fibre is favourable for acoustic insulation properties.

2.3.4 Disadvantages of Jute and Other Lignocellulosic Fibres

The major disadvantages of all the lignocellulosic fibres, including jute, are as follows:

- Affinity towards moisture.
- Lignin degradation sets in at around 200 °C.
- Low mildew resistance.

- Not all parts of the plant are used.

2.3.5 Morphology and Structure of Jute Fibre

The fibre when separated from the plant has a meshwork structure and the filaments or strands which comprise the mesh are composed of tiny fibres. The length of these fibres may vary from 0.7 to 6.0 mm and diameter from 0.005 to 0.025 mm.

The use of X-ray diffraction techniques has opened up a new dimension related to the molecular configuration and fine structure of jute fibre. The crystalline regions in the fibre structure are mainly responsible for rigidity, density and resistance to swelling. On the other hand, the amorphous regions are mainly concerned with softness, flexibility, dyeability and reversible extensibility. The degree of crystallinity and orientation of the jute fibre have been determined from X-ray diffraction photographs. It has been estimated that the percentage of cellulose crystallites in jute is approximately the same as in pure cellulose fibres [4-7].

Stronger swelling agents, such as alkaline solutions, are commonly employed in the preparation of textile fibres. Mild treatments [8] are similar to water in that they swell the fibre, allowing the release of internal strains and improved orientation of the crystallites. As such, small improvements in strength and elongation have been demonstrated for jute fibres by Sikdar and co-workers [8]. The use of strong alkali for mercerisation results in conversion of the cellulose crystal structure from amorphous. Cheek and Roussel [9] studied the effect of mercerisation on the mechanical properties of different natural fibres and found a significant drop in strength, and the ultimate elongation increased by 4-5 times compared with the virgin fibre.

2.3.6 Composition of Jute Fibre

Jute fibre contains 12-14% lignin, 21-24% hemi-cellulose and 58-63% α-cellulose besides some trace ingredients such as nitrogenous matter, fats, waxes and ash. Like wood fibre, the chemical composition of jute is not uniform. From staining tests [10] it is observed that the lignin is not uniformly distributed in the fibre. Astbury and co-workers [11] suggested a state of mixed crystallisation of cellulose, lignin, and hemicelluloses in jute and other lignified fibres.

Jute fibre itself can be considered to be a polymer composite. Jute, of all natural fibres, contains the highest proportion of stiff natural cellulose, approximately 75 wt% [12]. The lignin and hemicelluloses act as a matrix in a cellulose backbone. The approximate chemical composition of jute fibre has been reported by Macmillan [13].

Jute is a coarse fibre due to the presence of a substantial amount of lignin and a low wax content. Moreover, the presence of the cellulose backbone in a lignin matrix makes it the stiffest of the natural fibres and this is the basic reason for its technical difficulties as a textile fibre from the spinning point of view. However, this stiffness gives important design criteria for the development of fibre-reinforced composites. The tensile strength, modulus and elongation at break are 442 MPa, 56 GPa and 1.3% respectively [14]. Due to the low specific gravity of jute fibre, its specific properties are useful to the material scientist and comparable to those of glass fibre. The specific tensile strength is 303 MPa, about 22.5% that of glass fibre and the specific tensile modulus is 38.4 GPa which is about 35% more than that of glass fibre (**Table 2.3**). This is in agreement with the work done by Wells and co-workers [16]. The choice of jute as a reinforcing fibre is mainly due to its good specific mechanical properties, low cost and good availability. The long fibre and its flexibility also confer very good impact strength. Its less abrasive nature as compared to glass increases the machinery life.

In spite of all these advantages, the drawbacks of jute fibres are that they are not resistant to high temperature and they are susceptible to moisture absorption and mildew attack. For these reasons, chemical modification of the fibres is required in order to develop high performance composites.

2.4 Glass Fibre

2.4.1 Background

Glass fibre is a very popular reinforcing material in the composite industry. The reason for the widespread use of glass fibres in composites, both in the past and in the present, include availability, ease of processing, high tensile strength, and excellent insulating properties, in spite of its cost which is several times greater than the common agrofibres like jute. There are several applications where the very high cost and strength of glass fibre is neither justified nor required. This is one of the many reasons why, and how, agrofibres are making inroads into the composite industry. In spite of the dominating position and advantages of glass fibre in the composite industry, it has some drawbacks too. These are its high specific gravity, low tensile modulus, sensitivity to abrasion, low fatigue resistance and high hardness which causes excessive wear on moulding dies, mixers, extruders, cutting tools and so on.

2.4.2 Comparison with Jute Fibre

The comparative compositions and properties of E-glass and jute are given in **Tables 2.3** and **2.4**.

Table 2.3 Typical composition of glass and jute fibres

Constituents	E-glass fibre (wt%)	Constituents	Jute fibre (wt%)
Silica	54.5	α-Cellulose	60.70
Aluminium oxide	14.5	Lignin	12.50
Calcium oxide	17.5	Hemicellulose	23.14
Magnesium oxide	4.5	Fats and waxes	1.00
Sodium oxide	0.5	Ash	0.79
Boron oxide	8.5	Nitrogenous matter	1.87

Table 2.4 Typical properties of glass and jute fibres

Properties	E-glass	Jute
Specific gravity	2.54	1.46
Tensile strength (MPa)	3450	442
Specific tensile strength (MPa)	1358	303
Tensile modulus (GPa)	72.4	56
Specific tensile modulus (GPa)	28.5	38.4
Elongation at break (%)	4.8	1.3

2.5 Chemical Modification of Lignocellulosic Fibres

Lignocellulosic fibres are hygroscopic in nature, flammable and biodegradable. To develop composite materials with a long service life these drawbacks must be overcome by chemical modification. Several means by which this can be achieved are discussed below.

2.5.1 Acetals

The reaction between formaldehyde, acetaldehyde or other aldehydes with the hydroxyl groups of the lignocellulosic cell wall, forms an acetal linkage. The reaction is usually catalysed by strong acids.

$$LC\text{-}OH + HCHO \rightarrow LC\text{-}O\text{-}CH_2\text{-}OH$$

(hemiacetal)

$$\downarrow LC\text{-}OH$$

$$LC\text{-}O\text{-}CH_2\text{-}O\text{-}LC + H_2O$$

(acetal)

Where LC = lignocellulose

At a weight gain of 3.9% after acetal formation with formaldehyde, the equilibrium moisture content of pine fibre at 27 °C and 65% relative humidity, reduced to 4.2 from 12.0 % [17].

2.5.2 Esters

Among ester reactions anhydrides (such as acetic, phthalic, succinic, maleic and propionic) acid chlorides and many different types of isocynates form esters with reactive hydroxyl groups in lignocellulosic fibre cell walls. Of these reactions, acetylation by acetic anhydride has attracted commercial interest in the recent past for wood fibre modification [18].

Acetylation is a single-site reaction in which an acetyl group reacts with a hydroxyl group without self-polymerisation. The reactive hydroxyl sites in wood react with acetic anhydride forming esters, splitting off acetic acid as a by-product. Therefore, all the weight gain through acetylation can be directly related to units of reacted hydroxyl groups [19]. To attain maximum weight gain of fibres through acetylation, several parameters require prior attention such as the initial moisture content of the fibres, reaction time and temperature, catalyst and co-solvent. The acetylation reaction may be represented as follows:

$$LC\text{-}OH + CH_3\text{-}CO\text{-}O\text{-}CO\text{-}CH_3 \rightarrow LC\text{-}O\text{-}CO\text{-}CH_3 + CH_3COOH$$

2.5.3 Chemical Modification to Improve Dimensional Stability

Changes in dimensions, especially thickness, in the presence of water are a measure of dimensional stability. Water, being a polar liquid, possesses the ability to penetrate into lignocellulosic composites and wood and swell the cell walls by breaking interhydroxyl hydrogen bonds [20]. In the case of wood, this swelling is reversible, i.e., upon release of water the wood cell wall reverts to its original dimensions. In contrast, lignocellulosic composites or reconsitituted wood (in other words composites which are made under pressure), undergo swelling caused by the release of residual compressive stresses and this is known as irreversible swelling. Water sorption causes both reversible and irreversible swelling, and some of the reversible shrinkage occurs when the composite board dries. The dimensional instability of lignocellulosic composites has been the major reason for their restricted use. The mechanism of enhancing dimensional stability is based on blocking of lignocellulosic hydroxyl groups by the polymer or chemical resulting in composites which are more hydrophobic [21].

2.5.4 Modification of Fibre-matrix Interaction

Addition of filler or fibre to a polymer usually has a positive effect on the mechanical properties of the resulting composites, such as stiffness, but a negative effect on the elongation at break and impact strength. It is possible to improve these properties through modification of fibre-matrix interfaces. The basic role of such modifications is to reduce particle-particle interactions and to facilitate wetting [22].

There are three different types of surface modification:

- *Non-reactive type*: With this type of treatment, surfactants which are bonded to the fibre or filler surfaces do not react with the matrix. These surfactants can decrease the surface energy of the filler. Non-reactive surfactants change the particle-particle interaction, leading to improved dispersion and homogeneity and easier processing and usually also better impact properties. However, this kind of treatment can reduce the particle-matrix interaction and thus decrease the strength of the composite [23].

- *Reactive type*: Reactive surface treatment results in a chemical reaction of the added coupling agent, also known as the compatibiliser, with both the fibre and the matrix resulting in improved adhesion and, hence, improved mechanical properties [23]. In the case of lignocellulosic fibres, the compatibiliser (which is often a polymer) works by attaching itself to the fibre and leaving a polymeric 'tail' which is soluble in the matrix.

- *Formation of polymer interlayer on the surface*: Covering the filler with a polymer layer which is capable of diffusion into the matrix has been shown to be very effective in both stress transfer and in forming a thick diffuse interphase with acceptable deformability [22, 23].

2.6 Natural Fibre Composites

2.6.1 Background

The use of straw in mud huts from ancient times is probably the precursor of the modern man-made composites. The use of biobased composites has rapidly expanded in recent years and there is tremendous potential for future growth in this area. The reasons for this are mainly two-fold. One is to prevent deforestation by manufacturing wood substitutes and the other is to meet the demand for new materials, including composites, due to growth of the world population. The uses range from automotive interior components to geotextiles. A broad range of agro-based fibres is utilised as the main structural components or as fillers or reinforcing agents in these composite materials.

A composite can be defined as a material with two or more chemically distinct phases which, at the microscopic scale, are separated by a distinct interface (boundary). A composite can also be defined as a simpler way: it is a material which exhibits a combination of properties of two or more components held together by some type of mastic or matrix. The continuous phase is known as the matrix and the other component as the filler or reinforcement. In general, fibres are the principal load carrying constituents while the surrounding matrix keeps them in the desired position and, acts as a load transfer medium between them. So, from a theoretical point of view, fibres which have very good strength and modulus properties and which bond well with the matrix should give rise to good quality composite materials [24, 25].

The most common composites, used and seen by all of us, are decorative laminates and protective equipment such as helmets. Decorative laminates are composed of phenolic resin-impregnated Kraft paper as the base and a top layer of melamine resin coated paper with a design on it. The wood fibres present in the Kraft paper play the role of reinforcement and the resin is the binder. Chopped strand mat made of fibre glass, and unsaturated polyester resins, are the main constituents of protective helmets. Other examples of commercially accepted lignocellulosic-based composites include autotrims, pultruded profiles, medium density fibreboard, and railway coach building components like floor boards, roof linings and vertical panels.

Sustainable development is an important concept underlying many of today's renewable resource policies. While definitions of sustainable development vary, the essence of the concept is that we must sustain ecosystems and we must sustain people. This means that we must manage our ecosystems in ways that do not put them at risk, but we cannot abandon peoples' needs for fibre. Our goal is to use only resources which are sustainable, so we need a course of action to get us there. That course is ecosystem management, which is an approach to the management of natural resources that strives to maintain or restore the sustainability of ecosystems and to provide present and future generations with a continuous flow of multiple benefits in a manner which is in harmony with ecosystem sustainability [17, 26, 27].

2.6.2 Thermoplastic versus Thermoset Polymer Matrix

Polymers are divided into two broad categories: thermoplastics and thermosets. In a thermoplastic polymer, individual molecules are linear in structure with no chemical linking between them. They are held in place by weak secondary bonds (intermolecular forces), such as van der Waals forces and hydrogen bonds. With the application of heat and pressure, these intermolecular bonds in a solid thermoplastic polymer are temporarily broken, and the molecules can be moved relative to each other to flow into new positions: Upon cooling the molecules freeze in their new positions, restoring the secondary bonds between them and resulting in a new solid shape. Thus, a thermoplastic polymer can be heat-softened, melted, and reshaped (postformed) as many times as desired.

In a thermoset polymer, on the other hand, the molecules are chemically joined together by crosslinks, forming a rigid, three-dimensional network structure. Once these crosslinks are formed during the polymerisation reaction, also called a curing reaction, the thermoset polymer cannot be melted and reshaped (postformed) by the application of heat and pressure. However, if the number of crosslinks is low, it may still be possible to soften them at elevated temperatures.

Traditionally, thermoset polymers have been used as a matrix material for fibre-reinforced composites. Starting materials used in the polymerisation of thermoset polymers are usually low molecular weight liquid chemicals with very low viscosities. Fibres are either pulled through or immersed in these chemicals before the polymerisation reaction begins. Since the viscosity of the polymer at the time of fibre incorporation is very low, it is possible to achieve a good wet-out between the fibres and matrix without the aid of high temperature and pressure. Among other advantages of using thermoset polymers are their thermal stability and chemical resistance. They also exhibit much less creep and stress relaxation than thermoplastic polymers. The disadvantages [15] of thermosetting polymers are:

- Limited storage life at room temperature.
- Long fabrication time in the mould.
- Low strains to failure which also contribute to their low impact strengths.

The advantages of thermoplastics [15] are:

- High impact strength and fracture resistance.
- Higher strains to failure than thermoset polymers.
- Unlimited storage (shelf) life at room temperature.
- Shorter fabrication time.
- Postformability e.g., thermoforrning.
- Ease of repair by welding, solvent bonding, and so on.
- Ease of handling, e.g., no tackiness.

In spite of such distinct advantages, the development of thermoplastic matrices has been slower than that of thermoset matrices. Because of their high melt or solution viscosities, incorporation of continuous fibres into a thermoplastic matrix is difficult. Commercial engineering thermoplastic polymers, such as polycarbonate and nylons are of very limited interest in structural applications because they exhibit lower creep resistance and thermal stability than thermoset polymers [28, 29].

2.6.3 Natural Fibre Composites as Wood Substitutes

The most common form of nature made composite is wood. Continuous use of wood in increasing amounts for different applications such as furniture, packaging, paper pulp, fuel, and construction materials has led to drastic deforestation. This deforestation is, in turn, causing ecosystem imbalance. To stop this, there is an urgent need to develop wood substitutes using rapidly renewable resources as emphasised by Rowell and co-workers [17].

Both jute and wood are lignocellulosics and so far as chemical composition is concerned they are very close to each other, The main difference between them is that jute is flexible and wood is rigid. The very flexible nature of jute fibre allows the scientist to develop moulded composites with intricate shapes at lower machining and fabrication costs than for wood. The various jute composites developed by the Indian Jute Industries' Research Association (IJIRA), including the hybrid with glass using different thermosetting resins are described by Mitra and co-workers [30], and Ranganathan and co-workers [31].

Glass fibre, asbestos and cotton are the most frequently used reinforcing fibres in most of the composite industries. The high cost and very good mechanical properties of glass fibre composites are not often justified as a substitute for wood. The use of asbestos has been restricted due to its carcinogenic effect and cotton is also a costly proposition for its use as a reinforcing fibre for the development of wood substitute. Considering these facts, jute fibre has a future role to play as a low cost reinforcing fibre for the development of jute composites as wood substitutes.

2.6.4 Variables Influencing the Properties of Polymer Composites

The properties of composites are strongly influenced by factors such as proportions and properties of the matrix and reinforcements, the interface between these and the shape, size orientation, distribution of the reinforcement or filler and lastly the processing techniques [32, 33]. Some of the important parameters to generate a wide range of physical and chemical properties for composite materials include fibre orientation, stacking sequence and press cycle.

2.6.4.1 Aspect Ratio of the Fibre or Filler

The ratio of the length to the diameter of a fibre is known as the aspect ratio. It has been stated that the minimum relative fibre dimensions required for reinforcement, or critical aspect ratio, can be defined as [34]:

$$q_{crit} = \sigma_{fu} / 2\tau = l_f / d_f \qquad (2.1)$$

or

$$l_f = d_f \sigma_{fu} / 2\tau \qquad (2.2)$$

where:

q_{crit} = critical aspect ratio;

σ_{fu} = fibre ultimate tensile strength;

τ = interfacial fibre-matrix shear stress;

l_f = critical fibre length; and

d_f = fibre diameter.

Hence, for a typical jute fibre with σ_{fu} = 442 MPa and τ =5 MPa, and diameter = 15 µm, the minimum length required for reinforcement is 0.66 mm. This aspect ratio can be easily obtained from the agrofibres and that is why they are suitable candidates for reinforcement [35, 36].

2.6.4.2 Polymer-fibre Interface and Interphase

The interface regions strongly affect the end properties of the composites. For better strength there must be an excellent adhesion between the matrix and the reinforcement. In a composite two interfaces exist, one between the interphase and the fibre and another between the interphase and the polymer matrix.

2.6.4.3 Alignment or Distribution of the Reinforcement

Alignment or distribution of the reinforcement plays an important role for the development of performance driven composites. Unidirectional (i.e., parallel) alignment is known as anisotropy and gives the highest strength. Random arrangement of fibres (i.e., a mat) gives the lowest strength and is known as isotropic alignment. The square weave fabric is an example of bidirectional alignment and provides a strength that results from anisotropic and isotropic alignment [37].

2.6.4.4 Loading and Processing Techniques

Some processing techniques allow the reinforcements to be more carefully aligned or oriented. These result in products which are more anisotropic. During processing, reinforcements may be broken or damaged, resulting in reduced mechanical properties. Depending on the processing technique, particulate fillers and short fibres are more likely to have random rather than oriented placement in the matrix. This will result an isotropic composite [38].

Normally, the mechanical properties of the composite depend on the amount of reinforcing material added. Initially, there is a tendency to increase the mechanical properties with an increase in reinforcements, but often it reaches an optimum and then decreases. The reason may be attributed to the failure of reinforcement due to the presence of insufficient matrix at high levels of loading [39].

2.6.4.5 Wetting, Adhesion and Dispersion

Wetting, adhesion and dispersion are of great importance in achieving good interaction between the two phases. Wetting can be defined as the extent to which a liquid makes contact with a surface. In polymer composites, it is characterised by the degree of direct interfacial contact between the surfaces. Adhesion is the sticking together of two surfaces so that stresses can be transmitted between the phases and can be quantified by the amount of work required to pull the two surfaces apart. The degree of dispersion describes the degree of mixing in a colloidal system; good dispersion would correspond to colloidally stable individual particles and poor dispersion to filler agglomeration [26].

2.7 Short Natural Fibre-reinforced Thermoplastic Composites

2.7.1 Background

The art of making composites is mainly restricted to synthetic polymers and synthetic reinforcing fibres such as glass, carbon and nylon. The potential of natural fibre-based composites using cellulose, wood, jute, kenaf, hemp, sisal, pineapple, coir, and so on, as reinforcing fibres in a thermosetting resin matrix, has received considerable attention among scientists all over the world [16, 31, 40, 41] because some of these exhibit excellent specific properties and some of them have already found use as industrial products.

Composites based on thermoplastic resins are now becoming popular due to their processing advantages [42]. The mechanical properties of thermoplastics are often inadequate for specific applications and so they may be mixed with different man-made fibres (mainly glass) or other mineral fillers such as china clay, talc, calcium carbonate ($CaCO_3$) to form composites. On the other hand, natural plant fibres can be regenerated either annually or at regular intervals. These are lignocellulosic self-reinforced biopolymer fibres [43], which are non-abrasive, have low density, high strength, are abundantly available in tropical countries and are, therefore, of particular interest for use as reinforcing fibres in thermoplastic composites.

Among commodity thermoplastics, polypropylene (PP) possesses outstanding properties including low density, high Vicat softening point, good flex life, sterilisability, good surface hardness, scratch resistance, very good abrasion resistance and excellent electrical properties [39]. That is why polypropylene-based composites are popular and are of considerable commercial interest.

Natural fibres are predominantly polar due to the presence of various polar groups on their back-bones [44]. In contrast, PP is nonpolar [45], so to improve the interaction between the two otherwise incompatible surfaces, the use of a compatibiliser was suggested [46]. Grafting of dicarboxylic anhydrides on to polyolefins and their use as compatibilisers have been reported in the literature [47-50]. Comparative rheological studies on jute fibre and glass fibre-filled PP composite melts [51] revealed that both follow a power law relationship. Work on coir/polyvinyl chloride (PVC) and coir/PP [52] reported no remarkable increase in mechanical properties. The development of high density polyethylene impregnated jute cloth boards resulted in a 44-54% increase in tensile strength [53].

With the advent of different agrobased thermoplastic composites for different applications for which dynamic mechanical properties are particularly important (e.g., machinery, transport, buildings, domestic appliances, acoustic devices and so on), it is important to study the polymer composite structure to investigate the dynamic mechanical thermal properties, particularly the dynamic modulus and internal friction, over a wide range of temperatures [54-57].

As is known from the dynamic mechanical and thermal analysis principles when internal molecular motion occurs, the material responds in a viscoelastic manner and the strain response lags behind the stress. The tangent of this phase lag is expressed as tan δ and is also known as the damping factor. The damping properties of a material represent its capacity to reduce the transmission of vibration caused by mechanical disturbances to a structure. The dynamic mechanical and thermal properties are mostly dependent upon parameters such as amount of fibre [58, 59], the presence of additives such as compatibiliser, filler, or impact modifier [60], fibre orientation [61, 62] and the mode of testing. Over a wide range of temperatures, polymeric materials usually exhibit more than one relaxation region, or so-called transition [56]. In the majority of the studies the fibres are randomly placed and the aim is to compare different dynamic parameters at different levels of fibre loading within the experimental range. The effect of the compatibiliser has also been studied by researchers to obtain fundamental information regarding molecular interaction using the viscoelastic properties in dynamic conditions.

2.7.2 Preparation of Fibres

One of the important steps is to accommodate the fibres in conventional thermoplastic processing machinery. Natural lignocellulosic fibres are very long and very difficult to process in the conventional machinery due to the problem of entanglement and a tendency to agglomerate which results in a non-uniform distribution of the fibres.

To overcome this problem, the fibres are first chopped using a guillotine or similar cutting equipment to a length of around 100 mm. They are then reduced to a length of around 10 mm using a pulveriser at a speed of 500-700 rpm. Because of the very high speed, a substantial amount of frictional heat is developed. There is a high risk of fire during this operation, so the use of nitrogen during pulverisation is essential to avoid this unwanted problem.

This short fibre is then dried in an oven at 105 ± 5 °C to remove absorbed moisture from it. This dried short fibre is now ready for further compounding operations [63].

2.7.3 Compounding of Fibres

A common choice of compatibiliser for compounding of natural fibre-reinforced thermoplastic granules is maleated PP (maleic anhydride-modified PP). This compatibiliser forms a bridge between the natual fibre and the PP. The dose of this compatibiliser is 0-2% depending on the final mechanical properties required for a given application. The fibre loading varies from 30-60% depending on the mechanical properties required and the type of compounding machinery used. Impact modifiers are also used commercially as described by Rana and co-workers [64].

Natural fibre, PP, maleated PP, impact modifier, colour master batch, and other additives, are mixed in a ribbon blender in predetermined proportions and then fed into a co-rotating twin-screw extruder with mixing and compounding zones. Temperatures are restricted to below 200 °C, and preferably at 190 °C, to avoid charring. A vacuum is maintained in the barrel to remove volatiles formed at high temperature and to restrict the formation of bubbles in the granules.

The thread-like molten strands coming out from the extruder die are passed through water to solidify them and these are then cross cut using a circular cutter to produce granules loaded with short natural fibre. These granules are again dried in an oven before moulding.

Sometimes, high speed thermokinetic mixers are used for mixing purposes. In this case, the hot dough formed after mixing has to be cooled immediately by pressing in a cold hydraulic press between the flat platen. The sheets are fed into a granulator to break them up and the resulting granules are used for subsequent moulding operations.

2.7.4 Moulding

Injection moulding machines are most commonly used for thermoplastic moulding. In these machines, the molten mass is forced into the cold mould cavity by hydraulic pressure. Pack pressure and hold pressure are normally higher for natural fibre moulding than for mineral filler loaded material.

There are also manufacturing lines for thermoplastic-natural fibre sheet in which *in situ* compounding and calendering can be carried out. These sheets are thermoformed to mould them for autotrims.

2.7.5 Properties of Jute Fibre-reinforced Polypropylene Composites

In the experiment jute fibres were granulated using a granulator having a 4-mm screen. The calculated quantity of granulated jute fibre on a dry weight basis, compatibiliser and PP were mixed in a K-mixer. Jute fibre loadings were varied from 0 to 60 wt% and compatibiliser varied from 0 to 4 wt%. After mixing in the K-mixer, the dough was pressed in the cold hydraulic press and then transferred to the granulator to form granules suitable for injection moulding. Details of the experiments are described elsewhere [63, 64, 71].

2.7.5.1 Fibres and Other Reagents Used

- Jute fibre (Grade W-2, *Chorchorus capsularis*).
- Compatibiliser - Epolene G-3002, Eastman (maleated PP, acid number = 60 and Vicat softening point = 133 °C).
- PP Homopolymer - Fortilene, Solvay Polymers (melt flow index [MFI] = 12.0).

Jute/PP granules prepared during the experiment were dried at 105 °C for 4 h and then moulded using a Cincinnati Milacron 33T injection moulding machine to mould American Society for Testing and Materials (ASTM) test specimens for measuring the izod impact, flexural strength and tensile strength. Test specimens were moulded at 190 °C, 850 psi hold pressure and 1000 to 1650 psi pack pressure depending on the MFI. The MFI values, measured by MFI Tester (Keyness make), at 190 °C for different series are shown in **Table 2.5**.

Table 2.5 MFI values at different jute content (wt%)	
Sample identification	MFI (g/10 min)
PP	12.2
J300	2.6
J400	1.2
J500	0.05
J600	Not possible
J300 = Jute fibre, 30 wt%; compatibiliser, 0 wt% J400 = Jute fibre, 40 wt%; compatibiliser, 0 wt% J500 = Jute fibre, 50 wt%; compatibiliser, 0 wt% J600 = Jute fibre, 60 wt%; compatibiliser, 0 wt%	

2.7.5.2 Specific Gravity

The specific gravity (measured by ASTM D 792) of the composites varies from 1.02 (for the J300 series) to 1.10 (for the J600 series) as compared to 0.89 for PP, 1.23 for 40% glass fibre and 1.25 for 40% $CaCO_3$ loaded PP composites [65]. Due to the low specific gravity of jute/PP composites, their specific mechanical properties became comparable to those of glass/PP composites [65].

2.7.5.3 Water Absorption

Water absorption values after 24 h soaking in cold water and 2 h in boiling water, are shown in **Table 2.6**. Water absorption values were found to increase with increased fibre loading. Use of a compatibiliser, however, decreased water absorption at the same fibre loading. The reasons might be attributed to some of the hydrophilic -OH groups reacting with acid anhydride to form ester linkages and thereby giving lower water absorption values. Similar observations were also observed by Rowell and co-workers [66] during the acetylation of wood and Hedenberg and Gatenholm [49] during the conversion of plastic cellulose waste into composites.

2.7.5.4 Tensile Behaviour

The presence of a compatibiliser improves the tensile strength. It was observed that at 0% compatibiliser the tensile strengths of the composites were 33.65 to 36.17 MPa at fibre loadings of 30-60%. This suggests that there was a little stress transfer from the matrix to the fibres irrespective of the amount of fibre present. When only 1%

compatibiliser was added, the strengths increased phenomenally. Further increase in compatibiliser beyond 1% had very little effect. The effect was greater in composites with higher fibre contents, reaching up to 73.71 MPa for 60% fibre loading. The percentage maximum increases in the tensile strengths for the J300, J400, J500 and J 600 series were 50.15, 62.34, 87.48 and 119.05 respectively [63]. When compatibiliser was added, the sharp rise in strength was due to stress transfer from matrix to fibre *via* the compatibiliser [67]. The increase in fibre content restricted the mobility of the PP chain and this was reflected in the tensile modulus values though addition of compatibiliser in the same series did not alter the values appreciably. The moduli for the J300 to J600 series were 4.60, 6.20, 8.60 and 10.35 GPa respectively. A decrease in failure strain was observed with increase in fibre content due to restrictions posed by the fibre in the stress-strain curve. However, the addition of the compatibiliser helped in stress transfer from the matrix to the fibre. The reduction of the elongation at break with the addition of compatibiliser, suggested that there was a linkage at the interface between the fibre and the matrix formed by the compatibiliser.

Table 2.6 Specific gravity and percent water absorption (based on initial weight)

Test Parameter	PP	J300	J302	J400	J402	J500	J502	J600	J602
Specific gravity	0.89	1.02	1.05	1.05	1.05	1.08	1.08	1.10	1.10
2 h in boiling water absorption, %	0.1	0.93	0.77	1.5	1.16	2.12	1.47	3.06	2.22
24 h in cold water absorption, %	0.03	0.56	0.42	0.61	0.44	1.07	0.6	1.86	0.91

J300= Jute fibre, 30 wt%; compatibiliser, 0 wt%; J302 = Jute fibre, 30 wt%; compatibiliser, 2 wt%
J400 = Jute fibre, 40 wt%; compatibiliser, 0 wt%; J402 = Jute fibre, 40 wt%; compatibiliser, 2 wt%
J500 = Jute fibre, 50 wt%; compatibiliser, 0 wt%; J502 = Jute fibre, 50 wt%; compatibiliser, 2 wt%
J 600 = Jute fibre, 60 wt%; compatibiliser, 0 wt%; J602 = Jute fibre, 60 wt%; compatibiliser, 2 wt%

Reproduced with permission from A.K. Rana, A. Mandal, B.C. Mitra, R. Jacobson, R. Rowell and A.N. Banerjee, *Journal of Applied Polymer Science*, 1998, **69**, 329. ©1998, John Wiley & Sons, Inc. [63]

2.7.5.5 Flexural Behaviour

The changes in flexural strength and flexural modulus, in the presence of a compatibiliser, are significant. An increase in these values was observed and the percentage maximum increases in flexural strengths for the J300 to J600 series were 31.66, 44.09, 64.30 and 90.88 respectively [63]. At 0% compatibiliser, there was virtually no change in the flexural strength (56.48-61.29 MPa) with increased fibre loading from 30-60%, indicating therefore that there was little interfacial bonding for the transfer of the load from matrix to fibres. Yet, when compatibiliser was added, interfacial bonding had occurred between the fibres and the matrix. The optimal dose of compatibiliser was observed to be 2-3% . The trend in flexural modulus was identical to that in the tensile modulus described earlier, where compatibiliser had no effect. However, the effect of the fibre content was obvious.

2.7.5.6 Impact Behaviour

The effect of compatibiliser on impact strength was observed for both notched and un-notched samples. The notched sample values were found to increase from 26.9 J/m (for 30% fibre) to 33.73 J/m (for 60% fibre). The effect of compatibiliser content on notched impact strengths was negligible, with marginal improvement up to 2% compatibiliser beyond which deterioration in this property was observed and this had its most pronounced effect at 30% fibre content [64]. This could be attributed to the migration of too much compatibiliser around the fibres, causing self-entanglement between compatibiliser molecules rather than binding with the polymer matrix, resulting in slippage [67, 68].

The effect was quite different in the case of unnotched samples. The values at 0% compatibiliser were 70.7 J/m (60% fibres) and 195.6 J/m (30% fibre). The low value at high fibre content might be due to the presence of too many fibre ends within the body of the composites which could cause crack initiation and hence potential composite failure [69]. Addition of fibres also increased the probability of fibre agglomeration [68] which created regions of stress concentrations that required less energy to initiate a crack. Similar observations were also reported elsewhere [67, 69]. However in the presence of compatibiliser the fibres themselves could modify the microstructure of the surrounding matrix and thus there was a sharp increase in unnotched impact strength. It was apparent that even with 30% fibre content and 1% compatibiliser the unnotched impact strength was more or less equivalent to that of a composite with 60% fibre content. Unlike notched specimens no deterioration was observed up to 3% compatibiliser content. In this case, the compatibiliser migrated around the fibres and acted as a damper to the shock wave imparted during impact thereby transmitting it to the fibres evenly.

2.7.5.7 Tensile Energy Absorption

Tensile energy absorption (TEA) is the area under the stress-strain curve up to failure. TEA was found to decrease with increase in fibre loading. Increase in fibre loading restricted the mobility of the polymer chains and this caused a decrease in failure strain and TEA too [63, 67]. However, with the addition of compatibiliser, a sharp rise in TEA was observed. This was due to the better interaction at the jute-PP interface in the presence of the compatibiliser. This was observed up to 3% compatibiliser, but at 4% the energy absorption trend was found to be downward. This was probably due to the migration of too much compatibiliser around the fibre surfaces which caused slippage as described earlier.

2.7.5.8 Dynamic Mechanical Study

The dynamic mechanical properties were measured by using a Dupont DMA 983 analyser. The testing parameters were [71]:

Resonant frequency	Mode
12.5 × 3.00 mm	Sample size
10 × Thickness	Clamping distance
0.2 mm	Oscillation amplitude
5 °C/min	Heating rate
33 to 160 °C	Temperature range

It is found from the experimental results that the storage modulus (E′)of PP is much lower than that of the jute-PP system. With gradual addition of jute fibre, E′ values are also increased gradually. This is due to the reinforcement imparted by the fibre that allows stress transfer from the matrix to the fibre. With the addition of 2% compatibiliser, the E′ value is found to increase more than in the uncompatibilised system. This phenomenon is observed uniformly throughout the experiment. This may be due to the migration of compatibiliser to the fibre surface and linking with the matrix by its tail. Studies on the mechanical properties were also in agreement with this [63]. At 160 °C, it is found that the E′ value of PP is reduced to a very low level whereas compounds having jute fibre reinforcement have some strength depending on the fibre content (the E′ values of J600 and PP at 160 °C are 0.9677 and 0.06201 GPa respectively).

Results show that the variation of the shear storage modulus (G′) with temperature at different weight fractions of fibres were found to be very similar to that of the variation of E′ and the same explanation is true for this case also.

The variation of E′ at different compatibiliser levels (0-4%) with 50% weight fraction of fibre content corroborates the previous work. Earlier works on cellulose/lignocellulose-thermoplastic composites [49, 61, 62] have revealed that, during compounding, the compatibiliser migrates to the fibre to form a bond with it and the tail of the compatibiliser becomes entangled in the thermoplastic matrix. The same phenomenon is observed in our current experiment. The E′ values increased with an increase in compatibiliser content. This might be due to an increase in the bulkiness of the fibre due to the formation of bonds between the compatibiliser and the fibre. For the same reason, the variation of G′ with temperature at different compatibiliser levels also shows the same trend.

The improved interaction between fibre and polymer in the presence of a compatibiliser can also be understood when E′ is plotted against the weight fraction of jute fibre at 35 °C and 110 °C. This shows that, with an increase in fibre content the graphs become steeper for the compatibilised system. The modulus enhancement of both the control and compatibilised system with fibre weight fraction is also demonstrated by the relative flexural storage modulus (E'_c/E'_m where E'_c and E'_m are the storage moduli, of the composite and matrix respectively). At high fibre content the compatiblised system shows the higher value at a particular fibre content. At 110 °C this is more prominent than at 35 °C. This might be due to the fact that poor fibre-polymer interaction in the absence of compatibiliser becomes exaggerated at higher temperature.

Table 2.7 shows the variation of the modulus retention term [70] of the composites with weight fraction of the fibre. It is evident from the table that with increase in fibre content, the modulus retention values increase and further increase in the same value is observed with the addition of compatibiliser.

From the variation of tan δ with temperature for various weight fractions of fibre and compatibiliser it is observed that the introduction of fibres has reduced the peak height, i.e., tan δ is lowered with increase in fibre content. This phenomenon was also observed by earlier workers [72, 73]. The reason might be due to restriction of the mobility of polymer chains by the fibre. The other prominent effect is broadening of the transition region. This might be due to the inhibition of the relaxation process within the composites upon incorporation of the fibres.

If the applied mechanical energy (work) is not stored elastically, it must be lost and converted to heat through molecular friction, i.e., viscous dissipation, within the material. This is precisely the loss modulus [74]. It is seen that with gradual introduction of fibres the transitional peak is also gradually shifted to a higher region. From J300 to J600 a shift of 10 °C is observed. In the experimental range a tiny peak at 80 °C is observed for PP and at 110 °C a broad peak for J600 is also observed. This might be explained as a result of movement towards the glass transition of dry cellulose in the jute fibre, which is about 200 °C [75].

Table 2.7 Variation of modulus retention with weight fraction of jute fibre (control and 2% compatibiliser content)			
Sample ID	Jute fibre (wt%)	% Modulus retention	
		$E'_{110°C}/E'_{35°C}$	$E'_{160°C}/E'_{35°C}$
J600	60	53.50	18.94
J602	60	57.12	20.16
J500	50	59.80	16.08
J502	50	53.49	17.07
J400	40	49.50	14.28
J402	40	50.70	15.81
J300	30	43.55	13.74
J302	30	43.68	13.85
PP	0	25.03	3.77

J300 = Jute fibre, 30 wt%; compatibiliser, 0 wt%; J302 = Jute fibre, 30 wt%; compatibiliser, 2 wt%
J400 = Jute fibre, 40 wt%; compatibiliser, 0 wt%; J402 = Jute fibre, 40 wt%; compatibiliser, 2 wt%
J500 = Jute fibre, 50 wt%; compatibiliser, 0 wt%; J502 = Jute fibre, 50 wt%; compatibiliser, 2 wt%
J600 = Jute fibre, 60 wt%; compatibiliser, 0 wt%; J602 = Jute fibre, 60 wt%; compatibiliser, 2 wt%

Reproduced with permission from A.K. Rana, B.C. Mitra and A.N. Bannerjee, *Journal of Applied Polymer Science*, 1999, 71, 531. ©1998, John Wiley & Sons, Inc. [71]

2.7.5.9 Correlation of Mechanical Properties with other Factors

To understand the stiffness and strength of short fibre-reinforced thermoplastic between the nonpolar PP and polar natural fibre, several factors are very influential. These are dispersion, adhesion, fibre breakage and agglomeration, fibre distribution and orientation, volume or weight fraction of fibres, microstructure, and so on. To improve dispersion and adhesion the use of maleic anhydride grafted polyolefins as compatibilisers has been reported by earlier workers [49, 76, 77]. The formation of covalent linkages between maleic anhydride and the hydroxyl group of cellulose has been indicated by Hedenberg and Gatenholm [49] by infrared and environmental and stress cracking analysis. For better stress transfer efficiency of the interface, sufficient maleic anhydride group should be present in the compatibiliser for better interaction with the -OH group present on the fibre surface, and the compatibiliser also should have sufficient tail length for better entanglement with the matrix [78]. The K-mixer, used for mixing, caused substantial fibre breakage due to very high shear. The strengths obtained for composites were thus limited by this fibre breakage. Use of a compatibiliser did not alter the values of the moduli substantially, probably due to the formation of transcrystalline zone forming around the fibre in both the compatibilised and uncompatibilised systems. The complex fibre distribution and orientation also contributed to the high values of the moduli and the values increased with the fibre percentage. At high fibre loading, the possibility of fibre agglomeration hindered the plastic flow process and the resultant composite showed a lower TEA. In the presence of compatibiliser the fibres themselves could modify the microstructure of the surrounding matrix as was evident from the same studies.

2.7.5.10 Conclusions

The following conclusions may be drawn from the above discussion and these apply to both jute and to all the natural fibres:

- The sharp increase in mechanical properties and decrease in water absorption values after addition of compatibiliser could be attributed to linkage between the hydrophilic hydroxyl groups of jute and the carboxyl groups of the compatibiliser.
- All these results demonstrate that the role of jute fibre was not as a filler fibre but as a reinforcing fibre in a properly compatibilised system.
- This system produced a new range of low energy, low cost composites having interesting properties and should be given priority over costly and high energy synthetic reinforcing fibres wherever possible.
- The drawback of this system was that it had low impact strength as compared to 40% glass filled PP [29], but there is immense scope for improvement if an impact modifier is used.
- The jute-PP system might not be recommended for electrical applications due to its high water absorption values as compared to systems loaded with fillers such as glass, mica, $CaCO_3$ and talc.
- The E′ increased with increase in fibre content. With addition of compatibiliser, further increases in values were also observed.
- The transition, i.e., tan δ peak, shifted to a higher temperature as the amount of fibre loading increased.
- Marginal shifting of the transition was observed with increase in compatibiliser dose at a given level of fibre.
- Efficiency of the compatibilised system increased with an increase in the amount of fibre as compared to the uncompatibilised system.
- Fourier transform infrared studies indicated the presence of chemical linkages between the fibre and the polymer.

2.8 Long Natural Fibre-reinforced Thermosetting Composite Boards and Moulded Items

2.8.1 Choice of Thermosetting Resin

The following are the common thermosetting resins which have been used in composites. These have been studied in many research publications and their merits have been demonstrated:

- Phenol-formaldehyde (PF)
- Melamine-formaldehyde (MF)
- Urea-formaldehyde (UF)
- Melamine-urea-formaldehyde (MUF)
- Unsaturated polyester resin
- Polyurethane resin
- Epoxy resin

However, the choice of resin depends on the following factors:

- Specification of the final product: we have to judge whether the particular resin can meet the required criteria or not.
- Processing machinery available or to be acquired because the same machinery cannot be used for all the resins.

2.8.2 Compounding of the Resin Solution

The most popular and traditional thermosetting resins are formaldehyde-based resins (PF, UF and MF). The major advantages of these resins are water solubility, self-extinguishing, fire retardant, commercially available and easy to manufacture. The major disadvantage of these resins is that they release formaldehyde during their service life. Therefore, during compounding, it is better to use a formaldehyde scavenger so that formaldehyde will react with it instantly when released.

For cost reduction, creep resistance and to impart abrasion resistance, mineral fillers must be used for compounding. Fire retardant properties are very important for the transport sector and for this, chlorinated paraffin wax, boric acid and other fire retardants are used during compounding. Silicone emulsions are used for scratch and water resistance.

2.8.3 Impregnation of Fibre

Unlike thermoplastics, water-based thermosetting resins do not need reinforcing fibres to be prepared by size reduction and drying. However thermosetting resins like unsaturated polyester, where water retards the curing reaction, need to be dried before impregnation. To date the chemical modification of fibres has not been commercially viable, though this treatment has tremendous impact on the final properties of composites. Thermosetting resin impregnation normally needs the fibre to be in the continuous form. Long fibres in the form of a woven or nonwoven web are passed through a dryer, to drive out the moisture. The dried web is then dipped in the compounded resin solution. After impregnation, the web is squeezed between rollers, to drive out excess resin and then dried *in situ* in a vertical or horizontal dryer. After drying, the web is cut to the required length and this is known as a prepreg. The impregnation process is shown in **Figure 2.1**. In the prepreg, the resin remains in the semisolid form and not fully cured. One of the important aspects of impregnation is to impart the required amount of resin in the prepreg. This is done by adjusting the dipping length, temperature of the dipping bath and dryer, viscosity of the resin solution, squeeze roll pressure and finally, by the machine speed. Prepregs are checked for resin flow and resin content.

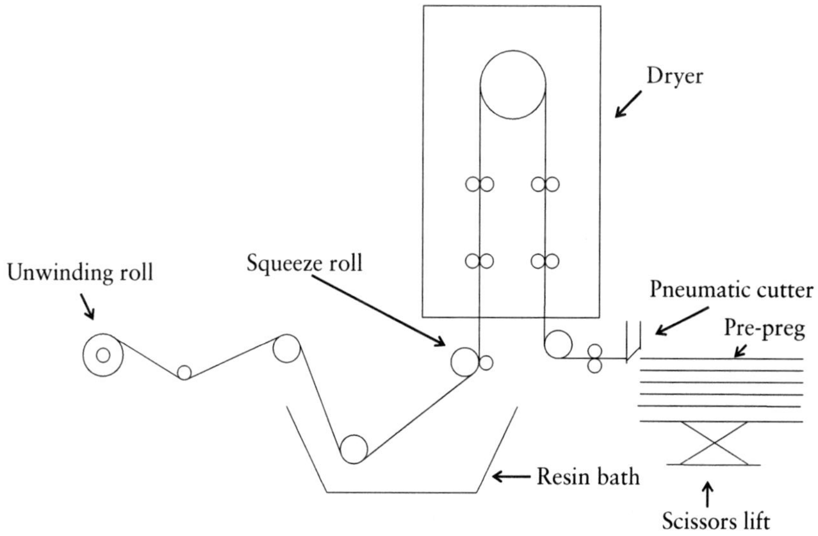

Figure 2.1 Impregnation of natural fibre web

2.8.4 Moulding

Normally jute boards are moulded in a multidaylight hydraulic press. The gap between the platen is known as 'daylight'. Between one and ten boards are moulded per daylight depending on the thickness of the board and the gap between the platen. Each board is first placed on a mother plate, known as a caul plate. All the caul plates are then simultaneously pushed pneumatically into the press and after completion of the heating and cooling cycle they are taken out from the press. Finally the boards are trimmed using a circular saw. A commercial multidaylight hydraulic press for composite moulding has a capacity of 3,000-5,000 tons, and has 5-10 daylights for 8 x 4 ft finished board. This process is also suitable for making doors, housing panels, ceilings, boards for furniture or other wood substitute products.

For intricate moulding, these prepregs are cut to fit the dimensions of the mould. Several such cut pieces are placed in the hot moulds and cured between male and female moulds under heat and pressure. After moulding, these are trimmed for finishing purposes. Different transport sector components like packing rings and autotrims, together with shoe components, and so on, are moulded by this method. A typical prepreg assembly is shown in **Figure 2.2**.

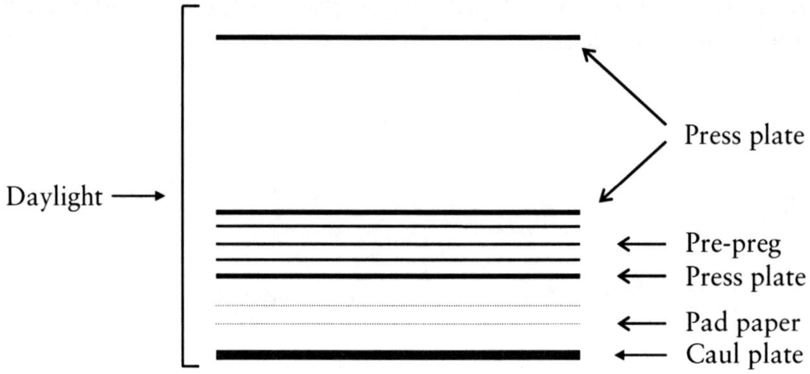

Figure 2.2 Prepreg assembly

2.8.5 Characteristic Properties

Characteristic properties of some jute composites are given below from the various experiments and plant scale trials done under the United Nations Development Programme at IJIRA, Kolkata, India.

Table 2.8 Typical physical and mechanical characteristics of jute-based decorative laminates

Characteristics	BIS 2046:1969	FC2	FC3
Bulk density (kg/m^3)	1300 - 1400	1250 – 1300	1220 – 1250
Flexural strength (MPa)	80 – 85	122.7	79.05
Flexural modulus (MPa)	-	12,540	7,233
Water absorption (%) maximum			
2 h cold soaking	-	2.01	2.19
24 h cold soaking	10.00	4-19	5.98
Thickness swelling (%) maximum			
2 h cold soaking	-	0.49	0.68
24h cold soaking	10.00	1.27	1.70
Boiling water test, 2 h			
Thickness swelling, %	10.00	4.75	8.53
Water absorption (%)	10.00	8.89	11.87

BIS: Bureau of Indian Standard
FC2: Fibre composite No.2
FC3: Fibre composite No.3
Reproduced with permission from Indian Jute Industries' Research Association, *Technical Brochure on Jute Composites and Technical Reports*, Kolkata, India, 1994-1998. ©1994-1998, IJIRA [79]

Table 2.9 Typical physical and mechanical characteristics of jute/MUF and jute/PF board

Characteristics	MUF board	PF board
Specific gravity	1.0 – 0.9	1.4 – 1.0
Tensile strength (MPa)	37 - 50	40 – 65
Tensile modulus (MPa), minimum	2500	3000
Flexural strength (MPa)	40 – 45	45 – 85
Screw withdrawal resistance, minimum (N)	200	400
Thermal conductivity, minimum (kcal cm/m^2/h/°C)	0.03	0.05 – 0.04
Water absorption (%) maximum		
2 h cold soaking	1.0 – 15.0	0.5 – 15.0
24h cold soaking	25.0 – 30.0	0.8 – 20.0
Thickness swelling (%) maximum		
2 h cold soaking	5.0 – 10.0	0.05 – 5.0
24 h cold soaking	15.0 – 20.0	0.09 – 8.0

Reproduced with permission from Indian Jute Industries' Research Association, *Technical Brochure on Jute Composites and Technical Reports*, Kolkata, India, 1994-1998. ©1994-1998, IJIRA [79]

Typical physical and mechanical characteristics of jute-based decorative laminates and boards are shown in **Tables 2.8** and **2.9**. These laminates and boards can be used for various applications and are ideal as wood substitutes.

Table 2.10 shows the physical and mechanical characteristics of a jute/polyester laminate and compared with masonite. This combination can be used for moulded panels for the transport sector.

Table 2.10 Typical physical and mechanical characteristics of jute/polyester laminate

Characteristics	Jute/polyester	Masonite
Thickness (mm)	1.5	2.5
24 h soak in cold water		
Water absorption (%)	10.50	25.45
Thickness swelling (%)	5.20	11.50
2 h in boiling water		
Water absorption (%)	13.20	50.00
Thickness swelling (%)	8.37	32.39
Tensile strength (MPa)		
Longitudinal	65.57	38.90
Transverse	27.89	26.66
Flexural strength (MPa)		
Longitudinal	85.00	50.21
Transverse	55.27	34.22

Reproduced with permission from Indian Jute Industries' Research Association, *Technical Brochure on Jute Composites and Technical Reports*, Kolkata, India, 1994-1998. ©1994-1998, IJIRA [79]

References

1. *http://www.naturalfibers2009.org*

2. W.D. Brouwer in *Common Fund for Commodities. Technical Paper Number 14. Alternative Applications for Sisal and Henequene*, Food and Agricultural Organization of the United Nations, Rome, Italy, and Common Fund for Commodities, Amsterdam, The Netherlands, 2001, p.75.

3. T.C. Ghosh in *Handbook on Jute*, Food and Agricultural Organization of the United Nations, Rome, Italy, 1983, paper 51.

4. M. Bannerjee and A.K. Roy, *Proceedings of the National Institute of Sciences of India*, 1944, **7**, 377.

5. P.K. Ray, *Journal of Applied Polymer Science*, 1969, **13**, 2593.

6. P.K. Ray, *Journal of Applied Polymer Science*, 1973, **17**, 951.

7. P.K. Ray, A.C. Chakraborty and S.B. Bandyopadhyay, *Journal of Applied Polymer Science*, 1976, **20**, 1765.

8. B. Sikdar, A.K. Mukhopadhyay and B.C. Mitra, *Indian Journal of Fibre and Textile Research*, 1993, **18**, 139.

9. L. Cheek and L. Roussel, *Textile Research Journal*, 1989, **59**, 478.

10. P.H. Hermans, *Textile Research Journal*, 1948, **18**, 9.

11. W.T. Astbury, A.G. Norman and R.D. Preston, *Nature*, 1944, **154**, 84.

12. S. Leannouth in *Fillers for Plastics*, Ed., W.C. Wake, Iliffe, London, UK, 1971, p.81.

13. W.G. Macmillan, *Textile Digest*, 1964, **25**, 169.

14. P. Ghosh in *International Seminar on Jute and Allied Fibres*, Calcutta, India, 1998, p.111.

15. P.K. Mallick in *Fiber Reinforced Composites - Materials, Manufacturing and Design*, Marcel Dekker, Inc., New York, NY, USA, 1988, Chapter 2.

16. H. Wells, D.H. Bowen, I. Macphail and P.K. Pal in *Proceedings of the 35th Annual Technology Conference*, Reinforced Plastics/Composites Institute, SPI, Section 1F, 1980.

17. Eds., R.M. Rowell, R.A. Young and J.K. Rowell in *Paper and Composites from Agro-based Resources*, CRC Lewis, New York, NY, USA, 1997, p.359.

18. R.M. Rowell, R.S. Lichtenberg and P. Larsson in *Pacific Rim Bio-based Composites Symposium*, Rotorua, New Zealand, FRI Bulletin No.176, 7-8th November, 1992.

19. R.M. Rowell, *Journal of Wood Science*, 1982, **15**, 172.

20. D.N.S. Han and W.C. Feist, *Journal of Wood Science*, 1981, **14**, 41.

21. M. Krzysik, J.A Youngquist, R.M. Muehl, R.M. Rowell, P. Chow and S.R. Shook in *Materials Interactions Relevant to Recycling of Wood-based Materials*, Eds., R.M. Rowell, T.L. Laufenberg and J.K. Rowell, Materials Research Society, Pittsburg, PA, USA, 1992, **266**, 73.

22. *Multicomponent Polymer Systems*, Eds., I.S. Miles and S. Rostami, Longman Scientific and Technical, New York, NY, USA, 1992.

23. E.L. Nielsen in *Mechanical Properties of Polymers and Composites*, Volume 2, Marcel Dekker, Inc., New York, NY, USA, 1974.

24. F.L. Matthews and R.D. Rawlings in *Composite Materials: Engineering and Science*, Chapman and Hall, London, UK, 1994.

25. B.D. Agarwal and L.J. Broutman in *Analysis and Performance of Fiber Composites*, John Wiley & Sons, Inc., New York, NY, USA, 1990.

26. D. Hull in *An Introduction to Composite Materials*, Cambridge University Press, London, UK, 1981.

27. D.G. Unger in *Perspective on Eco System Management, Symposium on Ecosystem Management*, The United States Department of Agriculture Forest Service, Burlington, VT, USA, 1994.

28. J.D. Muzzy and A.O. Kays, *Polymer Composites*, 1984, **5**, 169.

29. L.A. Goertler, *Polymer Composites*, 1983, **4**, 249.

30. B.C. Mitra, R.K. Basak and M. Sarkar, *Journal of Applied Polymer Science*, 1998, **67**, 1093.

31. S.R. Ranganathan, P.K. Pal, A.K. Rana and B.C. Mitra, inventors; assignee, Indian Jute Industries' Research Association, Kolkata; Indian Patent 170,305, 1990.

32. T.L. Richardson in *Industrial Plastics Theory and Application*, Delmer Publisher Inc., New York, NY, USA, 1989, Chapter 6.

33. *Handbook of Composites*, Ed., G. Lubin, Van Nostrand-Reinhold Co., New York, NY, USA, 1982.

34. B. Westerlind, M. Rigdahl, H. Hollmark and A. De Ruo, *Journal of Applied Polymer Science*, 1984, **29**, 175.

35. D.H. Page, F. El-Hosseiny, K. Winkler, and A.P.S. Lancaster, *TAPPI*, 1977, **60**, 114.

36. C.A. Cruz-Ramos in *Advances in Rheology*, Volume 3, Eds., B. Melia, A. Garcia-Rejon and C. Rangel-Nafaile, Elsevier, New York, NY, USA, 1985, p.733.

37. A.A. Berlin, S.A. Volfson, N.S. Emikolopian and S.S. Negmatov in *Principles of Polymer Composites*, Eds., G. Henrici-Olive and S. Olive, Springer-Verlag, Berlin, Germany, 1986, p.60.

38. R.G. Weatherhead in *FRP Technology:Fiber Reinforced Resin Systems*, Applied Science Publishers Ltd., London, UK, 1980, Chapter 4.

39. J.A. Brydson in *Plastics Materials*, 6th Edition, Butterworth Heinemann, London, UK, 1995, p.122.

40. A.G. Winfield, *Plastics and Rubber International*, 1979, **4**, 23.

41. K. Verma, S.R. Ananthakrishnan and S. Krishnamoorthy, *Composites*, 1989, **20**, 383.

42. J.T. Hoggat, S. Oken and E.E. House, *US Air Force Report AFWAL-TR- 80-3023*, April, 1980.

43. R.R. Atkinson in *Jute - Fiber to Yarn*, Temple Press Books Ltd., London, UK, 1964, p.31.

44. *International Encyclopedia of Composites*, Ed., S.M. Lee, Volume 4, VCH Publishers, New York, NY, USA, 1990.

45. *Encyclopedia of Polymer Science and Technology*, Eds., H.F. Mark, N.G. Gaylord and N.M. Bikales, Volume 11, Interscience Publishers, John Wiley and Sons. Inc., New York, NY, USA, 1969.

46. J. Stepek and H. Daoust in *Additives for Plastics: Polymers*, Springer-Verlag, New York, NY, USA, 1983, Chapter 4.

47. G. Samay, T. Nagy and J.L. White, *Journal of Applied Polymer Science*, 1995, **56**, 1423.

48. F. Hyndryckx, P.H. Dubois, M. Patin, R. Jerome, P.H. Teyssie and M. Garcia Marti, *Journal of Applied Polymer Science*, 1995, **56**, 1093.

49. P. Hedenberg and P. Gatenholm, *Journal of Applied Polymer Science*, 1995, **56**, 641.

50. A.R. Sanadi, C. Clemons, R.M. Rowell and R.A. Young, *Journal of Reinforced Plastics and Composites*, 1994, **13**, 54.

51. D. Basu, A.N. Banerjee and A. Misra, *Journal of Applied Polymer Science*, 1992, **46**, 1999.

52. O. Owolabi and T. Czvikovszky, *Journal of Applied Polymer Science*, 1988, **35**, 573.

53. A. Biswas, A.J. Hammond, N.L. Hancox and A. Rana, *Composites Manufacturing*, 1993, **4**, 53.

54. J.D. Ferry in *Viscoelastic Properties of Polymers*, Wiley, New York, NY, USA, 1961.

55. T. Murayama in *Dynamic Mechanical Analysis of Polymeric Materials*, 2nd Edition, Elsevier, Amsterdam, The Netherlands, 1978.

56. B.E. Read and G.D. Dean in *The Determination of Dynamic Properties of Polymers and Composites*, Adam Hilger, Bristol, UK, 1978.

57. L.E. Neilsen in *Mechanical Properties of Polymers and Composites*, Marcel Dekker, New York, NY, USA, 1975.

58. S.W. Tsai and H.T. Halpin in *Introduction to Composite Materials*, Technomic, Westport, CT, USA, 1980.

59. R.D. Adams and D.F. Short, *Journal of Physics D: Applied Physics*, 1973, **6**, 1032.

60. F. Kolarik, F. Lednicky and B Pukanszky, *International Conference on Composite Materials*, Volume 6, London, UK, 1987, p.452.

61. R.D. Adams in *Engineered Materials Handbook*, Volume 1, ASTM International, Metals Park, OH, USA, 1987, p.207.

62. D.F. Adams and D.R. Doner, *Journal of Composite Materials*, 1967, **1**, 4.

63. A.K. Rana, A. Mandal, B.C. Mitra, R. Jacobson, R. Rowell and A.N. Banerjee, *Journal of Applied Polymer Science*, 1998, **69**, 329.

64. A.K. Rana, A. Mandal and S. Bandyopadhyay, *Composites Science and Technology*, 2003, **63**, 801.

65. *Modern Plastics Encyclopaedia*, McGraw Hill, New York, NY, USA, 1993.

66. R.M. Rowell, A.M. Tillman and R. Simonson, *Journal of Wood Chemistry and Technology*, 1986, **6**, 427.

67. A.R. Sanadi, D.F. Caulfield, R.E. Jacobson and R.M. Rowell, *Industrial and Engineering Chemistry Research*, 1995, **34**, 1889.

68. L. Mascia in *The Role of Additives in Plastics*, Edward Arnold Ltd, London, UK, 1974, Chapter 3.

69. M.J. Folkes in *Short Fiber Reinforced Thermoplastics*, Ed., M.J. Devis, John Wiley and Sons Ltd., Hertfordshire, UK, 1982.

70. Y.P. Khanna, W.M. Wenner, R. Kumar and S. Kavesh, *Journal of Applied Polymer Science*, 1989, **38**,571.

71. A.K. Rana, B.C. Mitra and A.N. Bannerjee, *Journal of Applied Polymer Science*, 1999, **71**, 531.

72. N. Saha, A.N. Bannerjee and B.C. Mitra, *Journal of Applied Polymer Science*, 1996, **60**, 657.

73. N. Saha, and A.N. Banneijee, *Journal of Applied Polymer Science*, 1996, **62**, 1199.

74. S. L. Rosen in *Fundamental Principles of Polymeric Materials*, John Wiley & Sons, New York, NY, USA, 1982.

75. E. Alfthan, A de Ruvo and W. Brown, *Polymer*, 1973, **14**, 329.

76. J.M. Felix and P. Gatenholm, *Journal of Applied Polymer Science*, 1991, **42**, 609.

77. R.G. Raj, B.V. Kokta, D. Maldas and C. Daneault, *Journal of Applied Polymer Science*, 1989, **37**, 1089.

78. D.J. Olsen in the *Proceedings of the ANTEC '91*, Montreal, Canada, 1991, p.1886

79. Indian Jute Industries' Research Association, *Technical Brochure on Jute Composites and Technical Reports*, Kolkata, India, 1994-1998.

3 Cellulose-based Composite Systems for Biomedical Applications

Ana Baptista, Isabel Ferreira and João Paulo Borges

3.1 Cellulose-based Composites

Composite materials made from glass fibres or carbon fibres embedded into epoxy resins or unsaturated polyester typically exhibit excellent mechanical and thermal properties and are used in many applications, for instance in the aerospace and automotive fields. However, these materials are incinerated at the end of their life cycle, causing significant environmental issues [1]. Consequently, the increasing environmental awareness and ecological concerns have resulted in a renewed interest in natural-based and compostable materials, and therefore issues such as biodegradability and environmental safety are becoming important. Indeed, the concept of biocomposites made from cellulose-based feedstock appears to be an alternative route to achieve 'green' polymer composites.

Amongst the advantages of using cellulose in polymer composites, renewability, cheapness, high specific strength and modulus are the most important [2]. Taking account of these properties, the present chapter gives a widespread overview of the potential applications of cellulose-based composites in the medical field.

3.1.1 Biocomposites

In the materials science field, a composite is commonly defined as a material composed of two or more phases: a matrix (or a continuous phase) and at least one dispersed phase. The continuous phase is responsible for the main structure of the composite and acts as a support for the dispersed material(s). The dispersed phase is usually responsible for enhancing one or more properties of the matrix. Indeed, most composites show an enhancement of the mechanical properties of the matrix, such as stiffness and strength; however, other properties, such as thermal or electrical properties, density or bioactivity might also be improved. As a result, this synergism allows the preparation of materials with properties that are not exhibited by the individual constituent materials. However, there are some composites in which the dispersed phase and the matrix are composed of the same material [3].

Within the scope of this chapter, biocomposites are regarded as the combination of a dispersed phase (e.g., fibres, particles or biomolecules) with a polymer matrix in which at least one of these components is from a renewable source.

The recent developments in biocomposites research make it possible to foresee the transition from petroleum-based polymers (e.g., polyethylene (PE), polypropylene (PP)) to naturally derived biopolymers (e.g., cellulose and starch) [4]. As a result, natural fibres for composites are emerging as a feasible alternative to glass fibre, particularly in the automotive [5], packaging [6] and building fields [7].

Natural fibres can be classified according to their origin: bast fibres (jute, flax, and hemp), leaf fibres (sisal and pineapple), seed fibres (cotton and coir), and other types which include wood and roots. Besides their abundance in nature, natural fibres have many advantages such as low weight, cheapness, renewability, and they exhibit good mechanical properties.

Polymer composites reinforced by natural fibres have performances which are highly dependent on the chemical composition, structure, and physical and mechanical properties of the dispersed phase. **Table 3.1** shows the main composition and relevant mechanical properties of the most common plant fibres used for composite reinforcement.

Considering that the cellulose content (wt%) can differ from plant to plant, and even within different parts of the same plant, it is expected that the mechanical properties of these fibres increase with increasing cellulose content.

The basic structural unit of the plant cell wall is the cellulose microfibril. In most natural fibres these microfibrils orient themselves at an angle to the fibre axis called the microfibrillar angle. Plant fibres are more ductile if the microfibrils have a spiral orientation relative to the fibre axis. If the microfibrils are oriented parallel to the fibre axis the fibres will be more rigid, inflexible and have higher tensile strength. Consequently, the microfibrillar angle is a parameter that can define the stiffness of the fibres.

In a biocomposite the main function of the dispersed phase is to carry structural loads thereby improving the specific properties, whilst the shape, the surface appearance, environmental tolerance and durability are provided by the matrix, which can be from either renewable or non-renewable resources. Polyolefin thermoplastics such as PP and PE are examples of matrices used for structural applications. However, efforts have been made to replace those matrices by biopolymers such as starch, poly(lactic acid) (PLA) and polyhydroxyalkanoates [4].

Table 3.1 The main composition and relevant mechanical properties of the most common cellulose-based fibres

Plant fibre	Cellulose (wt%)	Hemicellulose (wt%)	Lignin (wt%)	Tensile strength (MPa)	Young's modulus (GPa)
Bamboo	26-43	30	21-31	140-230	11-17
Abaca	56-63	20-25	7-9	400	12
Hemp	68	15	10	690	-*
Flax	71	18-21	2	345-1035	28
Jute	61-71	14-20	12-13	393-773	27
Sisal	65	12	10	511-635	9-22

Adapted from O. Faruk, A.K. Bledzki, H-P. Fink and M. Sain, *Progress in Polymer Science*, 2012, 37, 1552 [4] and S.J. Eichhorn, *Journal of Materials Science*, 2001, **36**, 2107 [8]
*Not measured'

3.1.2 Cellulose

Cellulose-based products have been traditionally utilised by our society. However, more recently, cellulose has attracted considerable attention as one of the most well-known renewable and sustainable raw materials for obtaining environmentally friendly and biocompatible technological products. Biomass produced by photosynthetic organisms such as plants, algae, and some bacteria is made up of cellulose and for that reason it is the most abundant biopolymer on earth [9]. Therefore, the use of cellulose-based materials in composites has increased over the last few years because of their relative cheapness compared to conventional materials, their recyclability, and their mechanical properties.

3.1.2.1 Sources of Cellulose as a Raw Material

Wood pulp remains the most important raw material source for the processing of cellulose. The structure of wood is highly complex due to the presence of lignin, a three-dimensional (3D) polymer network that binds to carbohydrates (hemicellulose and cellulose) to form a tight and compact structure from which cellulose is isolated by large-scale chemical pulping, separation and purification processes [9]. Other sources of cellulose have been studied including cotton, potato tubers, sugar beet pulp, soybean stock and banana rachis where almost pure cellulose is available [10, 11].

Apart from plants, cellulose can also be produced by certain bacteria, algae and fungi. The cellulose obtained from algal species has a porous or sponge-like structure, which is substantially different from that obtained from plants. Bacteria, under special culturing conditions, can also produce a thick gel composed of cellulose microfibrils and water [12]. Because of their small structure, crystallinity, and reactivity these cellulose forms have attracted much research interest for the development of new materials and biomaterials.

3.1.2.2 Structure

Cellulose can be described as a polydisperse linear homopolymer composed of D-glucopyranose units linked with a β-1,4-glycosidic bond. The polymer chain contains free hydroxyl groups (-OH) at the C-2, C-3, and C-6 atoms. Based on the -OH groups and the oxygen atoms of both the pyranose ring and the glycosidic bond, ordered hydrogen bond networks can be found (**Figure 3.1**).

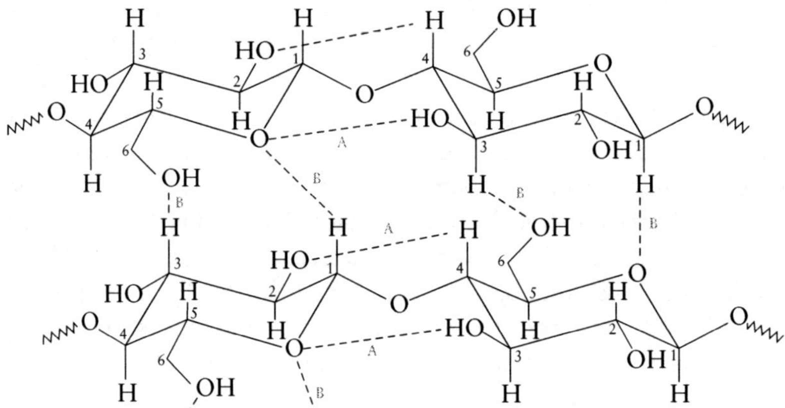

Figure 3.1 The structure and intra- (A), and interchain (B) hydrogen bonding patterns in cellulose

3.1.2.3 Cellulose Derivatives

Cellulose can be chemically modified to yield different derivatives. The hydrogen bonding patterns in cellulose are considered to be the main factor which determines its physical and chemical properties. The solubility, crystallinity and hydroxyl reactivity can be directly affected by intra- and intermolecular bond formation (**Figure 3.1**) [13]. Cellulose derivatives have been designed and fine-tuned to obtain certain desired properties and the chemical functionalisation of cellulose is done by changing the inherent hydrogen bonding network and by introducing different substituents. Indeed, the properties of cellulose derivatives are mainly determined by the choice of substituent group and the degree of substitution.

3.2 Applications of Cellulose-based Composites

Cellulose-based composites can be found in different fields of application, including automotive applications [14, 15], building [16], packaging [17] and medicine [18-23].

Biological tissues are essentially composite materials with particular mechanical properties that should be carefully considered during the design of innovative biomedical scaffolds made of composite biomaterials. From this point of view, cellulose-based composites have received special attention as suitable and inexpensive alternatives for a wide range of medical applications, including scaffolds for tissue engineering, wound healing and healthcare systems (**Table 3.2**). For that reason, the following subsection will give a detailed overview of the recent outcomes using cellulose-based composites for medical applications.

3.2.1 Medical Applications

Microbial cellulose has recently been used in the production of composites for the development of tissue engineered constructs due to its unique nanostructure that closely resembles the structure of native extracellular matrices (ECM) [18, 19].

Different approaches are found in the literature concerning the use of composites with microbial cellulose for medical applications. Wan and co-workers [18] report the development of a nanocomposite made of HA and bacterial cellulose (BC). HA is one of the bioceramics (biocompatible ceramics) which is frequently used for bone and dental tissue reconstitution. It has excellent biocompatibility with hard tissues, and high osteoconductivity and bioactivity. Many studies have been described using HA for tissue regeneration due to its ability to mimic the basic composition of bone. However, ceramics are by nature fragile materials and are therefore limited to nonstructural applications. By incorporating HA into BC, it will be possible to design a new HA/BC composite with high mechanical properties whilst keeping the good osteoconductivity and biodegradation of HA leading to a material with improved performance for tissue engineering and orthopaedic surgery.

Table 3.2 Summary of recent research on cellulose-based composites for medical applications

Biocomposite	Cellulose source	Dispersed phase	Applications	Relevant properties	Reference
Hydroxyapatite (HA)/chitosan/ carboxymethyl cellulose	Commercial	HA	Tissue engineering	Biological activity	[24]
Poly(3-hydroxubutyrate-co-4-hydroxubutyrate)/ cellulose	Microbial	Cellulose fibres	Tissue engineering	Improvement of mechanical properties (TS = 46MPa and E = 0.88 GPa)	[25]
Nanocrystalline cellulose (NCC)/ polyvinyl alcohol (PVA)	Commercial	Rod-shaped NCC	Tissue regeneration	Improvement of thermal properties	[26]
HA/cellulose	Microbial	HA	Bone tissue engineering	Biological activity	[27]
HA/cellulose	Microbial	HA	Cartilage and bone tissue engineering	Biological activity and improvement of mechanical properties	[18]
Cellulose/PVA	Microbial	Cellulose fibres	Tissue regeneration	Improvement of mechanical properties (TS = 0.2-0.6 MPa)	[19]
All-cellulose composite	Commercial	Cellulose nanowhiskers	Small grafts	Improvement of mechanical properties (TS = 11 MPa)	[28]
All-cellulose composite	Norway spruce	Cellulose fibres	Ligament or tendon substitute	Improvement of mechanical properties (TS= 25-30 MPa)	[21]
Laponite/cellulose	Microbial	Laponite (clay)	Drug delivery	Improvement of electrical properties and thermal stability	[29]

All-cellulose composite	Cotton linter pulp	Cellulose nanowhiskers	Drug delivery	Improvement of mechanical properties	[22]
HA/cellulose	Microbial	HA particles	Drug delivery	Biological activity	[30]
Cellulose acetate/ silver (Ag)	Commercial	Ag	Electrode for healthcare systems	Improvement of electrical properties	[31]
Cellulose/ montmorillonite (MMT)	Microbial	MMT (clay)	Wound healing	Improvement of mechanical properties (TS = 209.6 MPa), thermal stability and antibacterial activity	[20]
Cellulose acetate/ Ag	Commercial	Ag nanoparticles	Wound healing	Antibacterial activity	[32]
Lysostaphin (Lst)/ cellulose-based membrane	Commercial	Lst (enzyme)	Wound healing	Antibacterial activity	[23]
E = Young's modulus TS = Tensile strength					

The replacement of soft tissue is also an important area in which new materials would be very valuable. For instance, efforts have been made to create a biomaterial with properties which are adjusted for specific cardiovascular applications. One of the major difficulties in this application is to replace heart valves since their mechanical properties are quite different from those of natural ones. Indeed, tissue fatigue can be reached due to the inability of synthetic heart valves to stretch and relax with the aortic wall during the cardiac cycle. Common cardiovascular tissues are composite materials composed mainly of elastin and collagen, where elastin provides the initial elasticity and the collagen fibres contribute to the tissue stiffness. In view of this, Millom and Wan [19] propose an original concept composite made of BC and PVA to mimic the role of collagen and elastin, respectively, for possible use as a heart valve replacement. PVA when crosslinked by a low temperature thermal cycling process is one of the few materials that exhibit a stress-strain relationship similar to that of soft tissues, such as cardiovascular tissue. The combination of this property with the high elastic modulus and degree of crystallinity characteristic of BC fibres makes it possible to design a composite structure with mechanical properties similar to those expected for heart valves. This work has resulted in one type of PVA/BC composite that has tensile properties which are similar to those of a porcine aorta. Millom and

Wan have also found that specific composition and processing parameters should be chosen to create a custom-designed PVA-BC nanocomposite biomaterial in order to achieve the appropriate mechanical properties of the tissue to be regenerated.

Wound healing is another application where biocomposites can have a prominent role. Recently, Ul-Islam and co-workers [20] have opened up the possibility of incorporating clays with good antibacterial activity into a BC porous structure. MMT is a well-known clay that can be found in a wide range of applications in the medical field, for instance for cleaning and protection of the skin, and wound healing. Ul-Islam and co-workers have reported the preparation of BC-MMT nanocomposites through simple particle impregnation. This process consists of the attachment of MMT nanoparticles onto the surface of BC fibres followed by their penetration into the empty spaces of the BC matrix, resulting in a composite structure with improved mechanical properties and antibacterial activity. This work proved that it was possible to enhance the physico-mechanical properties of BC and to create an original composite with excellent properties for wound healing applications.

All-cellulose composites have also been introduced to the medical field. The dispersed phase and matrix used in these composites are made from the same material which brings benefits, such as recyclability and better interfacial adhesion.

An attempt to develop artificial ligaments and tendons with mechanical properties similar to, or better than, the natural ligaments or tendons using all-cellulose composites has been described by Mathew and co-workers [21]. In the human body the main function of the tendon is to transfer the force of the muscle contraction to the bones whereas ligaments stabilise the joints, preventing abnormal movements. In their work, Mathew and co-workers have proposed the preparation of a fibrous nanocomposite by a partial dissolution method. Firstly, a nanofibre network is produced by a mechanical fibrillation process and then, using an ionic liquid, the cellulose fibrous structure is partially dissolved. This method produces homogeneous and uniform structures leading to an enhancement of fibre-matrix interaction and, consequently, better mechanical properties.

The proposed all-cellulose nanocomposite demonstrates an excellent cytocompatibility and the ability to successfully support the adhesion and growth of ligament derived-fibroblasts which are fundamental requirements for their potential application as artificial ligaments.

Another approach reported by Wang and Cheng [22] proposed a suitable method to prepare nanowhisker-reinforced all-cellulose composite gels to develop high performance porous materials with potential biomedical applications, for instance as drug delivery systems. Cellulose nanowhiskers can develop hydrogen bonds, not only between themselves, but also with the host polymer matrix, creating a porous network. Indeed, the authors describe a thermally induced phase separation as the most appropriate and rapid method to prepare physically irreversible crosslinked cellulose nanowhisker/cellulose nanocomposite gels. Here, the cellulose nanowhiskers act as a 'bridge' to facilitate the crosslinking of cellulose chains, providing a positive reinforcement for the gel network, resulting in a nanocomposite with improved mechanical properties. Moreover, by analysing the controlled release of a model molecule such as bovine serum albumin from these nanocomposites, these authors demonstrated their great potential as drug delivery systems.

Nanostructured composite materials can also be obtained using electrospun cellulose-based membranes. Electrospinning is a simple and versatile technique for electrostatic fibre formation which utilises electrical forces to produce fibres with diameters ranging from 2 nm to several micrometres using polymer solutions from a vast range of materials, such as PLA [33], polyurethanes [34], silk fibroin [35], collagen [36], cellulose and its derivatives [37, 38], composites [39], and ceramics [40].

Nanofibrous membranes for biomedical applications often need to be functionalised to enhance their surface physicochemical properties, mechanical durability, biocompatibility, and cellular response. Electrospun nanofibres have an extremely large specific surface area and so can form 3D porous structures which can mimic natural ECM, thus providing a useful option for tissue regeneration, drug delivery, and wound dressings.

Beyond the improvement of the mechanical properties, electrospun composite fibres can include additional properties, for instance antibacterial activity. The incorporation of silver nanoparticles [32] or polypeptides [23] to kill bacterial pathogens into electrospun matrices has been reported. Son and co-workers [32] have described the production of electrospun nanocomposite membranes using cellulose acetate/silver nitrate solution. After membrane preparation, a rapid photoreduction of Ag^+ ions within the nanofibres was carried out by ultraviolet irradiation. The resulting membrane exhibited a strong antimicrobial activity which is necessary when considering wound dressing applications.

Miao [23] and co-workers reported the preparation of different electrospun cellulose-based membranes (cellulose, cellulose-chitosan, and cellulose-polymethyl methacrylate) followed by surface modification to generate enzyme-reactive fibres. Lst is an enzyme with specific bactericidal activity against *Staphylococcus aureus*, one of the most recognised bacteria responsible for hospital-acquired infections. The authors described the immobilisation of Lst onto electrospun fibres surface demonstrating a stable bioconjugation between them and these modified fibres showed high antimicrobial activity and biocompatibility.

The development of electrospun composite fibres opens new paths for the creation of novel, lightweight and flexible products. Regarding the powering of microelectromechanical systems, the battery system becomes a limiting factor for the lifetime and applicability of many biosensors. Even if some conventional batteries are biocompatible and have long lifetime, they will eventually require replacement or recharging. For short term applications, the conventional battery may provide a sufficient lifespan for a biomedical device, but for long term applications, it may be preferable to replace these batteries with alternative power sources, especially if the substitution or recharge procedure involves invasive surgery. For instance, a pacemaker is a common implantable system which requires an autonomous power source that functions completely independently of the outside world. Further development of this technology may be able to eliminate the costly and invasive surgery required to maintain the pacemaker, both decreasing medical costs and improving the quality of care for the patient.

Indeed, a new class of energy supply systems which is compatible with biomedical implantable devices is needed. In this context, our research team has recently proposed the development of a bio-battery (**Figures 3.2a** and **b**) comprising an ultrathin monolithic structure of an electrospun cellulose acetate (CA) membrane (**Figure 3.2c (1)**), over which thin metal film electrodes are deposited by thermal evaporation onto both surfaces (**Figure 3.2c (2)**) [31]. In fact, the polymeric matrix has an important role since it works not only as a separator but also as the support for the electrodes. When deposited, the metallic layers do not form a continuous film over the membrane surface. Instead, the metallic layers completely cover the fibres, thereby providing the required conductive properties but also allowing the preservation of the main properties of the membrane (light weight, flexibility, porosity, and large surface area).

Considering that harvesting energy directly from the environment is probably the most effective and promising approach for powering long-term biomedical devices, the proposed bio-battery can take advantage of the ion content of biological fluids, such as blood. This concept has been shown to work by using a Ag/cellulose acetate/aluminium nanofibre structure which demonstrated the ability to generate electrical energy from simulated body fluid showing a power density of 3 $\mu W/cm^2$. Considering that the typical power required for a pacemaker operation is around 1 μW, these results are quite promising [31]. Besides the supplying of low power consumption devices, biochemical monitoring systems and artificial mechanisms for human muscle stimulation can also be foreseen as potential fields of application for these kinds of implantable power sources.

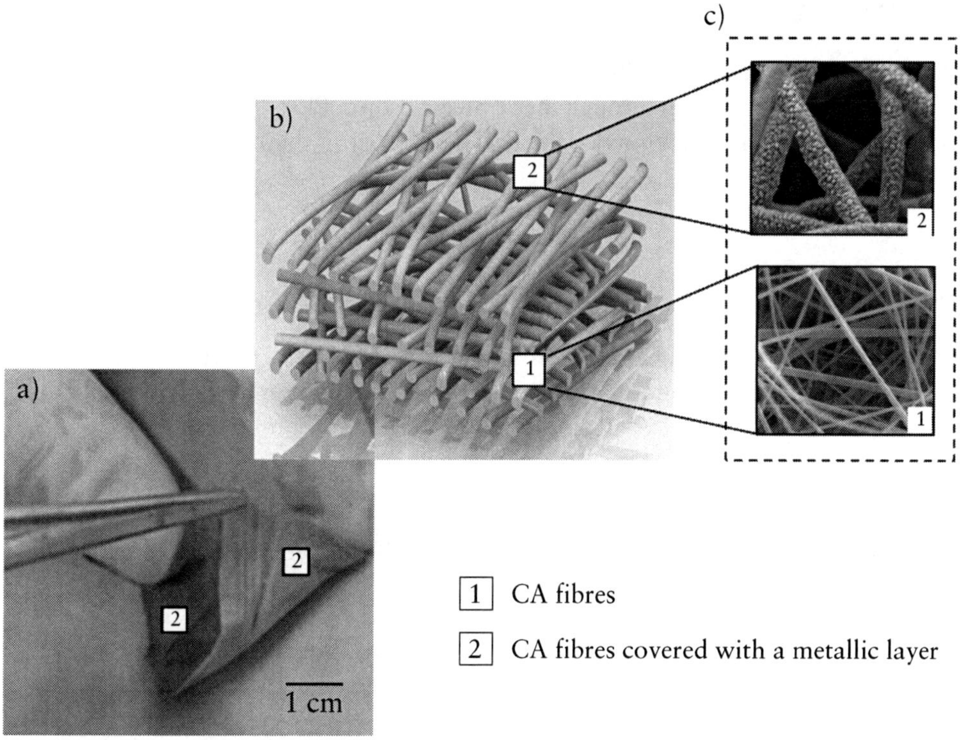

Figure 3.2 (a) Photograph of the flexible bio-battery; (b) schematic drawing of the bio-battery; and c) scanning electron microscope images of: (1) electrospun CA fibres and (2) electrospun CA fibres covered with a metallic layer.

3.3 Conclusions

Increasing environmental awareness and ecological concerns have renewed the interest in natural-based and compostable materials, and therefore issues such as biodegradability and environmental safety are becoming important. The concept of biocomposites made from cellulose-based feedstock appears to be an alternative route to achieve green polymer composites. From tissue engineering to biodevices, cellulose-based composites can be found in the most inspiring and challenging developments in the medical field.

Acknowledgements

The authors' work was partially supported by the Portuguese Science and Technology Foundation (FCT-MCTES) through the Strategic Project PEst-C/CTM/LA0025/2013-14. Ana Baptista also acknowledges FCT-MCTES for the doctoral grant SFRH/BD/69306/2010.

References

1. B.M. Cherian, A.L. Leão, S.F. de Souza, L.M.M. Costa, G.M. de Olyveira, M. Kottaisamy, E.R. Nagarajan and S. Thomas, *Carbohydrate Polymers*, 2011, **86**, 1790.

2. T. Nishino, I. Matsuda and K. Hirao, *Macromolecules*, 2004, **37**, 7683.

3. T. Huber, J. Müssig, O. Curnow, S. Pang, S. Bickerton and M.P. Staiger, *Journal of Materials Science*, 2011, **47**, 1171.

4. O. Faruk, A.K. Bledzki, H-P. Fink and M. Sain, *Progress in Polymer Science*, 2012, **37**, 1552.

5. J. Holbery and D. Houston, *Journal of the Minerals, Metals and Materials Society*, 2006, **58**, 80.

6. J. George, M.S. Sreekala and S. Thomas, *Polymer Engineering and Science*, 2001, **41**, 1471.

7. J. Reis, *Construction and Building Materials*, 2006, **20**, 673.

8. S.J. Eichhorn, *Journal of Materials Science*, 2001, **36**, 2107.

9. D. Klemm, B. Heublein, H.P. Fink and A. Bohn, *Angewandte Chemie*, 2005, **44**, 3358.

10. A. Dufresne, J-Y. Cavaillé and M.R. Vignon, *Journal of Applied Polymer Science*, 1996, **64**, 1185.

11. R. Zuluaga, J-L. Putaux, A. Restrepo, I. Mondragon and P. Gañán, *Cellulose*, 2007, **14**, 585.

12. M. Szymanska-Chargot, J. Cybulska and A. Zdunek, *Sensors*, 2011, **11**, 5543.

13. T. Kondo in *Polysaccharides: Structural Diversity and Functional Versatility*, Ed., S. Dumitriu, Marcel Dekker, New York, NY, USA, 2005, p.69.

14. A. Ashori, *Bioresource Technology*, 2008, **99**, 4661.

15. C. Alves, P.M.C. Ferrão, A.J. Silva, L.G. Reis, M. Freitas, L.B. Rodrigues and D.E. Alves, *Journal of Cleaner Production*, 2010, **18**, 313.

16. R. MacVicar, L.M. Matuana and J.J. Balatineez, *Cement and Concrete Composites*, 1999, **21**, 189.

17. A. Espert, F. Vilaplana and S. Karlsson, *Composites Part A: Applied Science and Manufacturing*, 2004, **35**, 1267.

18. Y.Z. Wan, Y. Huang, C.D. Yuan, S. Raman, Y. Zhu, H.J. Jiang, F. He and C. Gao, *Materials Science and Engineering: C*, 2007, **27**, 855.

19. L.E. Millom and W.K. Wan, *Journal of Biomedical Materials Research Part B: Applied Biomaterials*, 2006, **79**, 245.

20. M. Ul-Islam, T. Khan and J.K. Park, *Carbohydrate Polymers*, 2012, **89**, 1189.

21. A.P. Mathew, K. Oksman, D. Pierron and M-F. Harmand, *Carbohydrate Polymers*, 2012, **87**, 2291.

22. Y. Wang and L. Chen, *Carbohydrate Polymers*, 2011, **83**, 1937.

23. J. Miao, R.C. Pangule, E.E. Paskaleva, E.E. Hwang, R.S. Kane, R.J. Linhardt and J.S. Dordick, *Biomaterials*, 2011, **32**, 9557

24. L. Jiang, Y. Li, X. Wang, L. Zhang, J. Wen and M. Gong, *Carbohydrate Polymers*, 2008, **74**, 680.

25. C. Zhijiang, H. Chengwei and Y. Guang, *Carbohydrate Polymers*, 2012, **87**, 1073.

26. W. Li, J. Yue and S. Liu, *Ultrasonics Sonochemistry*, 2012, **19**, 479.

27. L. Hong, Y.L. Wang, S.R. Jia, Y. Huang, C. Gao and Y.Z. Wan, *Materials Letters*, 2006, **60**, 1710.

28. P. Pooyan, R. Tannenbaum and H. Garmestani, *Journal of the Mechanical Behavior of Biomedical Materials*, 2012, **7**, 50.

29. G.F. Perotti, H.S. Barud, Y. Messaddeq, S.J.L. Ribeiro and V.R.L. Constantino, *Polymer*, 2011, **52**, 157.

30. C. Tsioptsias and C. Panayiotou, *Carbohydrate Polymers*, 2008, **74**, 99.

31. A.C. Baptista, J.I. Martins, E. Fortunato, R. Martins, J.P. Borges and I. Ferreira, *Biosens Bioelectron*, 2011, **26**, 2742.

32. W.K. Son, J.H. Youk and W.H. Park, *Carbohydrate Polymers*, 2006, **65**, 430.

33. F. Yang, R. Murugan, S. Wang and S. Ramakrishna, *Biomaterials*, 2005, **26**, 2603.

34. H. Zhuo, J. Hu, S. Chen and L. Yeung, *Journal of Applied Polymer Science*, 2008, **109**, 406.

35. B-M. Min, G. Lee, S.H. Kim, Y.S. Nam, T.S. Lee and W.H. Park, *Biomaterials*, 2004, **25**, 1289.

36. L. Buttafoco, N.G. Kolkman, P. Engbers-Buijtenhuijs, A.A. Poot, P.J. Dijkstra, I. Vermes and J. Feijen, *Biomaterials*, 2006, **27**, 724.

37. J.P. Canejo, J.P. Borges, M.H. Godinho, P. Brogueira, P.I.C. Teixeira and E.M. Terentjev, *Advanced Materials*, 2008, **20**, 4821.

38. C-W. Kim, D-S. Kim, S-Y. Kang, M. Marquez and Y.L. Joo, *Polymer*, 2006, **47**, 5097.

39. A.F.R. Pimenta, A.C. Baptista, T. Carvalho, P. Brogueira, N.M.T. Lourenço, C.A.M. Afonso, S. Barreiros, P. Vidinha and J.P. Borges, *Materials Letters*, 2012, **83**, 161.

40. P.Q. Franco, C.F.C. João, J.C. Silva and J.P. Borges, *Materials Letters*, 2012, **67**, 233.

4 Primary and Secondary Processing of Biocomposites

Inderdeep Singh, Pramendra Kumar Bajpai and Vikas Dhawan

4.1 Introduction

Composite materials have been developed in response to the need for man-made materials with unusual combinations of properties which are not actually available in nature, such as, high specific strength and stiffness in one direction or a combination of high strength and toughness which are not available from any single material. Polymer matrix composites (PMC) are the most common and widely used composite materials. PMC offer several advantages over traditional metals and alloys, such as high strength to weight ratio, easy processing and the ability to produce near net shape products. Synthetic fibre-based polymer composites incorporate synthetic fibres (such as glass, carbon and aramid) as reinforcement in a thermoset or thermoplastic polymer matrix. Recently, PMC are emerging as viable alternatives to traditional materials in a wide range of fields such as automobile, sports, marine and aerospace applications. Although, these composites possess superior structural properties, their main drawback is that they are not biodegradable and therefore, these are a major threat to ecological systems. The synthetic fibres such as glass and carbon pose various health hazards and risks to human life. Requirements for lighter weight products and strict environmental rules and regulations have led to the development of natural fibre-reinforced biocomposites.

Natural fibre-reinforced biocomposites are light in weight, possess superior mechanical properties and, most importantly, these are environmentally friendly. These can be engineered according to the specific requirements and can be used for many different applications. The light weight nature of natural fibre-polymer composites has encouraged their use in the automobile industry because weight has a direct influence on the fuel efficiency of vehicles. Biocomposites have proved to be competitive or even better alternatives than synthetic fibre-reinforced composites. Biocomposites are either partially or fully biodegradable composites. In the case of partially biodegradable composites, natural fibres are used to reinforce petroleum-based synthetic polymers. These petroleum-based synthetic polymers are non-biodegradable in nature. Fully biodegradable composites use natural fibres to reinforce biodegradable polymers. These composites are also known as 'green' composites. Hybrid biocomposites have been developed to meet specific requirements by reinforcing with a combination of natural fibres in polymer matrix which may be a petroleum-based (thermoset or thermoplastic), or a biopolymer, or a combination of both [1]. Some examples of the constituents of biocomposites are shown in **Figure 4.1**.

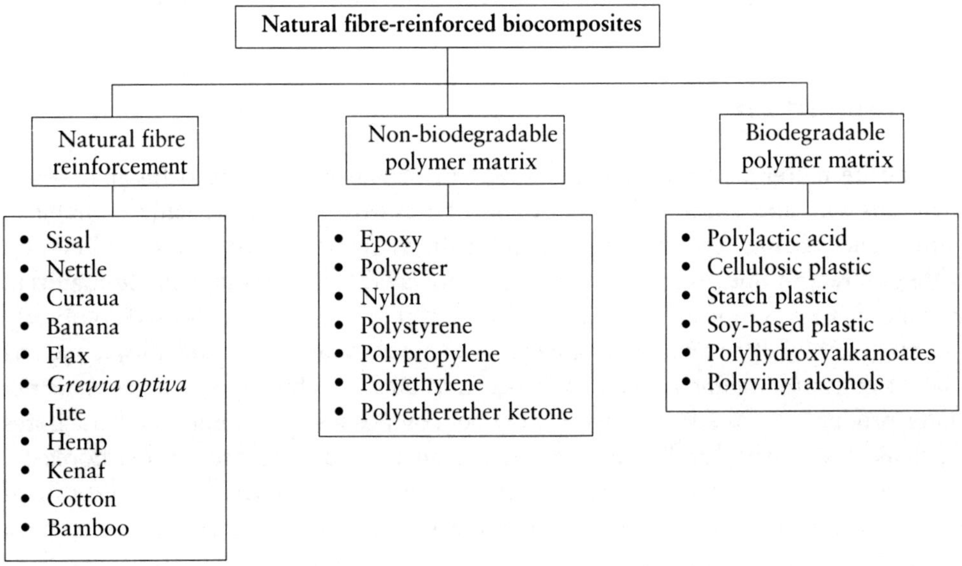

Figure 4.1 Constituents of biocomposites

The use of natural fibres as reinforcements in polymer composites has increased due to the enormous advantages associated with natural fibres such as low cost, their recyclability, light weight, ease of availability, their abundance in natural resources and good mechanical properties.

In view of the numerous advantages of biocomposites, it becomes necessary to focus on various aspects of the primary processing methods which are required to fabricate high performance biocomposites. Primary processing is the essential and the first step (after raw material preparation) to develop biocomposites and the suitability of the processes used determines the properties and performance of the final products. The following section provides a critical discussion of the various primary processing techniques used for the development of biocomposites, and the methods used to enhance their properties.

4.2 Primary Processing of Biocomposites: Challenges

Although biocomposites possess numerous favourable properties such as light weight, biodegradability, corrosion resistance, integral surface finish and good mechanical properties, it is important to understand the challenges which are associated with processing them. Some of these are discussed here:

- These materials (biocomposites) are used in many engineering applications, but it becomes a very confusing and difficult task to choose the best processing method and associated control parameters in order to achieve the properties which are required for a specific application.
- The properties of natural fibres are not standardised and vary according to the geographical regions in which the plants from which they are derived are grown The physical, mechanical and chemical properties of natural fibres and the polymer matrix are entirely different. The blending of these two components with their substantially different properties is the key to obtaining a composite with the desired properties.
- Generally, the adhesion efficiency between the natural fibre reinforcement and the matrix is poor. Several techniques may be used to improve the interfacial adhesion between the fibre and the matrix. These include chemical treatment or coating of the fibres and the addition of additives, fillers or a catalyst to the matrix. The interfacial zone is prone to failure and it is difficult but very important to avoid defects and subsequent failure at the interfacial region.
- Unlike conventional materials, composite materials are composed of entirely distinct components. Therefore, the tooling requirement is different from that for conventional materials because the tool comes into contact simultaneously with two types of materials which have different properties. One type of tooling which is good for the matrix material may at the same time damage the reinforcement, or *vice versa*.
- It is difficult to control the processing parameters and optimise the processing methods during the primary and secondary processing of composites because different conditions are required for the different constituents of the composite (fibre and matrix).

4.3 Processing Techniques for Biocomposites

Generally, the processing techniques used for the development of biocomposites are the same as those used for synthetic fibre-reinforced polymer composites. These processing techniques are broadly classified as open mould processes and closed mould processes. Hand lay–up, spray–up, filament winding and autoclave methods are open mould processes. Compression moulding, injection moulding and transfer moulding are closed mould processes [2].

4.3.1 Hand Lay-up Method

The hand lay-up method is the most frequently used process for the development of thermoset polymer-based biocomposites. This technique involves simple processing steps and minimum involvement of machinery. There are two mould plates: a top one and a bottom one. A mould release gel is sprayed onto the bottom plate to prevent the polymer from sticking to the mould surface. Reinforcement in the form of woven mats or chopped strand mats is cut as per the size of the mould and placed on the surface of the mould. Then, the thermosetting polymer in liquid form is mixed thoroughly in suitable proportions with a prescribed hardener (curing agent) and poured onto the surface of the mat that has already been placed in the mould. The polymer is spread uniformly with the help of a brush and a roller is used to remove the entrapped air and excess resin. Alternate fibre and polymer layers are then applied until the desired thickness of the composite is reached. Release gel is then sprayed on to the inner surface of the top mould plate and it is placed over the lay-up and pressure is applied. Thin plastic sheets may be used at the inner surface of the top and bottom mould plates to achieve a good surface finish of the composite part. After curing, either at room temperature or at some specified temperature, the mould is opened and the cured composite part is taken out and processed further. The cure time depends on the type of polymer used in the composite which is being processed. For example, for an epoxy-based system, the normal curing time at room temperature is 24-48 h. The hand lay-up method is mainly suitable for thermosetting polymer-based biocomposites. Capital requirement is less than for other composite processing techniques, but the production rate is lower and it is difficult to achieve a high volume fraction of reinforcement in the finished composites. The schematic of the hand lay-up process is shown in **Figure 4.2**.

Sisal fibre-reinforced unsaturated polyester biocomposite sheets were developed successfully using the hand lay-up technique. Firstly, the matrix of the composite was prepared by mixing unsaturated polyester resin with an accelerator and a curing agent. Chopped sisal fibres were then randomly laid in a stainless steel mould. A release agent (polyvinyl alcohol) had been applied onto the mould surface to make it easier to remove the cured composite part. The unsaturated polyester resin was carefully poured onto the fibres in the mould. The mould was then closed by screwing the cover plate onto the base as tightly as possible to force the resin to penetrate into the spaces between the sisal fibres, whilst the excess resin was forced out from between the cover plate and the base. The composite sheet was taken out after curing for 12 h at room temperature. [3].

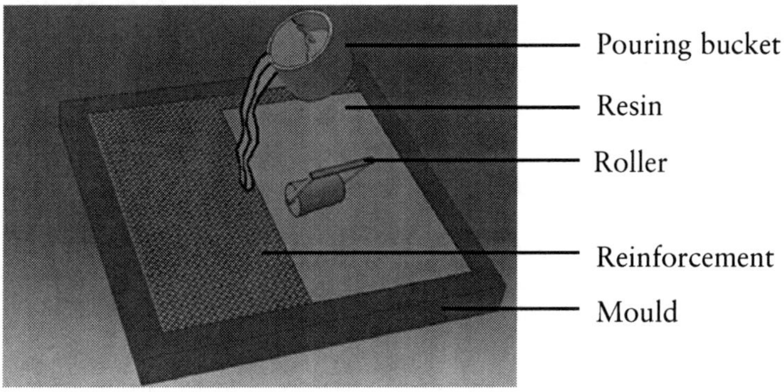

Figure 4.2 Hand lay-up method

4.3.2 Compression Moulding

Compression moulding is a closed moulding technique used for both thermoplastic and thermoset polymer composites. The schematic diagram of a compression moulding machine is shown in **Figure 4.3**. It has two mould plates: the lower mould plate is fixed and the upper mould plate is movable. The reinforcement (in the form of a mat, or long or short fibres) and the matrix are placed in a mould of the desired shape and the whole assembly is kept between the mould plates of the compression moulder. Temperature and pressure are set as appropriate for the constituents and processing technique for a definite period of time. As a result of the applied heat and pressure, the material in between the mould plates softens and acquires the shape of the mould cavity. Curing of the composite may be carried out either at room temperature or at some elevated temperature. After curing, the mould is opened and the composite product is removed for further processing. The temperature, pressure and time of exposure are critical in composite processing using the compression moulding technique. These three parameters should be optimised to achieve the desired quality of the composite part.

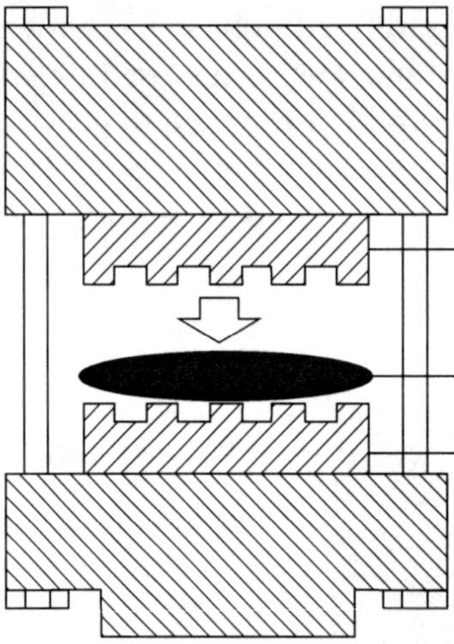

Figure 4.3 Compression moulding method

The compression moulding technique is suitable for all kinds of thermosets (such as epoxy, and polyester), thermoplastics (such as polypropylene (PP), nylon, polycarbonate, polyvinyl chloride and biodegradable polymers, e.g., poly(lactic acid) (PLA)) and various reinforcements (such as glass fibre, carbon fibre, aramid fibre, natural plant fibres (sisal, banana, nettle, hemp, flax and so on)). A wide variety of composite products including kitchen goods, automobile panels, vehicle roofs, bumpers, spoilers, and large liquid storage containers is produced by the compression moulding technique. This technique offers several advantages for composite processing such as high production rate, good surface finish of the composite part, high part uniformity and good flexibility in part design. However, this technique is not economical for making a small number of parts or for prototyping applications and it is a labour intensive process.

High strength manila hemp fibre-reinforced starch-based biodegradable resin composites have been produced with a hot pressing technique. The biodegradable resin was placed on the surface of manila hemp fibres and dried at 105 °C for 120 min in an oven. These preliminary composites were put in a metal mould and heated to 130 °C for 5 min with a flexible heater shaped to fit around the metallic mould and hot pressed at 10 MPa for 10 min. The volume fraction of manila hemp fibre in the specimens was varied from 30% to 70% during the processing [4]. Hemp fibre-reinforced PLA green composites with different fibre volume fractions (maximum 50%) were prepared using the hot pressing method. Before the raw material was placed into the mould, a mould release agent was used on the mould surface. The

PLA film and the short hemp fibre were placed into the mould layer by layer, and all the layers were of the same thickness. The mould assembly was then placed onto the heating panel of the hot press machine at a temperature of 170 °C and a pressure of 1.3 MPa for 10 min. When the temperature reached 170 °C the mould assembly was removed from the hot pressing machine and placed in another press machine. The pressure of 1.3 MPa was maintained at room temperature in the new press machine and then the composite plate was removed for further processing when the temperature of the composites reached room temperature [5].

Compression moulding through the film stacking process is a method which is commonly used to develop thermoplastic polymer-based laminated biocomposites. In this process, thermoplastic polymer pellets are converted into polymer thin films (generally 1 mm in thickness) by compression moulding at a temperature close to the melting point of the polymer and at a specified pressure for a contact time of a few minutes. Then, the films are allowed to cool under pressure and finally, they are removed from the mould. Thereafter, fibre mats (unidirectional long fibre mat, woven mat or randomly oriented short fibre mat) and polymer films are stacked alternately in a metal mould. At the top and bottom of the mould surface, a release gel (usually silica release gel) is applied to prevent the polymer films from sticking to the metallic plates. Sometimes, thin polymer sheets which have a melting point higher than the working temperature of the matrix can be used at the top and bottom of the mould to stop the resin from sticking to the mould surface and for easy removal. For example, when PP or PLA is used as the matrix material, Teflon (polytetrafluroethylene) sheets can be successfully used as Teflon has a higher melting point than that of PP and PLA. The whole assembly is then hot pressed at a specified temperature and at a high pressure for a few minutes. The pressure can be subsequently increased if desired and then the composite is cooled under pressure. The composite laminates are removed from the mould for further processing when the temperature reaches room temperature. The schematic diagram of composite fabrication through the film stacking technique is shown in **Figure 4.4**.

The film stacking method has been found to be effective for the processing of partially and fully biodegradable thermoplastic composite laminates using PLA and PP polymer matrices with nettle, *Grewia optiva* and sisal fibre mats as reinforcements. The finished laminated biocomposites have shown excellent mechanical (tensile, compressive, flexural and impact strength) and tribological characteristics (coefficient of friction and specific wear rate) [6, 7].

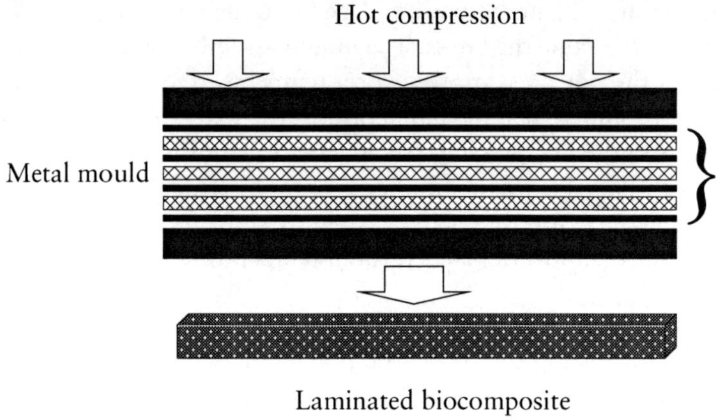

Figure 4.4 Schematic of film stacking method for composite fabrication

4.3.3 Extrusion and Injection Moulding Processes

For randomly oriented short fibre-reinforced biocomposites, extrusion and injection moulding processes are commonly used. Injection moulding is a closed moulding processs which is suitable for both thermoplastic and thermosetting polymer-based short fibre-reinforced composites. A schematic diagram of the injection moulding process is shown in **Figure 4.5**. The fibres and polymer are mixed manually and fed into the hopper as shown in this figure. Softening of the material takes place as it moves into the heated barrel. Due to the rotation of the screw, thorough mixing of polymer and fibres takes place and the material is pushed towards the converging section of the barrel where it is injected through a nozzle into the mould cavity at high pressure. Sometimes, a separate hopper to feed the natural fibres into the barrel is provided near the injection end of screw to avoid excessive damage to the fibres which may be caused by exposure to high temperatures and shearing action if they are in the barrel for a long time. If the extrusion system is used, the extruded fibre-matrix blend is separately processed into the desired shape using compression moulding. This system is used mainly for thermoplastic biocomposites. The technique is used for the processing of various composite parts, especially for automotive applications. The process is suitable for high production volume with low cost.

A variety of biocomposites have been developed using abaca, jute and flax fibres as reinforcement in a PP thermoplastic matrix using a mixer and the injection moulding process. All the natural fibres were dried at 80 °C in an air circulating oven for 24 h before mixing to remove the moisture present in the fibres. For each type of biocomposite processing, the natural fibres and PP polymer (in defined proportions) were placed into a hot mixer and heated up to the melting temperature of PP and then the hot agglomerated granules were transferred to a cool mixer to cool down to room temperature by the cold water which flows continuously through it. Then, the cold agglomerated granules were dried again. The composite specimens were prepared from the dried agglomerate by the injection moulding process at a barrel zone temperature of 150-180 °C, a mould temperature of 80 °C and under an injection pressure of 20 kN/mm^2 [8]. Fully biodegradable and green composites from waste materials have also been developed with desirable properties using the injection moulding technique. The PLA polymer and cellulose fibres (obtained from newspaper, magazine or kraft paper stock) were dried in a vacuum oven at 80 °C for 24 h to remove the moisture before processing. The dried polymer and fibres were mixed mechanically at 100 rpm with microcompounding moulding equipment at 183 °C for 10 min. The extruder had a screw of length 150 mm, and a length to diameter ratio of 18. The molten composite samples were transferred after extrusion through a preheated cylinder to a mini-injection moulder to obtain composite specimens [9].

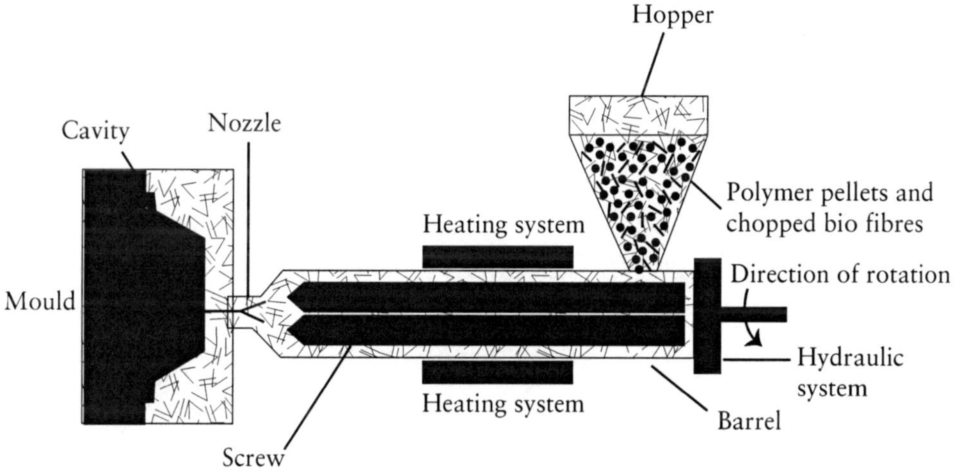

Figure 4.5 Injection moulding process

4.3.4 Resin Transfer Moulding

Resin transfer moulding is well-suited for biocomposite processing. It is a closed moulding technique used for large and complex composite products such as automobile parts. A schematic diagram of the resin transfer moulding process is shown in **Figure 4.6**. Initially, a release gel is applied to the mould surface for easy removal of the composite. Reinforcement in the form of a unidirectional mat, bidirectional woven mat or randomly oriented short or long fibre mat is placed on the surface of the lower half mould. The resin is transferred into the mould through ports. The uniformity of resin flow can be enhanced by using a catalyst as an accelerator. After curing, the mould is opened and the composite product is taken out. The viscosity of the resin is one of the critical parameters in the resin transfer moulding process which governs the injection time of the resin into the mould and defects like fibre wash, which arise during the processing of high viscosity resins. One of the main advantages of this technique is that a composite part produced by this method has a good surface finish on both sides of the product.

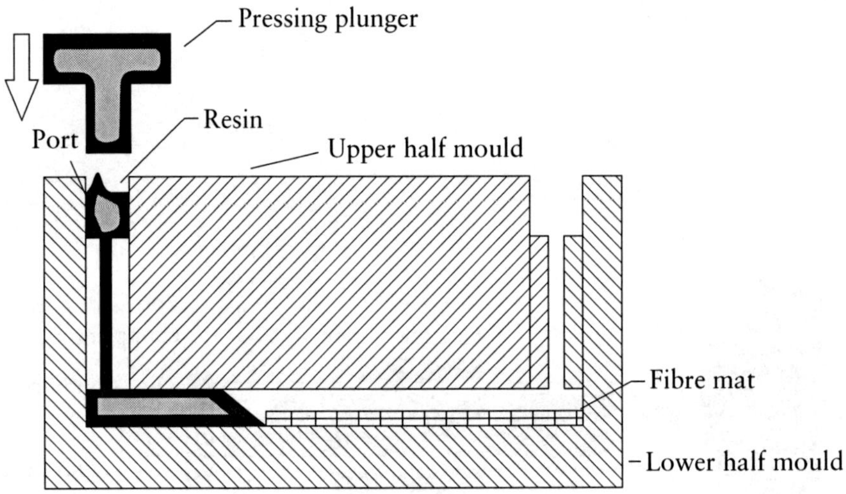

Figure 4.6 Resin transfer moulding process

The resin transfer moulding technique was used to produce hemp or kenaf natural fibre mat-reinforced polyester biocomposites. A manually operated mould with two inlet ports and a vent port made entirely of aluminum was used for composite fabrication. The temperature of the mould was maintained by circulating hot water flowing inside its upper and lower sections in a closed loop through a tank. Before starting the fabrication process, the surfaces of the mould were cleaned with a mould cleaner and mould release gel was applied to the surface of the mould. Natural fibre mats were cut as per the size of the mould and placed in the cavity. The mould was closed tightly and a vacuum of 725 mm of mercury was created in the cavity. Under these vacuum conditions, fibre mats already placed in the mould cavity were dried for 2 h by circulating water outside the mould at 55 °C. After that, cold water was circulated to cool the mould. After mixing with an initiator, the resin was injected into the mould at a constant pressure of 1.72×10^5 Pa. The vent port was closed when the resin was observed at the outlet. The resin was kept flowing at the inlet for 5 min more to make sure that the mould was completely filled. Then, the inlet ports were closed as well and hot water at constant temperature was circulated inside the mould. Finally, the composite part was cured under the same pressure mentioned above for 1 h. The resulting composites were of good quality and had acceptable mechanical properties, which showed the suitability of the processing method for these materials [10].

4.3.5 Prepregging of Biocomposites

The pre-impregnated lay-ups of fibres with resinous materials (generally thermosetting) are known as prepregs. The resinous material keeps the fibres in place. These prepregs are used to develop high performance composites. Prepregs can be unidirectional or bidirectional prepregs, depending upon the fibre positions. In the case of unidirectional prepregs, all the prepregs can be stacked in one orientation to achieve a composite laminate whose properties will be good in a particular direction. To process a composite laminate whose final properties will be almost the same in all directions, the prepregs are stacked in different orientations. These prepregs are available as tapes, cross-ply sheets and fabrics. Curing is carried out during or after shaping. Prepregs are fabricated with continuous filaments rather than chopped random fibres, which increases the strength and modulus. When prepregs are stored for further processing, a removable backing support is provided to prevent the layers from sticking together. These backings also provide an additional means of keeping the fibres in place. These prepregs can be developed from synthetic fibres (glass, carbon, aramid) and natural fibres (e.g., curaua, flax, hemp). Polyesters, phenolics, polyvinyl esters, polyamides and epoxy resins are used as resins for prepregging. Prepregs are used in a wide variety of applications such as aerospace components, home interiors, sports goods, medical applications, rocket engine nozzles, automotive body parts and fishing rods.

There was a noticeable improvement in the mechanical properties of fully biodegradable, green composites fabricated with ramie fibres and modified soy flour resin when the prepregging process was used, compared with other primary processing techniques. To fabricate composite sheets, the ramie fibres were soaked in precured soy resins and then squeezed by hand to remove the excess resin. This soaking/squeezing sequence was repeated several times to obtain complete impregnation of the fibres with the resin. The wet fibre bundles were hand-laid on Teflon-coated glass plates by aligning them layer by layer and then drying them at 45 °C. Then, these prepreg composite samples were cured by hot pressing at 120 °C for 50 min under 7 MPa of pressure [11].

4.3.6 Pultrusion

Pultrusion is a continuous process for biocomposite fabrication. Reinforcement in the form of continuous rovings or fibre mats is passed through a resin tank. In the resin tank, the fibres are dipped thoroughly so that they become completely wetted. These resin-saturated fibres are moved to a hot die in which they are moulded into the shape of the die. The composite is also heat cured in the die. The cured composite profile is pulled from the hot die with the help of gripper and finally, cut using a cutter which is positioned after the pulling mechanism in the pultrusion system. The process is mainly suitable for products with constant cross-sectional areas such as solid rods, tubing, and long, flat sheets. A schematic diagram of a pultrusion system is shown in **Figure 4.7**.

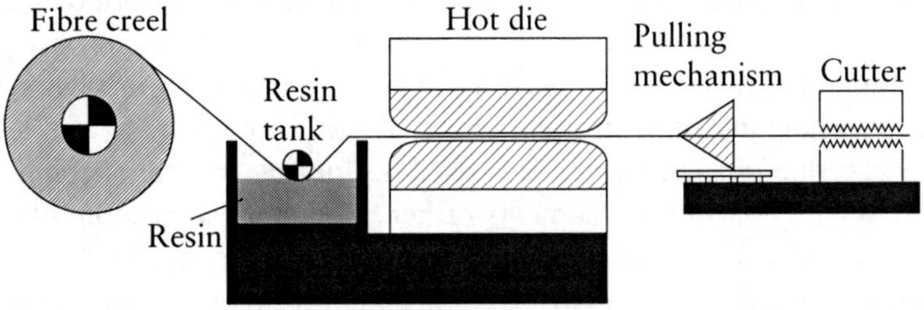

Figure 4.7 Pultrusion system

The pultrusion process was successfully used to develop hemp/wool-reinforced thermoset polymer biocomposite rods using three different types of resins (polyester, polyurethane and vinyl ester). The constituents of the biocomposite in each case were 5% wool, 30% hemp fibre, and 65% resin by weight. The composite rods with a diameter of about 4.8 mm were cured at a temperature range of 120-190 °C depending on the resin used. The resulting biocomposites showed much higher tensile strength than some of the common plastics at a comparable density (but lower than that of glass fibre-reinforced composites). Pultrusion has been suggested as a good choice of method for manufacturing biocomposites for applications such as construction boards and insulation boards [12].

4.3.7 Sheet Moulding Compounds

The sheet moulding compound (SMC) process is one of the main processing methods used for synthetic fibre- (glass fibre)-reinforced polymer composites. In this technique, continuous SMC sheets are fabricated with short fibres impregnated with a resin system. Continuous fibre roving is chopped into short fibres which fall onto uncured resin poured onto a continuously moving belt. The basic scheme of the SMC technique is shown in **Figure 4.8**. Sheet moulding compounds in the form of sheets are commonly used for compression moulding. Commonly available SMC sheets contain either randomly oriented short fibres or a combination of unidirectional fibres with randomly oriented short fibres. The production rate of this technique is high and the cost is low. Biocomposites have also been processed successfully with a modified SMC technique known as the biocomposite sheet moulding compound panel manufacturing process, which yields continuous production of biocomposites on a large scale. Biocomposites, as well as hybrid biocomposites, were developed with various natural fibres including big bluestem grass, jute, and industrial hemp. The processed low cost biocomposites were suggested for use in various panel applications such as for housing and transportation after testing of their mechanical and thermal properties [13]. **Table 4.1** summarises the constituents of various biocomposites and their processing techniques.

4.4 Applications of Biocomposites

The aim of the automotive industry is to always reduce the weight of automobile parts without sacrificing their performance, because weight has a direct impact on the fuel consumption of the vehicle. The automotive industry is one which has made substantial use of biocomposites and has developed various new biocomposite-based components. Biocomposites have also been introduced in the fields of automobiles, furniture and many others. According to a study, wood-based plastic composites now cover a diverse field of application areas such as automobile parts (the field of greatest usage), construction, marine applications, electrical components, consumer products, appliances and aerospace applications (the field of lowest usage) [26]. Jute fibre-reinforced polymer composites are used in many applications such as automobiles, building components, furniture and other domestic as well as electrical appliances. Banana fibre-reinforced epoxy composites are suitable for fabrication of telephone stands and other household applications [27]. Structural panels and beams were successfully manufactured as part of a project to develop an all-natural composite roof for housing applications using soybean oil-based resin and natural fibres (flax, cellulose, pulp, recycled paper, chicken feathers) and vacuum-assisted resin transfer moulding technology. Physical and chemical investigations and mechanical testing of the beams yielded good results in line with the desired structural performance [28]. Kenaf fibre-reinforced PLA matrix composites have been used for spare tyre covers, circuit boards and other applications [29]. Studies have shown that natural fibres such as cotton, kenaf, and flax have the ability to form a good bond with thermoplastic polymer binders and the resulting composites are suitable for automotive applications [30].

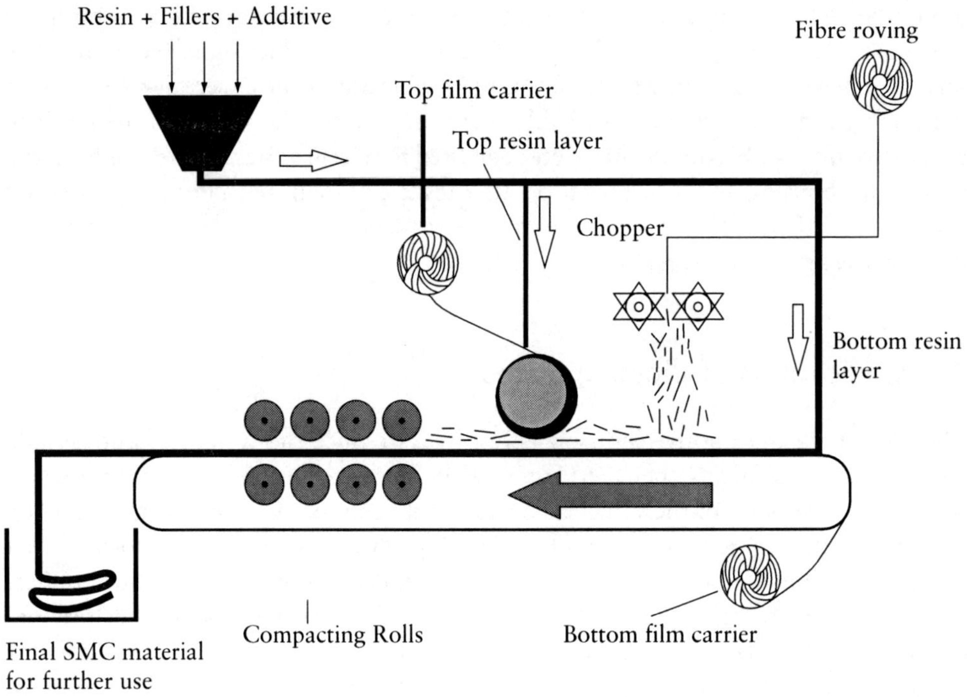

Figure 4.8 Sheet moulding compound manufacturing process

Table 4.1 Processing of various biocomposites			
Composite		Processing technique	Reference
Fibre	Matrix		
Raw and plasma-treated jute fibre	Unsaturated polyester resins	Hand lay-up	[14]
Flax and hemp fibres	PP, polyethylene, PLA, polyhydroxybutyrate	Twin screw extrusion	[15]
Kenaf	PLA	Wet impregnation method	[16]
Cuphea, *Lesquerella* and milkweed	PLA	Twin screw extrusion + injection moulding	[17]
Kenaf	PLA	Hot pressing	[18]
Glass fibres and curaua fibres	Orthophthalic polyester resin	Hand lay-up	[19]
Poplar (*Populus deltoides*) in particle form	PP	Twin screw extrusion + injection moulding	[20]
Alkali-treated hemp fibres	PLA	Hot compression	[21]
Kenaf fibres	High density polyethylene	Melt compounding (internal mixing) + compression moulding	[22]
Kenaf	PLA	Carding, punching followed by hot-pressing	[23]
Cotton, hemp, kenaf and lyocell fibres	PLA	Roller carding + compression moulding	[24]
Jute fibres	PP	Single screw extrusion + injection moulding	[25]

4.5 Secondary Processing of Biocomposites

4.5.1 The Need for Secondary Processing

Natural fibres provide a viable and abundantly available substitute for expensive and non-renewable synthetic fibres. Automotive components are continually tasked to meet higher specifications in terms of increased vehicle efficiency, lower fuel consumption, and an improved performance at lower cost. Now, many automotive parts are fabricated using natural fibre-reinforced polymer composites. Both, structural and interior components have incorporated polymeric composite materials due to the advances in the polymer processing industry and the development of novel, low cost light weight thermoplastic composites [31].

Except for simple composite parts which are produced to near net shape, the processing of polymer composites generally involves two steps: primary processing and secondary processing. The standard primary processing methods to produce a composite product are discussed in detail in the previous section of this chapter. To produce a composite-based automotive part, various secondary processes (machining operations) may be required to shape the composite parts. Of these machining operations, hole making is the most important process because this is necessary in order to assemble most composite parts to form the final product. Since adhesive bonding of polymer composites is not always a feasible option to obtain the structural integrity of the assembly, mechanical joining (using fasteners or rivets) is an important method for the assembly of composite parts. For mechanical fastening, drilling is one of the frequently used methods of making holes in order to assemble the polymer composite parts. There are various techniques for making holes in polymer composites, but conventional drilling is the established one and it is commonly used worlwide. The drilling operation is discussed in detail in the following section.

4.5.2 Drilling of Biocomposites

The effect of the drilling operation is entirely different in polymer composites as compared to drilling in metals because, in composites, the tool has to cut distinct phases simultaneously. Drilling of biocomposites depends on the properties of the fibre and the matrix, as well as their response to this machining process. The anisotropic nature of the composite materials results in their completely different behaviour from metals towards this machining operation. The diverse properties of the fibre and the matrix, combined with fibre orientation, have a significant effect on the machining process. Drilling of biocomposites causes different types of damage within the composite and at its surface. Drilling-induced damage (such as matrix burning and deformation, delamination, fibre pull-out) has a severe impact on the structural integrity of composite parts. Drilling process parameters such as cutting speed, feed rate, tool material and tool geometry, are some of the main factors which govern the quality of drilled holes. Optimisation of the operating conditions, tool material selection, tool design, unconventional methods of hole making and analytical modelling are some of the important research initiatives which have been undertaken to optimise the drilling process in order to obtain damage-free holes.

4.5.2.1 Input parameters

4.5.2.1.1 Cutting Speed and Feed Rate

When drilling biocomposites, the cutting speed and feed rate have to be optimised to have minimum drilling forces (thrust force and torque) and minimum drilling-induced damage. Generally, drilling of biocomposites shows that feed rate plays a significant role in their machining behaviour. The drilling force increases as the feed rate increases. The cutting speed has less effect on the drilling force. Modelling and optimisation of the process parameters (spindle speed, feed rate and drill diameter) using the Taguchi approach whilst drilling a coir fibre-reinforced polyester biocomposite showed that a drill bit with a diameter of 6 mm, spindle speed of 600 rpm and feed rate of 0.3 mm/rev gave the minimum value of thrust force, torque and tool wear. The optimum set of conditions resulted in minimum power consumption, minimum cutting forces, minimum tool wear, maximum material removal rate and enhanced hole quality and accuracy [32].

4.5.2.1.2 Selection of Drill Bit

Drill geometry has a significant effect on drilling forces and drilling-induced damage. Different types of drill point geometries such as trepanning tool, JO drill (single spindle drill), split point, Brad and Spur point, straight shank and U shape, have been proposed for drilling polymer composites under different drilling conditions. Modifications in the drill geometry have resulted in better quality holes, the ability to drill a greater number of holes and a reduction in the cutting forces. Drill geometry modifications such as change in the helix angle, rake angle or point angle and so on, have a substantial effect on the drilling forces and the amount of damage around the drilled hole. The drill bit diameter also has a significant effect on these two parameters. However, in most cases, the hole diameter required for a given part is fixed. Therefore, the drill geometry has to be optimised in order to minimise the damage associated with drilling.

Selection of the drill bit material is also an important factor when drilling biocomposites. A high speed steel (HSS) drill bit is used frequently for drilling biocomposites but, to increase the tool life and the quality of the drilled hole, other types of drill bit material such as HSS–cobalt and carbide can also be used. The drilling behaviour (thrust force and delamination extension) of a sisal fibre-reinforced epoxy plate has been studied under two different feed rates of 0.05 mm/rev (low) and 0.20 mm/rev (high) with two different drill point geometries (twist and brad type drill) at constant cutting speed. The variation of thrust force due to an increase in the feed rate was more evident for the twist drill. Thrust force analysis showed that Brad drills are recommended for the drilling of sisal-epoxy biocomposites, under the experimental conditions described. Variation of delamination due to the feed rate was more evident for the twist drill in sisal-epoxy biocomposite laminates. The results indicate the importance of selecting the appropriate drill geometry for the type of composite laminate to be drilled [33].

4.5.2.2 Output Parameters

4.5.2.2.1 Drilling Forces

During the drilling operation, the thrust force and torque signals are recorded by a dynamometer attached to the drilling set-up. The signals can be processed by a data acquisition system and recorded on computer. With the help of signals, the numerical values of maximum, minimum and average thrust force and torque can be obtained. The drilling behaviour of three different types of natural fibre-reinforced polyester composites (banana, sisal and roselle fibre-reinforced composites) has been studied. The thrust force and torque increased with increase in drill diameter and feed rate for all types of composites. A major reduction in thrust force was observed when the hole was predrilled to 0.4 mm and above. The thrust force and torque increased with an increase in the fibre volume fraction. The increase in torque was much smaller than that in thrust force, with an increase in the cutting speed. The average torque was higher with a multifacet drill at low cutting speed but lower at high cutting speed than that of the twist drill [34].

The machining behaviour of a nettle fibre-reinforced PP thermoplastic polymer biocomposite laminate was studied. The drilling process was carried out at different cutting speeds (900, 1800 and 2800 rpm) and feed rates (0.05, 0.12 and 0.19 mm/rev) using three different drill bits (twist, JO drill and parabolic) of diameter 8 mm. A typical graph of thrust force signals recorded during the drilling of nettle/PP biocomposite laminates with a twist drill is shown in **Figure 4.9.** It can be seen from the plot that, as the drill bit touches the biocomposite laminate, the drilling forces increase rapidly. After this rapid increase, a constant value zone is obtained which is known as the full engagement zone. After that, the thrust force value rapidly decreases and becomes zero as the drill bit comes out of the composite laminate. The twist drill was found to be suitable for drilling PP-nettle biocomposite laminates under the experimental conditions used.

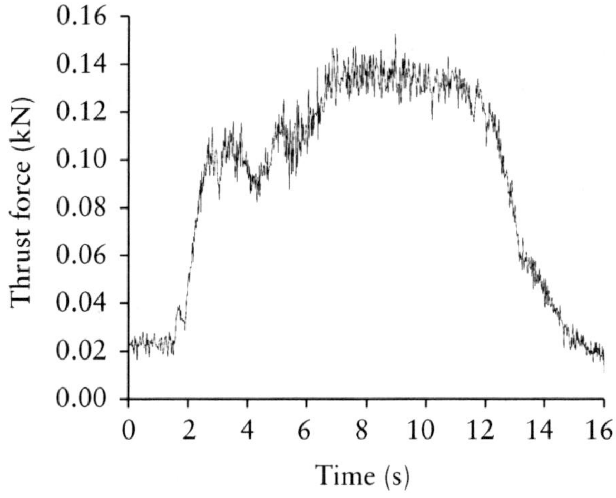

Figure 4.9 Variation of thrust force with twist drill (cutting speed: 900 rpm and feed rate: 0.05 mm/rev)

The drilling forces depend on the type and properties of the materials used to prepare a biocomposite and the sublaminate thickness. There are various numerical tools and techniques which have been used successfully to predict the critical drilling forces. The optimisation of process variables to minimise the damage to the composite caused by drilling and tool wear has been attempted. The machinability characteristics of a coir and glass fibre-reinforced polyester hybrid composite were investigated in terms of drilling using different process parameters. A regression model was developed for correlating the interactions of drilling parameters such as the drill bit diameter, spindle speed and feed rate, and their effects on responses such as thrust force, torque, and tool wear. Feed rate was found to be the most significant parameter in the analysis of thrust force and torque in drilling these hybrid composites. It has been established that a drill bit of diameter 8 mm with cutting speed of 1503 rpm and feed rate of 0.2 mm/rev generates the minimum value of thrust force, torque and tool wear during the drilling of a coir and glass hybrid composite [35].

4.5.2.2.2 Drilling-induced Damage

The estimation of drilling-induced damage has been an area of utmost importance in the machining of polymer composites because the damage generated during the drilling operation is responsible for the failure of the composite parts during operation. The nature of the drilling-induced damage depends upon the drilling process parameters, constituents of the composites, nature of the polymer matrix (thermoset or thermoplastic polymer, brittle or ductile polymer), characteristics of the reinforcements, tool geometry and material, stacking sequence and environmental operating conditions. Delamination is the most common type of damage that occurs during the drilling of polymer composite laminates. Delamination occurs both when the drill is entering the composite laminate and also when it is being removed from it. Delamination during entry of the drill into the laminate is known as the peel-up type and that caused during exit is known as push-down delamination [36]. Push-down delamination can be avoided by the use of a backing plate during drilling of a composite plate. Apart from delamination, there are many other types of damage such as matrix burning, matrix cracking, hole ovality, fibre pull-out, debonding and uncut fibres which occur during the drilling operation in composite laminates. The performance of any biocomposite part is significantly affected by drilling-induced damage, which also results in poor quality and surface finish. Preventive and quality control measures are essential to detect the damage incurred during drilling in order to ensure the high quality and performance of the finished biocomposite parts. The studies on drilling-induced damage have been qualitative as well as quantitative in nature. The quantitative evaluation of damage around the drilled hole can be done by various techniques such as examination using a microscope, X-ray non-destructive inspection, computer vision inspection, non-destructive dye penetrant testing and ultrasonic C-scans.

The delaminated region around the drilled holes in sisal/epoxy biocomposite laminates after drilling has been evaluated using digital scanning and enhanced digital radiography methods. It has been established from the calculation of area of damage based on delamination factor criteria that delamination in sisal/epoxy biocomposite laminates is always higher, than in glass/epoxy polymer composites because the mechanical resistance of the former is lower than that of the latter [33].

4.6 Concluding Remarks

With the growing concern for the environment and ecosystems, there is now a challenge for scientists and engineers to develop technologies which are environmentally friendly. Natural fibre-reinforced biocomposites form one such class of engineering materials, which not only possess superior properties but are also *'green'*. The conceptualisation and development of green composites has reached a stage where they can act as successful alternatives to conventional engineering materials. In order to further enlarge the application spectrum of these materials, there is an urgent need to develop high quality and cost-effective processing techniques. The present chapter highlights the important processing techniques which have been employed to produce parts made of natural fibre-reinforced composites. Although these techniques are similar in principle to those used for the processing of synthetic fibre-reinforced composites, there are many important differences in terms of process variables, tool/die design and the operating environments. Secondary processing in terms of machining has also been discussed in order to familiarise the reader with the complete processing cycle, starting from the primary processing and ending with the final product.

References

1. A.K. Mohanty, M. Misra and L.T. Drzal, *Journal of Polymers and the Environment*, 2002, **10**, 1/2, 19.

2. P.K. Bajpai, I. Singh and J. Madaan, *Journal of Thermoplastic Composite Materials*, 2012, DOI: 10.1177/0892705712439571.

3. S. Sangthonga, T. Pongprayoona and N. Yanumet, *Composites: Part A*, 2009, **40**, 687.

4. S. Ochi, *Composites: Part A*, 2006, **37**, 1879.

5. R. Hu and J.K. Lim, *Journal of Composite Materials*, 2007, **41**, 1655.

6. P.K. Bajpai, I. Singh and J. Madaan, *Materials and Design*, 2012, **35**, 596.

7. P.K. Bajpai, I. Singh and J. Madaan, *Proceedings of the Institution of Mechanical Engineers, Part J: Journal of Engineering Tribology*, 2012, **227**, 385.

8. A.K. Bledzki, A.A. Mamun and O. Faruk, *eXPRESS Polymer Letters*, 2007, **1**, 11, 755.

9. M.S. Huda, A.K. Mohanty, L.T. Drzal, E. Schut and M. Misra, *Journal of Materials Science*, 2005, **40**, 4221.

10. D. Rouisona, M. Sainb and M. Couturiera, *Composites Science and Technology*, 2004, **64**, 629.

11. J.T. Kim and A.N. Netravali, *Journal of Agriculture and Food Chemistry*, 2010, **58**, 5400.

12. X. Peng, M. Fan, J. Hartley and M. Al-Zubaidy, *Journal of Composite Materials*, 2011, **46**, 2, 237.

13. G. Mehta, A.K. Mohanty, K. Thayer, M. Misra and L.T. Drzal, *Journal of Polymers and the Environment*, 2005, **13**, 2, 169.

14. E. Sinha and S. Panigrahi, *Journal of Composite Materials*, 2009, **43**, 17, 1791.

15. I. Taha and G. Ziegmann, *Journal of Composite Materials*, 2006, **40**, 21, 1933.

16. T. Nishino, K. Hirao, M. Kotera, K. Nakamae and H. Inagaki, *Composite Science and Technology*, 2003, **63**, 1281.

17. V.L. Finkenstadt, C.K. Liu, R. Evangelista, L.S. Liu, S.C. Cermakc, M.H. Evangelista and J.L. Willett, *Industrial Crops and Products*, 2007, **26**, 36.

18. S. Ochi, *Mechachnics of Materials*, 2008, **40**, 446.

19. R.V. Silva, E.M.F. Aquino, L.P.S. Rodrigues and A.R.F. Barros, *Journal of Reinforced Plastics and Composites*, 2009, **28**, 15, 1857.

20. A. Nourbakhsh, A. Karegarfard, A. Ashori and A. Nourbakhsh, *Journal of Thermoplastic Composite Materials*, 2010, **23**, 169.

21. R. Hu and J.K. Lim, *Journal of Composite Materials*, 2007, **41**, 1655.

22. H. Ismail, A.H. Abdullah and A.A. Bakar, *Journal of Reinforced Plastics and Composites*, 2010, **29**, 2489.

23. B.H. Lee, H.S. Kim, S. Lee, H.J. Kim and J.R. Dorgan, *Composite Science and Technology*, 2009, **69**, 2573.

24. N. Graupner, A.S. Herrmann and J. Müssig, *Composites: Part A*, 2009, **40**, 810.

25. Md.R. Rahman, M. Hasan, Md.M. Huque and Md.N. Islam, *Journal of Reinforced Plastics and Composites*, 2010, **29**, 445.

26. A. Ashori, *Bioresource Technology*, 2008, **99**, 4661.

27. S.M. Sapuana and M.A. Maleque, *Materials and Design*, 2005, **26**, 65.

28. M.A. Dweiba, B. Hu, A. O'Donnell, H.W. Shentonb and R.P. Wool, *Composite Structures*, 2004, **63**, 147.

29. R. Nakamura, K. Goda, J. Noda and J. Ohgi, *eXPRESS Polymer Letters*, 2009, **3**, 19.

30. M.G. Kamath, G.S. Bhat, D.V. Parikh and D. Mueller, *International Nonwoven Journal*, 2005, **14**, 34.

31. J. Holbery and D. Houston, *Journal of the Minerals, Metals and Materials Society*, 2006, **58**, 80.

32. S. Jayabal and U. Natarajan, *Bulletin of Materials Science*, 2011, **34**, 7, 1563.

33. L.M. Durão, D.J.S. Gonçalves, J.M.R.S. Tavares, V.H.C. de Albuquerque, T.H. Panzera, L.J. Silva, A.A. Vieira and A.M. Baptista, *Materials Science Forum*, 2013, **730-732**, 301.

34. D. Chandramohan and K. Marimuthu, *International Journal of Advanced Engineering Research and Studies*, 2011, **1**, 1, 134.

35. S. Jayabal, U. Natarajan and U. Sekar, *International Journal of Advanced Manufacturing Technology*, 2011, **55**, 263.

36. H. Ho-Cheng and C.K.H. Dharan, *Transactions of the ASME: Journal of Engineering for Industry*, 1990, **112**, 236.

5 Composites of Thermosetting Polymers Reinforced with Natural Fibres

Daniella R. Mulinari, Camila C. Barros and George J.M. Rocha

5.1 Introduction

Composite materials consist of two different materials that are combined. In engineering, the definition can be narrowed down to a combination of two or more distinct materials into one with the intention of suppressing undesirable constituent properties in favour of the desirable ones [1]. One constituent is called the reinforcing phase (or dispersed phase) and the one in which it is embedded is called the matrix (or continuous phase). The reinforcing phase material may be in the form of fibres, particles or flakes [2].

Composites are classified by the geometry of the reinforcement (particles, flakes and fibres) or by the type of matrix (polymer, metal, ceramic and carbon) [2]. The use of short natural fibres as reinforcements in thermosetting polymer matrices will be discussed in this chapter.

Among the short natural fibre-reinforced polymer matrices combine acceptable mechanical properties with a low density. Such composites offer a number of advantages, which include low cost, abundance of renewable natural resources and biodegradability [3-6].

The utilisation of natural fibres for processing novel composites has attracted growing interest due to their eco-friendly characteristics and renewable nature, especially in tropical regions [7-10]. However, chemical treatment of the fibre surface is needed to decrease the hydrophilicity and roughness of the fibres, because the polymer matrices are hydrophobic [11-16].

5.2 Natural Fibres

Natural fibres can be classified according to whether they are obtained from a vegetal, animal or mineral source. All vegetal fibres are composed mainly of cellulose, hemicellulose and lignin, whilst animal fibres consist largely of proteins (hair, silk, and wool). Vegetal fibres are subdivided into classes such as: bast fibres (including jute, flax, hemp, kenaf), leaf fibres (such as pineapple, sisal, henequen, cuaraua, banana pseudostem), seed or fruit fibres (such as coir, cotton and oil palm) and grasses and reeds (such as bagasse and sape) [17].

Vegetal fibres are obtained by direct extraction from plants or from the use of industrial wastes as, for example, sugarcane bagasse (which, together with ethanol, is a by-product from sugar mills) and the husks of the coconut fruit and banana fruit [18, 19]. After obtaining the fibres, some type of surface treatment is usually carried out in order to improve their compatibility with polymer matrices when they are used as reinforcement in composites.

The chemical composition of natural fibres (also known as lignocellulosic fibres) varies, depending upon their type. The main components of natural fibres are cellulose, hemicellulose, lignin, pectins, and waxes [17].

Cellulose is the most abundant organic material on earth and, together with lignin, hemicellulose and pectin, is the major component of plant cell walls, with an annual production of 50 billion tonnes in Brazil. It is a long chain linear homopolymer with the empirical chemical formula $(C_6H_{10}O_5)n$, where *n* varies from a minimum of 200 to values greater than 7000 [20]. Its repeating unit, called cellobiose, is composed of two glucose molecules linked by β-(1→4) glycosidic bonds, containing six hydroxyl groups and their presence in the cellulose molecule leads to a large amount of intra- and intermolecular hydrogen bonding [20]. It has been reported [20] that the degree of crystallinity of cellulose varies according to its origin and the way in which it has been processed Cellulose from cotton, for example, has more ordered chains, with a crystallinity of about 70%, while the cellulose from trees has a crystallinity index of around 40%.

After cellulose, hemicellulose is the most commonly occurring organic compound in the biosphere. Along with cellulose, pectin and glycoproteins, they form the cell walls of plant cells. They are polysaccharides with lower molecular weights than that of cellulose and are interspersed with cellulose microfibrils, preventing them from touching each other and giving flexibility to the cell wall structure. Hemicelluloses are mainly divided into pentosans (composed of xylose and arabinose) and hexosans (composed of galactose, mannose and glucose), with the general formulae $(C_5H_8O_4)_n$ and $(C_6H_{10}O_5)_n$ respectively, where *n* is the degree of polymerisation [20]. Because of their open structure containing many hydroxyl groups, hemicelluloses are partly soluble in water and are hygroscopic [21].

Lignin is a macromolecular compound with a polyphenolic structure which cannot be converted into sugars by fermentation. This resin is an amorphous material with a hydrophobic, highly branched three-dimensional structure, which acts as a cement between the cellulose fibrils and is present in all layers of the cell wall of the plant. However, it is concentrated mainly in the secondary cell wall and occurs in association with cellulose and hemicellulose. This biopolymer is formed from three different phenyl ether units and the proportion of these compounds results in different types of lignin. It is insoluble in water and its architecture, as well as the chemical complexity of lignin, not only hampers its isolation from plants but also their plasticisation by economically viable processes [21].

5.3 Composites Containing Short Natural Fibres

The practice of choosing an appropriate manufacturing method is usually based on the actual part size and geometry required, the unit count, the precursor material (initial state of the composite material), the selected components of the composite, i.e., the reinforcement and the matrix, and the cost [1, 2].

The composite types can be broadly divided into composites made from short fibres (aspect ratio less than 100) and continuous fibres. The three most common mass production processes for manufacturing short fibre composites are injection moulding, compression moulding and extrusion. These processes were adopted from the polymer processing industry that had developed the equipment to produce parts in high volumes from polymers [1, 2].

The fibres can be cut or chopped and compounded in an extruder with any polymer to from a pellet consisting of short fibres or if pultruded, the pellet will consist of aligned fibres [1, 2]. The first generation of composites used chopped or short fibres pre-impregnated with thermoplastic polymer matrices in the form of pellets. The pellets are usually a few centimetres in length and a few millimetres in diameter. These pellets are fed into the hopper of either an injection moulding machine or an extruder. The polymer contained in the pellets melts inside the barrel of the machine because of the heating and viscous dissipation due to the shearing of the polymer against the barrel and the screw. The short fibres suspended in the molten polymer are forced either (a) into a mould cavity along with the polymer in the case of discontinuous operations, such as injection moulding, or (b) through a die in the case of continuous operations, such as extrusion. Thus, the existing machines used for polymer processing could be easily adapted for fibre-reinforced polymers. The advantages of these processes are stiffer and stronger components with lower degrees of warpage and shrinkage. The disadvantage is that there is no control over the fibre orientation and fibre attrition in the composites [1, 2].

5.4 Thermosetting Polymers Reinforced with Natural Fibres

The use of natural fibres has been studied as an alternative to synthetic fibre reinforcement in thermosetting polymers in order to decrease the cost of materials without significantly changing the mechanical properties of the composites [22-24].

Thermosetting polymers are used as a matrix for the production of composites and these can be reinforced using many different particulate fillers and fibres. Thermosetting resins are only 50 to 500 times more viscous than water and can impregnate the empty spaces between the fibres readily. Thermosetting resins are viscous fluids which, before the cure reaction, impregnate the filler particles or fibres, promoting high distribution of reinforcement and excellent homogeneity of the composites [22-24]. After the cure reaction, the polymeric resin has a crosslinked structure that physically anchors the reinforcement in the matrix, leading to an excellent combination between the properties of the matrix and the reinforcement [22-24]. However, thermosetting composites have some disadvantages such as the fact that thermomechanical reprocessing (and therefore extrusion and injection moulding) cannnot be used. Recycling of thermosetting composites is also a technological challenge.

Recently, thermosetting composites have been preferentially used for specific applications which use only small quantities for high-cost components such as electronic devices, and for the aeronautic industry.

5.4.1 Coconut Fibre-reinforced Polyester Resin

Mulinari and co-workers [25] studied the fatigue behaviour of polyester resin matrix composites reinforced with coconut fibres which had been pretreated with alkali solution. Firstly, coconut fibres were extracted from the exocarp, ground in a mill, sieved and then dried at 80 °C for 24 h. To remove the soluble extractives and to facilitate adhesion between the fibres and the matrix, the untreated coconut fibres were modified by pretreatment with alkali solution 1% (w/v). The effects of the surface modification of the fibres'were evaluated by scanning electron microscopy (SEM), thermal analysis and Fourier-transform infrared (FTIR) spectroscopy.

The alkali treatment promotes the removal of partially amorphous constituents such as hemicelluloses, lignin, waxes and oils which are soluble in alkaline solution, and therefore reduces the level of fibre aggregation, and makes the surface rougher [26-30]. During the alkali treatment, the OH groups present in the fibres react with sodium hydroxide (NaOH) according to **Equation 5.1:**

$$\text{Fibre-OH} + \text{NaOH} \rightarrow \text{Fibre-O-Na} + H_2O \quad (5.1)$$

The chemical composition (cellulose, hemicelluloses and lignin) of the original starting coconut fibre (untreated) and pretreated fibre (cellulignin) was determined according to the method used by Gouveia and co-workers [31]. Composites of unsaturated polyester resin and pretreated coconut fibres were prepared in a glass mould (250 × 250 mm). They were prepared by the compression moulding technique and the components were mixed manually. The concentration of fibres added was 10 wt% of the mass of the pure polymer. After curing for about 24 h, the material was removed from the mould and was subjected to tensile and fatigue tests.

To determine the fatigue behaviour of the composites it was necessary to analyse the results of the tensile tests. The form and dimensions of the samples under tension and tension-tension fatigue were defined were carried out according to American Society for Testing and Materials (ASTM) standard D3039. A sample was tested for each applied tension, under ambient temperature, at 6 Hz frequency and load ratio R = 0.1. The values of maximum tension for each test were defined taking as basis the tensile strength of the material around 16 MPa to obtain the S (stress)-N (number of cycles) curve in the band between 10^3 and 10^5 load cycles. The fractured surfaces were examined by SEM.

SEM is an excellent technique for examining the surface morphology of fibres. **Figure 5.1** shows SEM micrographs of untreated coconut fibres and **Figure 5.2** shows SEM micrographs of pretreated coconut fibres. Examination of the untreated fibres shows a large amount of debris adhering to the surface of the fibre bundles, because they are coated with amorphous material (**Figure 5.1**).

After treatment of the coconut fibres it was observed that wax, pectin, lignin and hemicelluloses had been removed from the surface of the fibres (**Figure 5.2**). It was also verified that the superficial layer (parenchymatous cells) had been removed, increasing the contact area. As a consequence an increase in the roughness of the fibre surfaces was observed, which can increase the adhesion between the fibres and the matrix.

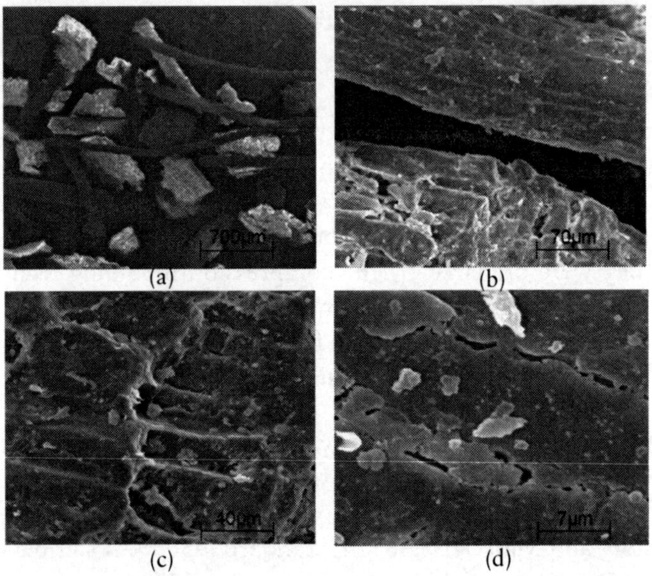

Figure 5.1 SEM of untreated coconut fibres at different magnifications (a, b, c and d)

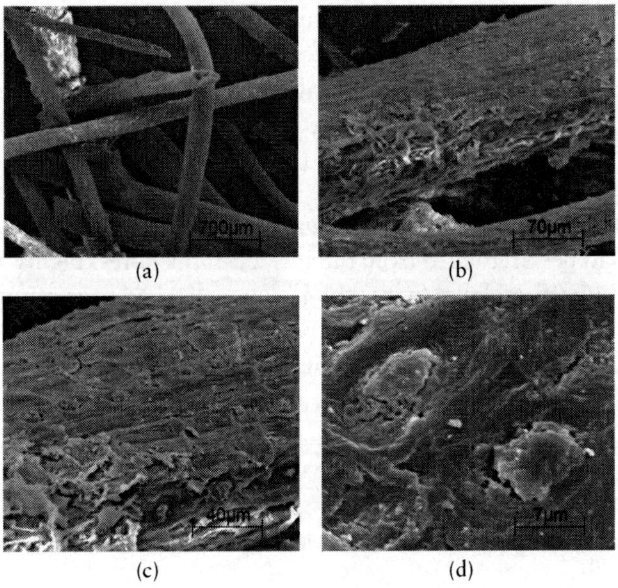

Figure 5.2 SEM of pretreated coconut fibres at different magnifications (a, b, c and d)

Figure 5.3 shows thermogravimetric analysis (TGA) curves of treated and untreated coconut fibres.

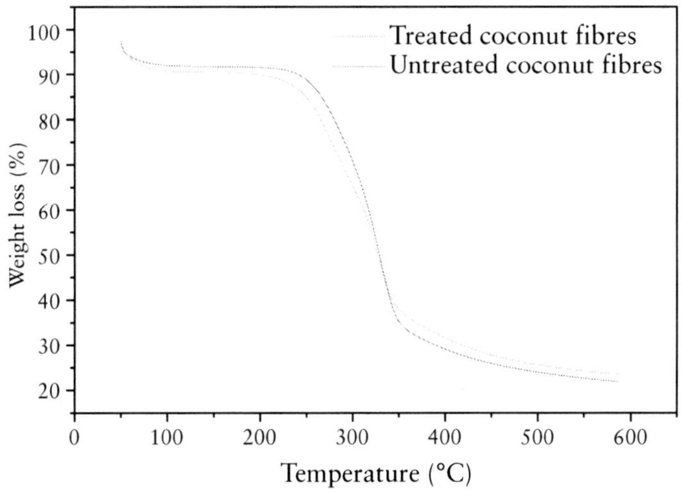

Figure 5.3 TGA curves of coconut fibres

Analysing **Figure 5.3**, it was observed that treated coconut fibres show higher thermal stability than untreated fibres. Both types of fibre are stable up to the temperature range of 220 to 300 °C but above 380 °C, decomposition of all the samples occurs.

Figure 5.4 shows the infrared spectra of coconut fibres. The most visible differences between the spectra of untreated and treated coconut fibres are the modifications of the signal at 2885 cm^{-1} and 1732 cm^{-1}, characteristics of the stretching of symmetrical CH groups and stretching of unconjugated CO groups present in polysaccharides and xylans. Considering the first region, the ratio between intensity of the CH stretching band (approximately 2900 cm^{-1}) is lower in the spectrum of the treated coconut fibres compared to the untreated coconut fibres.

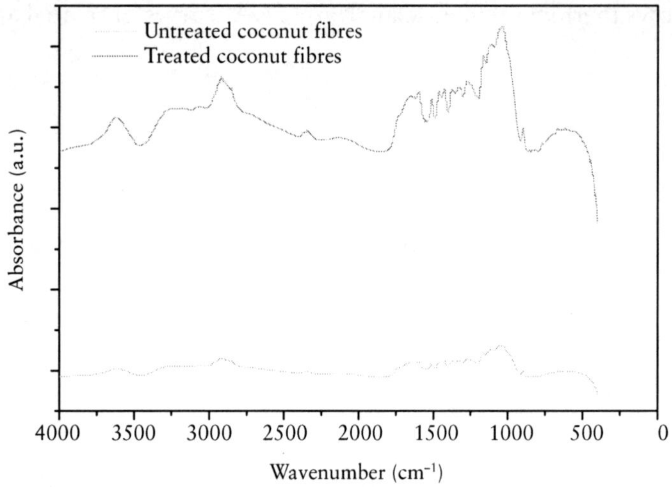

Figure 5.4 FTIR spectra of the coconut fibres (wavenumber 400 to 4000 cm^{-1})

Figure 5.5 shows the S-N curve obtained from fatigue tests.

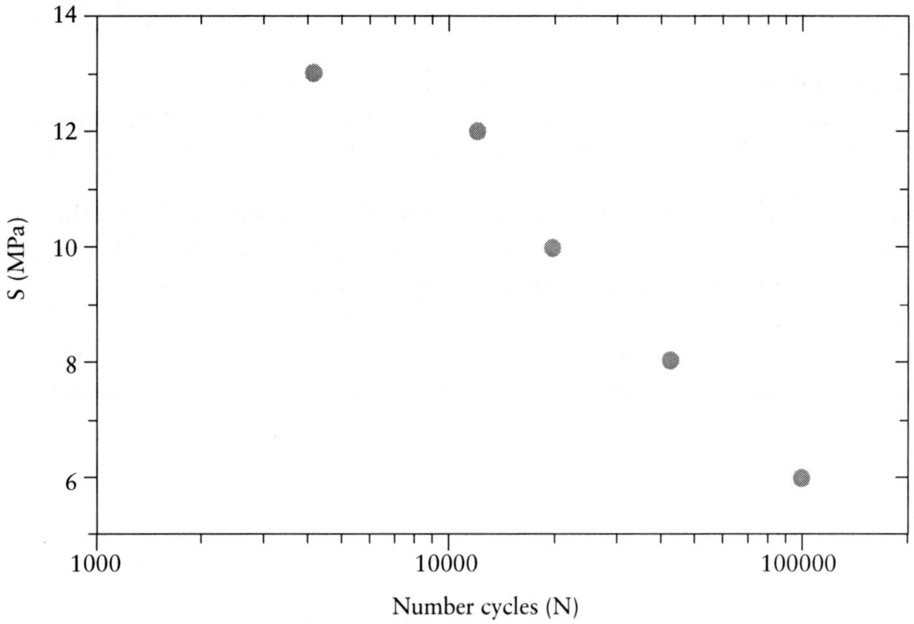

Figure 5.5 Fatigue tests: S-N curve of composites

Analysing **Figure 5.5**, it was observed that there was a relationship between the maximum tension applied and the number of cycles before rupture occurred. High performance composite materials show a relationship between tensile strength and fatigue strength. However, this relationship decreases when natural fibres are used. The coconut fibre-reinforced polyester resin studied showed a relationship between the fatigue strength after 10^5 cycles and a tensile strength of 6 MPa. The number cycles was low due to poor fibre-matrix interaction. Improving the interaction will increase the strength and, consequently, the number cycles.

Figure 5.6 shows the fracture region of the composites, and demonstrates the failure mechanism.

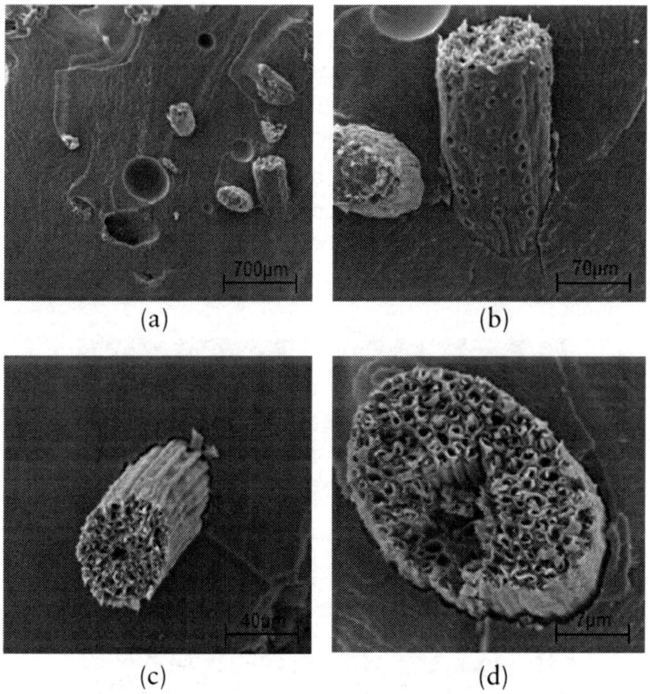

Figure 5.6 SEM of fracture surface of composites at different magnifications (a, b, c and d)

Analysing **Figure 5.6**, poor dispersion of fibres in the matrix, fractured fibres and the presence of pull-out can be seen. It is therefore necessary to improve fibre surface treatments in order to improve the mechanical properties of the composites.

5.4.2 Sugarcane Bagasse Fibre-reinforced Polyester Resin

Rodrigues and co-workers [32] studied the chemical modification of sugarcane bagasse fibres by esterification *via* an anhydride system prior to their use as reinforcement in a polyester matrix. Sugarcane bagasse fibres were esterified by treating them for 5 h with acetic anhydride, toluene, acetic acid and percloric acid. The chemical composition of the fibres was determined according to the method used by Gouveia and co-workers [31]. The modification was also evaluated using SEM, X-ray diffractometry, TGA and FTIR spectroscopy techniques. The modified fibres were then mixed with the polyester resin so that they formed 5 wt% of the total composition, and the material was compression moulded. After that, the composites were left to cure at room temperature for 24 h and machined into specimens according to the ASTM D-3039 specification.

Figure 5.7 presents SEM micrographs of unmodified sugarcane bagasse fibres, which show a large amount of extractives. However, modification of the sugarcane bagasse fibres removed the extractives from the surface of the fibres as can be observed in **Figure 5.8**. It was verified also that with the elimination of superficial layer the contact area increased. As a consequence, an increase in the roughness of fibres was observed, which can contribute to an increase of the interfacial bonding between the fibres and the matrix.

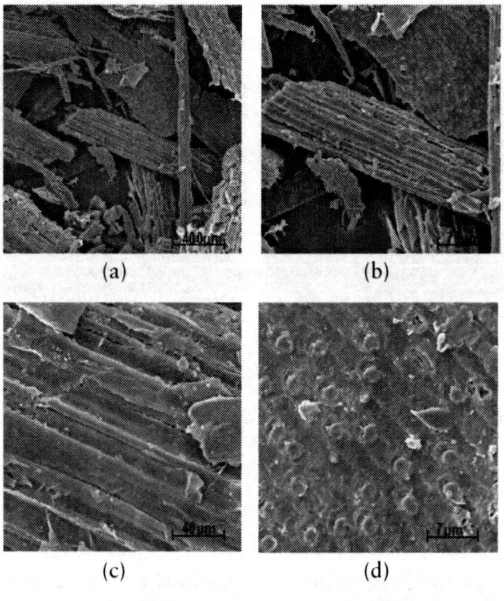

Figure 5.7 SEM of the unmodified sugarcane bagasse fibres at different magnifications (a, b, c and d)

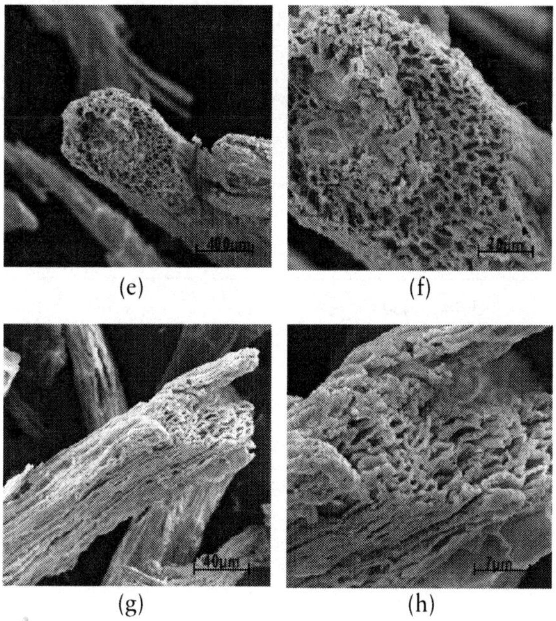

Figure 5.8 SEM of the modified sugarcane bagasse fibres at different magnifications (a, b, c and d)

Figure 5.9 shows the X-ray diffractogram of the sugarcane bagasse fibres. It presents two peaks, which are well-defined. X-ray diffraction peaks for both materials can be attributed to disordered regions. The materials tested are semicrystalline. The spectrum corresponding to the unmodified fibres shows diffraction peaks at the following 2θ angles: 15.9° and 22.0°. In the spectrum for the modified fibres, the diffraction peaks can be observed at 15.9° and 22.2°. The presence of these peaks at 15° and 22° shows that the fibre surface has been modified. The different intensities of these peaks indicates an increase of the interplanar distance compared to the modified fibres. This behaviour occurs due to the generation of disorder when the fibres are modified. The patterns for both materials are similar, however the modified fibres are less crystalline than the modified ones.

The mechanical properties of the composite and pure polyester are summarised in **Table 5.1**. The pure polymer showed higher tensile strength values compared to the composites reinforced with sugarcane bagasse fibres (unmodified and modified). However, a minimal difference in tensile strength was observed between the composites. In contrast, both composites showed higher tensile modulus values than the pure polymer. This occurred because of the good interaction between the fibres and the matrix. The presence of unmodified fibres led to an increase in tensile modulus of 71.5% compared to that of the polyester resin. However, composites reinforced with modified fibres showed a tensile modulus which was 13% higher than that of the composites which contained unmodified fibres.

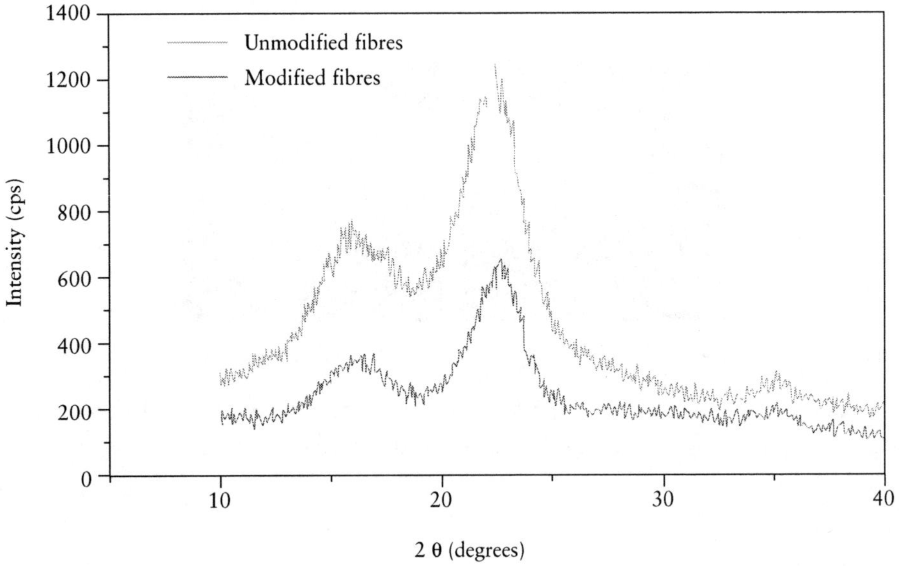

Figure 5.9 X-ray diffractogram of the sugarcane bagasse fibres (modified and unmodified)

Table 5.1 Mechanical properties of the sugarcane bagasse-reinforced polyester composites

Materials	Tensile strength (MPa)	Tensile modulus (MPa)
Polyester resin	19.7 ± 0.6	894.3 ± 29.5
Unmodified sugarcane bagasse fibres (5 wt%)/polyester	11 ± 0.001	1358.1 ± 0.02
Modified sugarcane bagasse fibres (5 wt%)/ polyester	12.5 ± 2.2	1534.1 ± 86.6

5.4.3 Banana Pseudostem Fibre-reinforced Polyester Resin

Paiva and co-workers [33] evaluated the mechanical behaviour of the fibres from banana pseudostem reinforced polyester composites. The biomass used in this work was obtained from bunches of bananas (*Musa sapientum*). As soon as obtained, the material was cut into small pieces and subsequently washed with water until neutral in pH and dried at 50 ± 5 °C (untreated material). Some of the material was submitted to dilute acid pretreatment in a 350 L stainless steel reactor, under these conditions: 1.0% (w/v) sulphuric acid solution in a 1:10 biomass:acid ratio, at 120 °C for 10 min. After reaction the resulting solid material (cellulignin) was separated by centrifugation, washed with water until the pH was neutral, and dried at 50 ± 5 °C until the moisture content was 50%. The modification of the fibre surface was analysed by SEM and TGA techniques.

The chemical composition (cellulose, hemicelluloses and lignin) of the original starting material (untreated) and pretreated (cellulignin) was determined according to methods described by Gouveia and co-workers [29].

Composites of unsaturated polyester resin and pretreated coconut fibres were prepared in a glass mould (250 × 250 mm). The composites were prepared by compression moulding. The components of the polyester resin were mixed manually and then the fibres were added. The concentration of fibres added was 5 wt% of the final mass of the finished composite. After curing for approximately 24 h, the material was removed from the mould and was submitted to tensile testing. Samples of the polyester matrix which did not contain banana pseudostem were also prepared. All the processes were performed at room temperature.

The composition of the fibres is shown in **Table 5.2**. The fibres from banana pseudostem showed a higher ash content than other lignocellulosic materials, such as sugarcane bagasse, for example [34, 35]. It has been reported that banana fibres, and the fibres from most of the plants that are used for cellulose production, have a high ash content and high solubility of the raw material in cold water, hot water and in NaOH solution (1% w/v) [36, 37].

Table 5.2 also shows that, after pretreatment with sulfuricacid (1.0% w/v), at 120 °C for 10 min, almost 60% of the hemicellulose fraction had been removed, increasing the exposure of cellulosic fibres.

The morphology of the untreated fibres can be observed in **Figure 5.10**. These fibres show a superficial layer with a high percentage of extractives. A superficial layer of parenchymatous cells can also be seen.

Table 5.2 Chemical composition and mass balance of raw banana pseudostem and after pretreatment

Components	Raw banana pseudostem (%)	Banana pseudostem, pretreated (%)	Solubilised components after pretreatment (%)	Values of chemical characterisation of banana pseudostem, adjusted by yield from pretreatment step (%)
Yield of reaction after pretreatment	-	-	-	33.0
Cellulose	46.9 ± 0.9	55.8 ± 0.3	29.6	7.1
Hemicellulose	17.3 ± 0.4	12.0 ± 0.5	58.9	16.0
Lignin	18.7 ± 0.3	27.1 ± 0.2	14.2	3.6
Ash	9.2 ± 0.0	6.0 ± 0.1	-	-
Extractives	4.9 ± 0.0	-	-	-
Total	97.0 ± 1.6	100.9 ± 1.1		

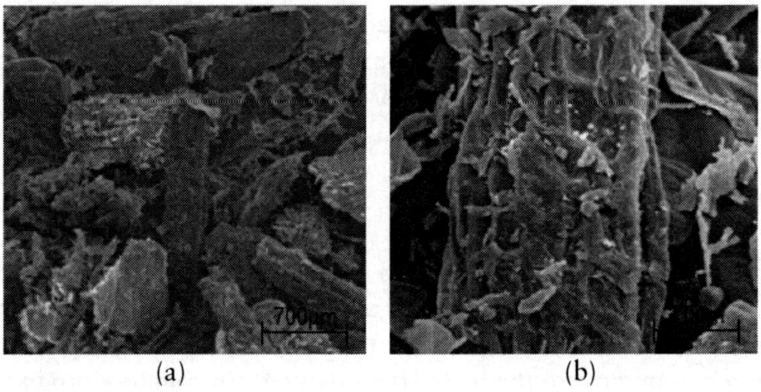

Figure 5.10 Morphology of the untreated banana pseudostem fibres at different magnifications (a and b)

After the pretreatment of banana pseudostem fibres, it was observed that the extractives had been removed from the surface of the fibres. This can be seen in **Figure 5.11.** It was verified also that with the elimination of the superficial layer the contact area increased. As a consequence, an increase in the roughness of the fibres was observed, which contributes to the increase of the adhesion between fibre and matrix. These observations have also been obtained with other treated fibres, such as sisal, flax,which have similar structures.

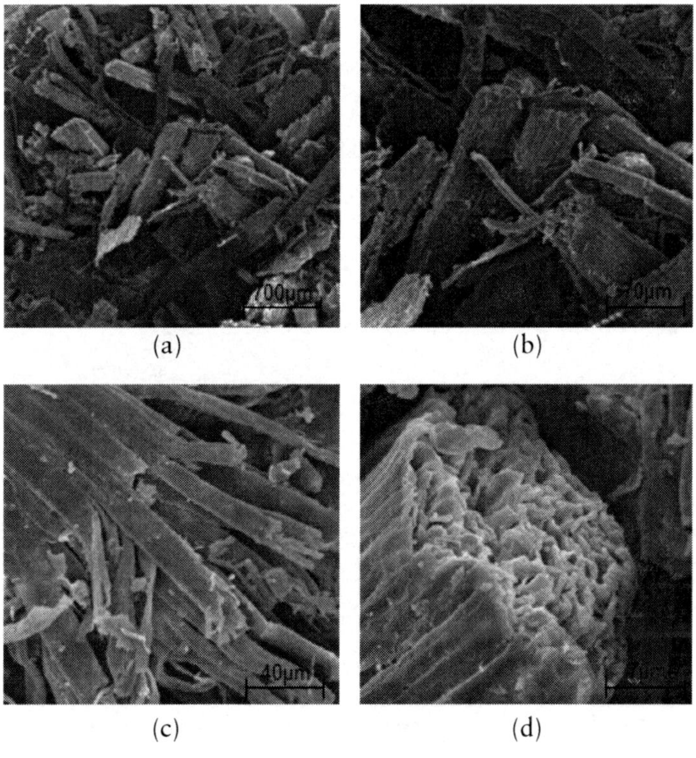

Figure 5.11 Morphology of the pretreated banana pseudostem fibres at different magnifications (a, b, c and d)

Figure 5.12 shows TGA curves of pretreated and untreated banana pseudostem fibres. From the TGA curves it was possible to observe that the weight loss of both pretreated and untreated banana fibres started at at 100 °C. This corresponds to the temperature at which water vapour and bound water are removed from the samples. Further weight loss of the pretreated and untreated banana fibres at higher temperatures is attributed to thermal degradation of the fibres.

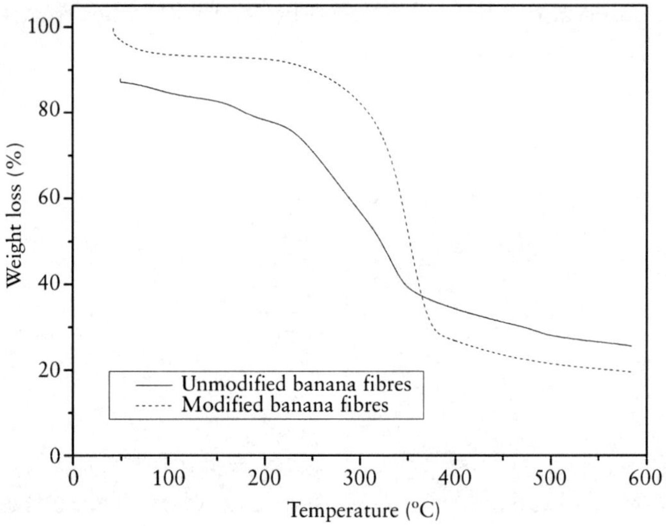

Figure 5.12 TGA curves of banana pseudostem fibres

The tensile properties of the pretreated banana fibre-reinforced polyester composite and the pure polyester are summarised in **Table 5.3**.

Table 5.3 Mechanical properties of banana fibre-reinforced polyester composites

Materials	Tensile strength (MPa)	Tensile modulus (MPa)
Polyester resin	19.7 ± 0.6	894.3 ± 29.5
Pretreated banana fibres (5 wt%)/polyester	12.1 ± 2.2	1339.7 ± 55.2

Pretreated banana fibre-reinforced polyester composites showed higher tensile strength and modulus results compared to the pure polyester. This occurred as a result of good interaction between the fibres and the matrix. The addition of the fibres contributed to the modulus increase, which exhibited an increase of 50%, compared to that of the polyester resin.

References

1. S.G. Advani and E.M. Sozer in *Process Modeling in Composites Manufacturing*, Taylor and Francis Group, CRC, Boca Raton, FL, USA and London, UK, 2010, p.1.

2. A.K. Kaw in *Mechanics of Composite Materials*, Taylor and Francis Group, CRC, Boca Raton, FL, USA and London, UK, 2005, p.1.

3. B. Mano, J.R. Araújo, M.A.S. Spinacé and M.A. De Paoli, *Composites Science and Technology*, 2010, **70**, 29.

4. C. Alves, P.M.C. Ferrão, A.J. Silva, L.G. Reis, M. Freitas, L.B. Rodrigues and D.E. Alves, *Journal of Cleaner Production*, 2010, **18**, 313.

5. K.J. Wong, S. Zahi, K.O. Low and C.C. Lim, *Materials and Design*, 2010, **31**, 4147.

6. U. Nirmal, B.F. Yousif, D. Rilling and P.V. Brevern, *Wear*, 2010, **268**, 1354.

7. M.M. Ibrahim, A. Dufresne, W.K. El-Zawawy and F.A. Agblevor, *Carbohydrate Polymers*, 2010, **81**, 811.

8. J.D. Megiatto, Jr, E.C. Ramires and E. Frollini, *Industrial Crops and Products*, 2010, **31**, 178.

9. I.M. De Rosa, C. Santulli and F. Sarasini, *Materials and Design*, 2010, **31**, 2397.

10. E.C. Ramires, J.D. Megiatto, Jr, C. Gardrat, C. Castellan and E. Frollini, *Bioresource Technology*, 2010, **101**, 1998.

11. A.I.S. Brígida, V.M.A. Calado, L.R.B. Gonçalves and M.A.Z. Coelho, *Carbohydrate Polymers*, 2010, **79**, 832.

12. K.C.C. Carvalho, D.R. Mulinari, H.J.C. Voorwald and M.O.H. Cioffi, *BioResources*, 2010, **5**, 2, 1143.

13. A.K. Bledzki, A.A. Mamun, A. Jaszkiewicz and K. Erdmann, *Composites Science and Technology*, 2010, **70**, 854.

14. D.R. Mulinari, H.J.C. Voorwald, M.O.H. Cioffi, G.J.M. Rocha, and M.L.C.P. Da Silva, *BioResources*, 2010, **5**, 2, 661.

15. H.U. Zaman, A. Khan, K.A. Khan, T. Huq, M.A. Khan, M. Shahruzzaman, M.M. Rahman, M. Al-Mamun and P. Poddar, *Fibres and Polymers*, 2010, **11**, 2, 258.

16. M. Krouit, M.N. Belgacem and J. Bras, *Composites: Part A*, 2010, **41**, 703.

17. A.K. Mohanty, M. Misra and L.T. Drzal in *Natural Fibers, Biopolymers, and Biocomposites*, Taylor and Francis Group, CRC, Boston, MA, USA, 2005, p.39.

18. F.T. Wallenberger and N. Weston in *Natural Fibers, Plastics and Composites*, Kluwer Academic Publishers, New York, NY, USA, 2004, p.249.

19. F. Willems and P. Moens in *Green Composites: Properties, Design and Life*, Nova Science Publishers, New York, NY, USA, 2009, Chapter 3.

20. K.K. Chawla in *Fibrous Materials*, Cambridge University Press, New York, NY, USA, 1998, p.115.

21. D. Fengel and G. Wegener in *Wood – Chemistry, Ultrastructure, Reactions*, Walter de Gruyter, New York, NY, USA, 1989, p.257.

22. D.G. Hepworth, D.M. Bruce, J.F.V. Vincent and G. Jeronimids, *Journal of Materials Science*, 2000, **35**, 293.

23. H. Kaddami, A. Dufresce, B. Khelifi, A. Bendahou, M. Taourirte, M. Raihane, N. Issartel, H. Sautereau, J. Gérard and N. Sami, *Composites: Part A*, 2006, **37**, 1413.

24. J.M. Raquez, M. Deléglise, M.F. Lacarmpe and P. Krawczak, *Progress in Polymer Science*, 2010, **35**, 487.

25. D.R. Mulinari, C.A.R.P. Baptista, J.V.C. Souza and H.J.C. Voorwald, *Procedia Engineering*, 2011, **10**, 2074.

26. H. Gu, *Materials and Design*, 2009, **30**, 3931.

27. E.M.S. Sanchez, C.S. Cavani, C.V. Leal and C.G. Sanchez, *Polímeros*, 2010, **20**, 194.

28. E.M. Teixeira, T.J. Bondancia, K.B.R. Teodoro, A.C. Corrêa, J.M. Marconcini and L.H.C. Mattoso, *Industrial Crops and Products*, 2011, **33**, 63.

29. B.L.L. Sipião, R.L.M. Paiva, S.A.S. Goulart and D.R. Mulinari, *Procedia Engineering*, 2011, **10**, 2028.

30. P.C. Miléo, D.R. Mulinari, C.A.R.P. Baptista, G.J.M. Rocha and A.R. Gonçalves, *Procedia Engineering*, 2011, **10**, 2068.

31. E.R. Gouveia, R.T. Nascimento and G.J.M. Rocha, *Química Nova*, 2009, **32**, 1500.

32. E.F. Rodrigues, T.F. Maia and D.R. Mulinari, *Procedia Engineering*, 2011, **10**, 2348.

33. R.L.M. Paiva, P.S. Souza, S.A.S. Goulart and D.R. Mulinari in *Second International Symposium Frontiers in Polymer Science*, Lyon, France, 2011.

34. C. Martin, B. Alriksson, A. Sjöde, T.N. Nilvebran and L.J. Jönsson, *Applied Biochemistry and Biotechnology*, 2007, **134**, 339.

35. V.F.N. Silva, P.V. Arruda, M.G.A. Felipe, A.R. Gonçalves and G.J.M. Rocha, *Journal of Industrial Microbiology and Biotechnology*, 2011, **38**, 809.

36. A.G. Silva, *Utilização do Pseudocaule de Bananeira para Produção de Celulose e Papel*, Universidade Federal de Viçosa, Viçosa, Portugal, 1998. [Dissertação (Mestrado em Ciência Florestal)]. [In Portuguese]

37. N. Cordeiro, M.N. Belgacem, I.C. Torres and J.C.V.P. Moura, *Industrial Crops and Products*, 2004, **19**, 147.

6 The Potential Use of Biomass as Reinforcement in Biocomposites

Manju Kumari Thakur, Vijay Kumar Thakur and Amar Singh Singha

6.1 Introduction

Himachal Pradesh is one of the hilliest states in India and is located in the Himalayan region. It has many kinds of vegetation and is full of different kinds of natural resources [1-3]. It produces a huge amount of biomass yearly which for the most part is wasted [4, 5]. The mountainous topography and lack of suitable roads mean that there is no proper utilisation of the natural waste biomass and excess biomass waste is anticipated to increase in the future as people are using the waste biomass to a lesser extent than in the past, when they used to make lots of household items from it especially from natural cellulosic fibres [6, 7].

In order to conserve energy and protect our environment, considerable attention should be given to the effective utilisation of waste biomass [8, 9]. One of the most important applications of waste biomass is its use to reinforce polymeric matrices [10, 11]. The incorporation of waste biomass as reinforcement in polymer composites brings a number of benefits especially environmental friendliness, high mechanical properties such as tensile strength, relatively high modulus of elasticity and the technology for the easy extraction of fibres from waste biomass for use in polymer composites is well-developed [12-14]. Waste biomass, especially lignocellulosic fibres, is becoming currently one of the indispensable components of polymer composites [15, 16]. It is frequently used as a reinforcement in polymer matrices and biomass-based composites are finding applications in many different areas such as furniture for the garden and house, parts for cars and technical devices [17-20]. In the case of polymer composites the efficiency of lignocellulosic fibres as reinforcements depends on the fibre/matrix interface and the ability to transfer stress from the matrix to the fibres [21, 22]. Depending upon the polymer matrix (natural or synthetic) the bond between the lignocellulosic fibres and matrices may be chemical, physical or a combination of the two [23, 24]. The chemical bonds in polymer composites are thought to be hydrogen bonds and/or hydroxyl bridges [25-31]. The hydrogen bonds may form between cellulosic fibres or between fibres and the polymer matrix. On the other hand in thermoplastic polymer matrices, physical bonding could occur during the preparation of composites

Among various fibres, *Saccaharum cilliare* (SC) fibres are obtained from a grass which grows abundantly in the hilly regions of Himachal Pradesh. SC fibres are lignocellulosic natural fibres obtained from this grass, which would otherwise not be used, and so are renewable resources which have the potential to be directly or indirectly used for a number of applications such as in green composites. SC fibres consist mainly of three major components, i.e., cellulose, hemicelluloses and lignin. Each of these components plays an important role in determining the strength of the materials. Lignocellulose, which is the prime constituent of biomass, is a linear polymer chain which is formed by joining anhydroglucose units into glucan chains [32-37]. The anhydroglucose units in cellulose are linked together by β-(1,4)-glycosidic bonds. This chapter focuses primarily on the synthesis and characterisation of lignocellulosic fibre-reinforced composites and gives a brief description of the mechanical properties of the composites.

6.2 Experimental

6.2.1 Materials and Methods

Phenol, formaldehyde solution and sodium hydroxide, used in the synthesis of polymers were kindly supplied by Qualigens Ltd., and were used as received. Phenol-formaldehyde (PF) resin was used as a novel thermosetting polymer matrix for preparing biocomposites. SC fibres were collected locally from the Himalayan region and were purified using a standard method as reported in some of the earlier studies [1]. After this the fibres were converted into long fibre form.

6.2.2 Fabrication of Biomass-based Biocomposites

For the fabrication of biomass-based biocomposites, the polymer matrix (PF) resin was synthesised by the standard method reported in the literature [1, 32, 33, 36, 37]. Subsequently, dried long SC fibres were mixed thoroughly with PF polymer resin at different loadings (10, 20, 30 and 40 wt%) using a mechanical stirrer. The above mixtures were then poured into specially made moulds. The surfaces of the moulds were coated on the inside with oleic acid to avoid adhesion of the biocomposite mixture of PF resin and the reinforcing cellulosic SC fibres. The coating on the surface allows the easy removal of the biocomposites from the mould. The mixture was then spread evenly on the surface of the mould. Composite sheets of size 150 × 150 × 5.0 mm were prepared by the compression moulding technique.

6.2.3 Mechanical Characterisation of Biomass-based Biocomposites

Mechanical characterisation of the biomass-based biocomposites prepared using long *Saccaharum cilliare* fibres was carried out by studying the properties such as tensile strength, flexural strength, compressive strength, and wear resistance. The properties were determined on specimens cut from the prepared green biocomposite materials. Tensile, compressive and flexural strength measurements were carried out on a computerised universal testing machine whilst wear testing was done on a wear and friction monitor. The specimens with dimensions of 100 × 10 × 5 mm were used for analysis. The tensile, compression, flexural and wear tests were conducted in accordance with American Society for Testing and Materials (ASTM) D 3039, ASTM D 3410, ASTM D 790 and ASTM D 3702 methods respectively under standard laboratory conditions [1, 32, 33, 36, 37].

6.2.4 Morphological and Thermal Behaviour of Biomass-based Biocomposites

In order to evaluate changes in the surface morphology of the long *Saccaharum cilliare* fibre-reinforced biocomposites, the surfaces were analysed by scanning electron microscopy (SEM). The excitation energy used was 5 keV. To achieve good electrical conductivity, carbon sputtering of all the samples was carried out, followed by sputtering with a gold-palladium mixture before examination. SEM micrographs of the biocomposite samples with different loadings (10-40 wt%) showed the morphology of the long SC fibre-reinforced green biocomposites prepared using a compression moulding technique. The thermogravimetric properties of the green polymer biocomposites were studied in a nitrogen atmosphere on a thermal analyser (Perkin Elmer) at a heating rate of 10 °C/min.

6.3 Results and Discucssion

A polymer composite is a material that is formed by two or more components which are combined as a macroscopic structural unit, with one component serving as a continuous matrix, and other as the reinforcement [1, 32, 33, 36, 37]. Among various kinds of polymer composites, biomass-based polymer composites are rapidly emerging as one of the materials for which the number of applications is growing rapidly especially in the automotive and the building industries. These composites are generally manufactured by dispersing the biomass in different forms, such as in particle form or short or long fibre form in polymer matrices to form a composite material using various processing techniques such as extrusion, compression or injection moulding. Polymer composites were first commercially prepared from PF and these were later developed with different modifications, depending upon the applications. In the present work the PF polymer resin which was subsequently used as the matrix was synthesised by the condensation polymerisation of phenol and formaldehyde [1, 36, 37]. The synthesised polymer matrix contains hydroxyl groups and posses good compatibility with the reinforcing SC fibres, as these fibres also contain hydroxyl groups. Because suitable functional groups are present on both the polymer matrix and the reinforcement, there are enough active sites for crosslinking during the precuring and curing processes. The resulting mechanical properties of SC fibre-reinforced polymer composites depend upon (a) the extent of bonding beetween the SC fibres and the polymer matrix, and (b) the effective load transfer from the polymer resin matrix to the SC fibres used as reinforcement [1, 32, 33, 36, 37]. It has been observed that the high degree of bonding between the polymer resin matrix and SC fibres as reinforcement facilitates the load transfer from the matrix to these fibres, which contributes to the higher mechanical properties. However SC fibre loadings beyond 30 wt% result in decreased mechanical properties due to the agglomeration of fibres at higher loading [1, 32, 33, 36, 37].

6.3.1 Mechanical Properties of Saccaharum Cilliare Fibre-reinforced Bicomposites

Figure 6.1a depicts the tensile strength for biocomposites with different SC fibre contents in long fibre form. The tensile strength of the biocomposites has been found to gradually increase with increasing SC fibre contents. It has been observed that an approximately linear relationship exists between the tensile strength and the fibre content over the fibre loadings studied. Composites containing 30 wt% SC fibres in the polymer matrix showed maximum tensile strength, followed by those which contained 40, 20 and 10 wt% SC fibres. The failure of the SC fibre-reinforced biocomposites at a particular loading during tensile strength testing, has been attributed to the breaking of the SC fibre reinforcement at the weakest point, followed by further cracks under the applied load which is transferred to the remaining intact reinforcement, leading to complete rupture of the composites [1, 32, 33, 36, 37]. **Figure 6.1b** shows the compressive strength results for the SC fibre-reinforced biocomposites. These results have been found to follow the tensile strength results. The failure of SC fibre-reinforced biocomposites during the compression test has been found to occur due to the sudden and dramatic buckling of the of SC [1, 32, 33, 36, 37]. The primary mode of failure in the SC fibre-reinforced composites with 10-40 wt% loading under compressive test can be due to the buckling of columns or microbuckling, which was caused by debonding and microcracking of the matrix. **Figure 6.1c** shows the flexural strength of SC fibre-reinforced biocomposites loaded with long SC fibres from 10-40 wt% loading. The flexural strength gradually increases with increasing SC fibre content and above 30 wt%, an increase of fibre content is likely to result in little improvement of flexural strength following the same trends as obtained in the tensile strength and compressive strength tests [1, 32, 33, 36, 37]. Wear resistance test results (**Figure 6.1d**) demonstrate that SC fibre used as long fibre reinforcement significantly improved the wear resistance as compared to the parent polymer matrix. It has been observed that the maximum wear resistance behaviour is shown by green composites with 30 wt% loading followed by 40, 20 and 10 wt% loading.

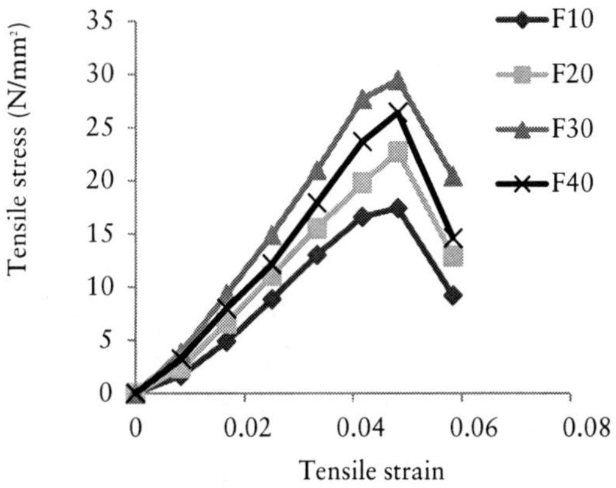

(a)

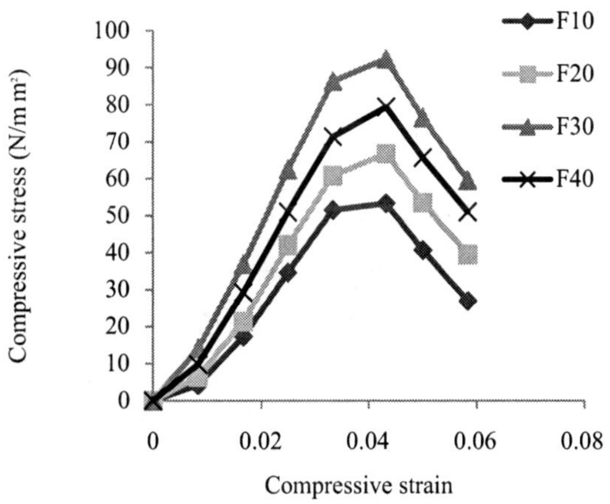

(b)

Figure 6.1 Tensile, compressive and flexural stress-strain and wear resistance results for *Saccaharum cilliare* fibre-reinforced biocomposites with different fibre loadings (F).

6.3.2 Morphological and Thermal Characterisation of the Biocomposites

The morphological results (**Figure 6.2**) as shown in the SEM micrographs of the polymer biocomposites with different SC fibre loadings. It has been observed that there is good intimate mixing of the optimised fibre loading (30 wt%) with the polymer matrix in the synthesised bicomposites. **Table 6.1** shows the thermogravimetric analysis (TGA) results of SC fibre, polymer resin matrix and the SC fibre-reinforced composite with 30 wt% loading. The biocomposites have been found to show intermediate behaviour between the thermal stability of the SC fibre and the polymer matrix. The behaviour is consistent with earlier reported results [1, 32, 33, 36].

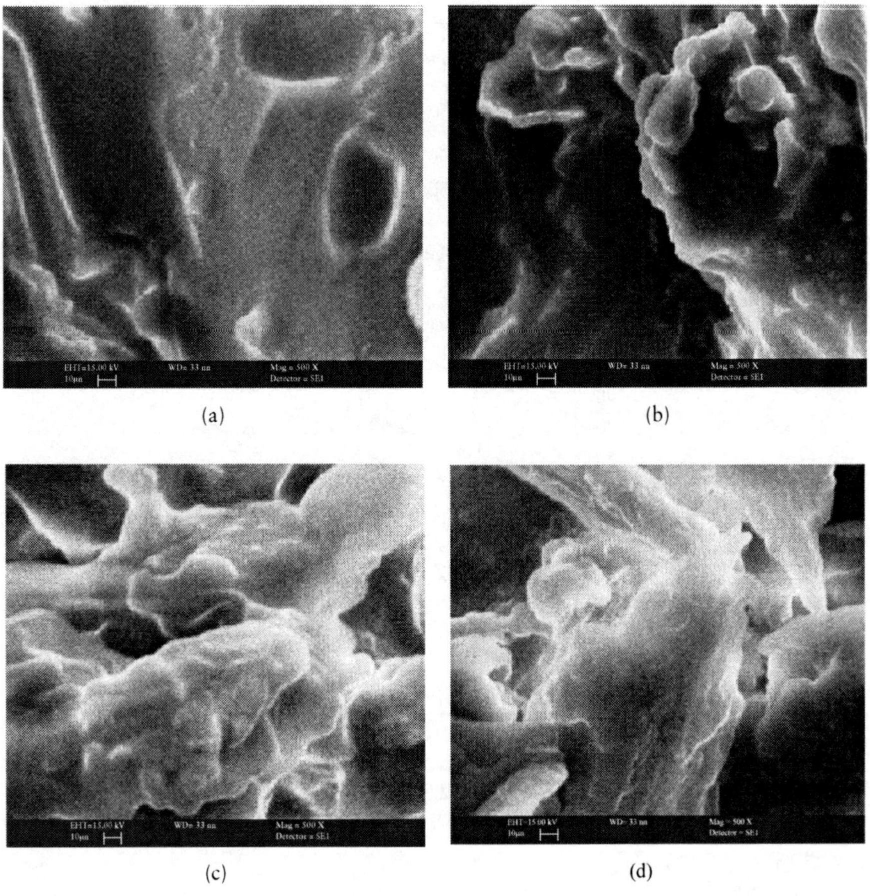

Figure 6.2 Scanning electron micrographs of cellulosic fibre-reinforced green composites with 10-40 wt% fibre loading

Table 6.1 TGA of PF polymer resin, SC fibre and biocomposites with optimum loading (30 wt%)

Sample specification	Sample name	IDT (°C)	Weight loss (%)	FDT (°C)	Weight loss (%)	Final residue (wt%)
1	SC fibre	210	14.6	479	83.7	16.3
2	PF polymer resin	404	17.3	1180	53.5	46.5
3	Composite	341	28.2	991	66.4	33.6

FDT: Final decomposition temperature
IDT: Initial decomposition temperature

6.4 Conclusions

SC fibres are an important biomass of the Himalyan region, which are considered as a waste product. These fibres were traditionally used by people of earlier civilisations to prepare different materials, such as ropes, carpets, baskets and so on. The present study successfully demonstrates that it is possible to use this waste biomass as a low cost environmental friendly reinforcement in biocomposites. The mechanical properties of the polymer resin matrix were significantly enhanced by reinforcement with SC fibres. These fibres are potentially very suitable candidates to replace the traditional toxic synthetic fibres for the synthesis of various kinds of biocomposites. These SC fibres can be used in the future for the fabrication of green materials which can be used in various fields including the automotive and building industries.

Acknowledgements

The authors wish to thank their respective institutes for providing the necessary facilities to accomplish the present research work.

References

1. V.K. Thakur, A.S. Singha and M.K. Thakur, *Advances in Polymer Technology*, 2013, **32**, S1, E427.

2. V.K. Thakur, A.S. Singha and M.K. Thakur, *Journal of Polymers and the Environment*, 2012, **20**, 1, 164.

3. V.K. Thakur, A.S. Singha and M.K. Thakur, *Journal of Polymers and the Environment*, 2012, **20**, 2, 412.

4. C.K. Williams and M.A. Hillmyer, *Polymer Reviews*, 2008, **48**, 1, 1.

5. A. Gandini, *Macromolecules*, 2008, **41**, 24, 9491.

6. V.K. Thakur, A.S. Singha and M.K. Thakur, *International Journal of Polymer Analysis and Characterization*, 2012, 17, 7, 547.

7. L. Yu, K. Dean and L. Li, *Progress in Polymer Science*, 2006, **31**, 6, 576.

8. V.K. Thakur, A.S. Singha and M.K. Thakur, *International Journal of Polymer Analysis and Characterization*, 2012, **17**, 6, 401.

9. V.K. Thakur, A.S. Singha and M.K. Thakur, *International Journal of Polymer Analysis and Characterization*, 2012, **17**, 2, 133.

10. V.K. Thakur, A.S. Singha and M.K. Thakur, *International Journal of Polymer Analysis and Characterization*, 2012, **17**, 1, 48.

11. V.K. Thakur, A.S. Singha and B.N. Mishra, *Journal of Applied Polymer Science*, 2011, **122**, 1, 532.

12. M. Sanchez-Garcia and J. Lagaron, *Cellulose*, 2010, **17**, 5, 987.

13. Y.X. Wang, H.F. Tian and L.N. Zhang, *Carbohydrate Polymers*, 2010, **80**, 3, 665.

14. H.D.X. Cao and Ch.M. Li., *Biomacromolecules*, 2007, **8**, 899.

15. V.K. Thakur and A.S. Singha, *International Journal of Polymer Analysis and Characterization*, 2011, **16**, 6, 390.

16. V.K. Thakur and A.S. Singha, *Bulletin of Material Science*, 2010, **33**, 3, 257.

17. V.K. Thakur and A.S. Singha, *Polymer-Plastics Technology and Engineering*, 2010, **49**, 11, 1101.

18. V.K. Thakur and A.S. Singha, *Polymer-Plastics Technology and Engineering*, 2010, **49**, 7, 694.

19. N. Lavoine, I. Desloges, A. Dufresne and J. Bras, *Carbohydrate Polymers*, 2012, **90**, 2, 735.

20. V.K. Thakur and A.S. Singha, *Iranian Polymer Journal*, 2010, **19**, 1, 3.

21. A.L. Goffin, J.M. Raquez, E. Duquesne, G. Siqueira, Y. Habibi, A. Dufresne and P. Dubois, *Biomacromolecules*, 2011, **12**, 7, 2456.

22. Y.X. Wang and L.Y. Chen, *Carbohydrate Polymers*, 2011, **83**, 4, 1937.

23. M.N. Angles and A. Dufresne, *Macromolecules*, 2001, **34**, 9, 2921.

24. A.P. Mathew and A. Dufresne, *Biomacromolecules*, 2002, **3**, 3, 609.

25. A.P. Mathew, W. Thielemans and A. Dufresne, *Journal of Applied Polymer Science*, 2008, **109**, 6, 4065.

26. M.J. John and S. Thomas, *Carbohydrate Polymers*, 2008, **71**, 3, 343.

27. A.S. Singha and V.K. Thakur, *Journal of Polymer Materials*, 2009, **26**, 1, 81.

28. A.L. Goffin, J.M. Raquez, E. Duquesne, G. Siqueira, Y. Habibi, A. Dufresne and P. Dubois, *Polymer*, 2011, **52**, 7, 1532.

29. V.K. Thakur and A.S. Singha in *Nanotechnology in Polymers*, Eds., V.K. Thakur and A.S. Singha, Studium Press LLC, Houston, TX, USA, 2012, Chapter 10.

30. A.P. Kumar and R.P. Singh, *Bioresource Technology*, 2008, **99**, 18, 8803.

31. A.S. Singha and V.K. Thakur, *E-Journal of Chemistry*, 2008, **5**, 1, 1055.

32. A.S. Singha and V.K. Thakur, *International Journal of Polymer Analysis and Characterization*, 2010, **15**, 2, 87.

33. Y.S. Lu, L.H. Weng and X.D. Cao, *Carbohydrate Polymers*, 2006, **63**, 2, 198.

34. A. Alemdar and M. Sain, *Composites Science and Technology*, 2008, **68**, 2, 557.

35. A.S. Singha and V.K. Thakur, *Polymer Composite*, 2010, **31**, 3, 459.

36. A.S. Singha and V.K. Thakur in *Green Polymer Materials*, Studium Press LLC, Houston, TX, USA, 2011.

7 Recent Progress in Polymer Natural Fibre Composites Made from Recycled Plastics

Shu-Kai Yeh, Christopher C. Yen and Johann Bödecker

7.1 Introduction

'Biocomposites', or more specifically natural fibre polymer composites, are defined as any combination of natural fibres from plants and polymers for the purpose of this publication. Strictly speaking, any material matrix with biofibres is a biocomposite including, but not limited to, premixed composites and manual combinations of a resin with a biobased substrate. As the demand for biocomposites is on the rise new variations emerge and old technologies resurface. The global natural fibre composites market reached 2.1 billion USD in 2010, and the compound annual growth rate has been 15% in last five years. By 2016, the natural fibre composite market is expected to reach 3.8 billion USD [1].

However, the history of the use of plant fibres to enhance other materials stretches further back; three thousand years ago a mixture of of twitch grass and clay was the first biocomposite that was used to build clay huts. Natural fillers have even been known to be used in concrete as early as Roman times. The first high performance integration into polymers was achieved in 1939 in Britain for the purpose of manufacturing parts of the Spitfire fighter with flax and phenolic resin. Ford's 1942 prototype car famously used a hemp polymer composite, and the famous East German Trabant car used a cotton and polyester composite [1]. It is interesting to see that there are almost no intermediate steps between the clay huts of our ancestors and the aeronautical applications during the Second World War. The commercial use of polymers only becomes established in the early 1940s and develops hand in hand with that of composites (not only biocomposites). In 1942, military applications of polyester and glass fibre are already on the way. It takes an amazing 50 years before such composites are used in aeronautics again but the technology has not changed much. Obviously, polymer science and manufacturing techniques have evolved in leaps and bounds and now biofibres have been identified as suitable for use in biocomposites.

This chapter deals specifically with composites of recycled polymers and plant fibres.

The reason that this is a noteworthy subcategory is the fact that the main force behind the recent swift rise of biocomposites is their lower embodied energy, so using recycled polymers to complement renewable fibres is the next logical step. In the following section we will argue that the developments of biocomposites are highly dependent on non-scientific factors like: government regulations and market demands. Case studies of existing companies that market their composites successfully will help to demonstrate the importance of these factors and present the 'hot' research areas that will receive attention and funding to develop new materials in the future.

7.2 Background on Natural Fibres

Natural fibres usually contain cellulose, hemicellulose, pectin, lignin, water-soluble ingredients and wax [2]. The actual composition of these materials in different natural fibres varies from species to species. Here, water-soluble constituents and wax are considered as extractives since cellulose, hemicellulose, and lignin are considered to be the basic components with regard to physical properties [2]. Generally, cellulose is the main component of natural fibres and it occupies more than 50 wt% [3]. The chemical structure of cellulose is shown in **Figure 7.1**. The hydroxyl groups on the cellulose make natural fibres hydrophilic. These hydrophilic hydroxyl groups absorb water molecules easily and swell the fibres. The theoretical swelling pressure for natural fibres can be as high as 1630 atm [4]. Thus, coupling agents or surface modifications are needed in order to stop the fibres from absorbing water into the polymer matrix.

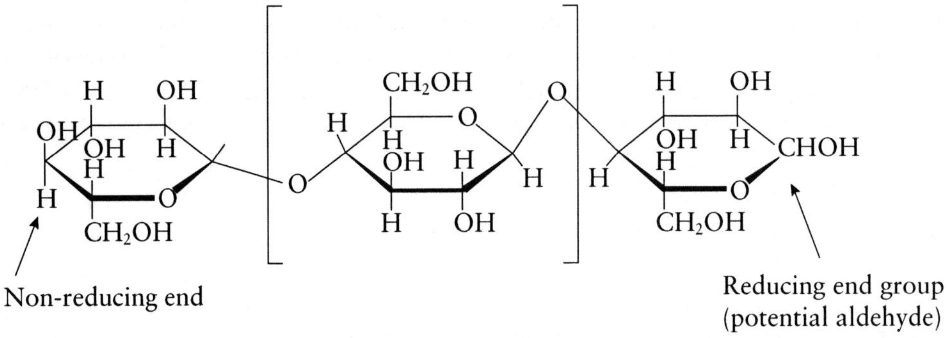

Figure 7.1 Chemical structure of cellulose [2]. Here, 'n' indicates the number of repeating units

The moduli of cellulose-based materials are very different from each other. For example, the Young's modulus of solid wood is around 10 GPa. With the use of an appropriate pulping process the modulus of a single pulp fibre can be as high as 40 GPa. Microfibrils can be separated from pulp fibres after mechanical disintegration and hydrolysis. The modulus of microfibrils is around 70 GPa. Finally, the modulus of cellulose nanocrystals is obtained by theoretical calculation, and this can be as high as 250 GPa [2]. The modulus values of wood fibres range from 10 to 90 GPa and other natural fibres can have moduli as high as 128 GPa [2, 3]. From these values, we can say that it is reasonable to compound natural fibres with plastics to increase the modulus of resins of interest because the modulus of natural fibres is much higher than that of most plastics. The high moduli of natural fibres make these good candidates to be reinforcements for plastics. Additionally, the price of natural fibres is usually much lower than that of polymer matrices and they provide other advantages such as biodegradability. Although the mechanical properties of natural fibres are usually lower than those of E-glass fibres, the low density of natural fibres offers more flexibility in composite structure design. The specific modulus or specific strength (modulus or strength divided by density) is usually higher and the current price of natural fibres is also much lower than that of E-glass fibres. Not only does the biodegradability of natural fibres comply with the green concept, producing natural fibres also consumes about 30-40% less energy than producing glass fibres. Life cycle analysis assessment has shown that making natural fibre-reinforced composites consumes about 40-60% less energy compared to glass fibre-reinforced composites [5].

As mentioned earlier, the properties of natural fibres vary from species to species and region to region. Usually, they contain approximately 6-22 wt% of water and need to be dried before processing [3]. This can cause problems in the utilisation and processing of natural fibre composites (NFRC). The main problems related to the processing of NFRC are [6]:

- Region and species variations in the quality of raw material.
- Compatibility limitations due to compounding hydrophilic natural fibres with hydrophobic matrix.
- Limited thermal stability during processing. Natural fibres cannot be processed above 200 °C or they will burn and change the appearance and mechanical properties of the composite.
- Shape deviation of the component caused by the swelling of natural fibres.

Among the problems stated above, the quality variation and thermal instabilities are intrinsic problems associated with natural materials. Nevertheless, the compatibility of natural fibres with thermoplastics and the shape deviation caused by the swelling of natural fibres can be solved by appropriate modification of the surface of the fibres. In this way, the strength of NFRC can be improved, and the rate of water absorption of the composites can be reduced.

There are many review articles that summarise the mechanical properties, moisture content, price, content and surface modification of natural fibres [2, 3, 6-10]. Among these articles, Bledzki and co-workers were the first to review the mechanical properties, surface properties, surface modification methods and prices of natural fibres [2, 6]. Later on, the same research group published another review article that focuses on using natural fibres for automotive applications [10]. The major advantage of applying natural fibres in automobiles is their biodegradability which could significantly reduce the non-recyclable portion of automobiles. The recent review article written by Hota and co-workers summarises the properties of natural fibres from most of the published work and integrates the information into useful tables and charts [3]. The article highlights the applications of using NFRC in building and construction of infrastructures. It also provides information from the viewpoint of economics and environmental impact. Due to the limitations of this chapter, the compiled properties of natural fibres are not listed here but they can be found in references [3] and [11].

Unless governments provide more incentives, the cost of growing and harvesting natural fibres in developed countries will not be able to compete with natural fibres grown in developing countries. In the past, those fibres were considered as waste but now, growing them can create new jobs. Developing countries can not only provide natural fibres, they can also manufacture the NFRC. Since natural fibres are hydrophilic and thermoplastic polymers are hydrophobic, surface modifications of fibres or additives such as coupling agents are needed. However, due to the limitations of this chapter, this will not be discussed here. More detailed information can be found in reference [8].

In summary, the high modulus and low price of natural fibres make them good candidates as plastic reinforcements. However, the thermal instability of natural fibres places an upper limit on the processing temperature. Other intrinsic properties such as quality variation and shape irregularity also limit the applications of natural fibres. The most important problem, however, is the hydrophilic nature of natural fibres. Due to this, water uptake swells natural fibres and affects the dimensional stability and the fungus susceptibility of the composite.

7.3 Fibre Markets

7.3.1 Classification by Origin

Natural fibres used in NFRC can be classified into three categories in terms of their production:

- Fibres that are specifically grown for use in biocomposites.
- Fibres which are by-products after plants have been used for different applications.
- Waste fibres which are currently too expensive to dispose of because there is not sufficient demand for them in the NFRC industry.

It is not necessarily true that the less desirable fibres are the less abundant ones (quality and quantity are not inversely correlated) but some abundant fibres of inferior quality make their way into biocomposites. For example the supply of coconut fibres is plentiful as the production of coconut oil is a prospering industry and, coincidentally, coconut oil is also a raw material for some polymers.

In addition to availability, shipping distance, storage considerations and geographic distance from the production area are the main considerations when using natural fibres. The problem can be overcome by proper supply chains that fall into place as the demand scales up.

7.3.2 Classification by Biological Features

Plant cells naturally have rigid outer walls, giving them very different mechanical properties from animal cells Plant fibres can be separated into three categories, i.e., those which are obtained from the leaf, stem or fruit/nut of the plant. The properties are different depending on the origin of the fibre. Generally speaking, the most desirable structural features for tensile strength can be found in the stem and leaf fibres. However, fruit and nut fibres tend to be more affordable and therefore have to be considered. The use of biofibres is strongly related to their cost effectiveness as displayed in **Table 7.1** [12].

Because of the big price difference between natural fibres and synthetic polymers currently, the main applications of biofibres are not for NFRC but for fibre boards and insulation panels that are manufactured without polymers. Arguably the stage is set for a trend towards NFRC.

Table 7.1 Cost effectiveness of natural fibre and other reinforcing materials						
Fibre	Carbon	Steel	Glass	Sisal	Jute	Coir
Cost (USD/kg)	200	30	3.25	0.36	0.3	0.25
Modulus/cost (GPa/USD)	2	6.7	21.5	41.7	43.3	20
Reproduced with permission fromA.K. Mohanty, M. Misra and L.T. Drzal, *Journal of Polymers and the Environment*, 2002, 10, 1-2, 19. © 2002, Springer						

7.4 Introduction to Recycled Plastics and Waste Flow Analysis

Plastics are commonly used in everyday life because they are lightweight, durable and low cost materials. However, the non-biodegradability of plastics causes severe environmental problems. Based on their properties; they can be classified as thermoplastic materials and thermoset materials. Plastic waste usually makes up 7% of the solid waste stream. Of this, thermoplastic materials contribute to 80% of the polymer consumption in Europe [13]. Thermoplastic materials are considered to be recyclable materials since they can be melted and reprocessed by extrusion. Plastic recycling could significantly reduce the amount of solid waste. In addition, most plastics are made from petroleum. It is estimated that around 4% of annual petroleum production is converted to the feedstock for plastics and it will take an equal amount of fossil fuel to make plastics since making plastic materials requires energy. Therefore, the price of plastic materials increases with the price of oil [14]. The composition of plastics usage in Europe is shown in **Figure 7.2**.

Depending on the applications, packaging materials account for 35-37% of plastics usage, which is the single largest use [15]. Other than that, plastic materials are also used in building and construction, furniture and housewares, the automotive and transport industries, and agriculture [14]. Since the life cycle of packaging materials is considerably shorter than other applications, the main stream of plastic recycling is focused on that of these materials. The annual consumption of plastic packaging materials per person is as high as 40 kg and it keeps growing at an annual rate of 3%. If those materials could be recycled and recovered, energy consumption could be reduced significantly. The major plastic packaging materials are PET (drinks bottles), polyolefins (milk containers and plastic bags), PVC (labels) and PS (packaging foam).

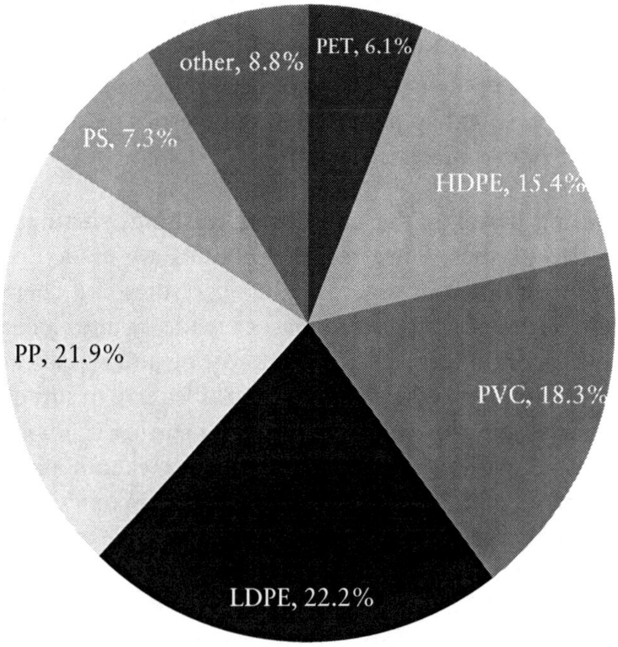

Figure 7.2 Plastics usage in 15 European Union countries in 2002 (HDPE: high density polyethylene; LDPE: low density polyethylene; PET: polyethylene terephthalate, PP: polypropylene; PS: polystyrene; and PVC: poly(vinyl chloride)). Adapted from J. Hopewell, R. Dvorak and E. Kosior, *Philosophical Transactions of The Royal Society B: Biological Sciences*, 2009, **364**, 1526, 2115 [14]

According to American Society for Testing and Materials standard D7209, there are four major methods of recycling and recovering plastics: re-extrusion (primary recycling), mechanical recycling (secondary), chemical recycling (tertiary) and energy recovery (quaternary). Chemical recycling is defined as converting polymers back into monomers and energy recovery means the incineration of plastics for power generation. Since chemical recycling is an evolving technology, and since capital investment for a chemical plant is considerably higher than in a plastics processing facility, it would be fairly difficult to convince the current plastic recycling business to move from mechanical recycling to chemical recycling. The incineration of plastics causes severe air pollution problems. The combustion of plastics are known to generate volatile organic compounds, smoke, particulate-bound heavy metals, polycyclic aromatic hydrocarbons, polychlorinated dibenzofurans and dioxins. A reactor design which will lead to complete combustion of plastics must be used [13]. For these reasons tertiary and quaternary recycling will not be discussed in detail in this chapter.

Re-extrusion is the reprocessing of industrial plastic scrap and rejected parts by extrusion. The products made by the re-extrusion process are similar to those made of virgin plastics since the polymers have not been used and contaminated by other materials. Re-extrusion is widely practiced in the plastics industry since it can save a significant amount of raw material cost.

Mechanical recycling involves the collection, washing, sorting, shredding and extrusion of recycled plastics. Since recycled plastics are made up of all kinds of different polymers with different processing temperatures and chemical structures, the mechanical properties of the recycled plastics made by mechanical recycling will be inferior to those of virgin plastics. Thus, the sorting and classification of recycled plastics is very important. Nevertheless, it is very challenging to sort different plastics. A typical solid separation technique such as density sorting is not very helpful since the densities of plastics overlap with each other. Other technologies such as Fourier transform near infrared and X-ray fluorescence spectroscopies and optical colour sorting systems using charge-coupled device cameras are able to identify the chemical structure of different plastics and thus sort the polymers [13, 14]. Unfortunately, no matter what kind of technology is applied, the recycled plastics made from the mechanical recycling process are usually contaminated with a certain amount of impurities. These impurities will act either as nucleating agents or stress concentrators and have a detrimental effect on the mechanical properties and appearance of the plastics. Therefore, the applications of recycled plastics are limited and the price of recycled plastics is usually lower than that of virgin polymers.

Since mechanically recycled plastics usually contain impurities, it is usual to compound 20-50% fillers or reinforcements with them in order to mask the effects of the impurities. The properties of recycled plastics containing a high concentration of fillers can be compared to those of a virgin polymer that contains the same level of fillers. Wood/plastic composites are good examples that adopt this strategy [16]. The mechanical properties of recycled plastics mixed with other resins can also be manipulated by adding elastomers or compatibilisers which will significantly improve the quality of the recycled plastics. Detailed information can be found in reference [17]. In the following paragraphs, the plastic waste flow analysis and the concept of compounding different natural fibres with recycled plastics will be reviewed.

There are two big factors that primarily influence the percentage of post-consumer polymers that end up being recycled. One is the willingness of the consumer to separate their waste (and therefore, implicitly, the incentives he receives for doing so) as this leads to a purer waste stream, e.g., thermoplastics only and so on. On the other hand there are the municipal and national collection points, recycling facilities, where it is necessary to reach a critical amount of any one given polymer in order to make it worthwhile recycling it. Therefore, the purity and volume of the waste stream are crucial factors and they need to be stable, or at least predictable, over time in order for an economically viable recycling infrastructure to develop. **Figure 7.3** gives a good idea as to which materials are predominant in the waste-stream:

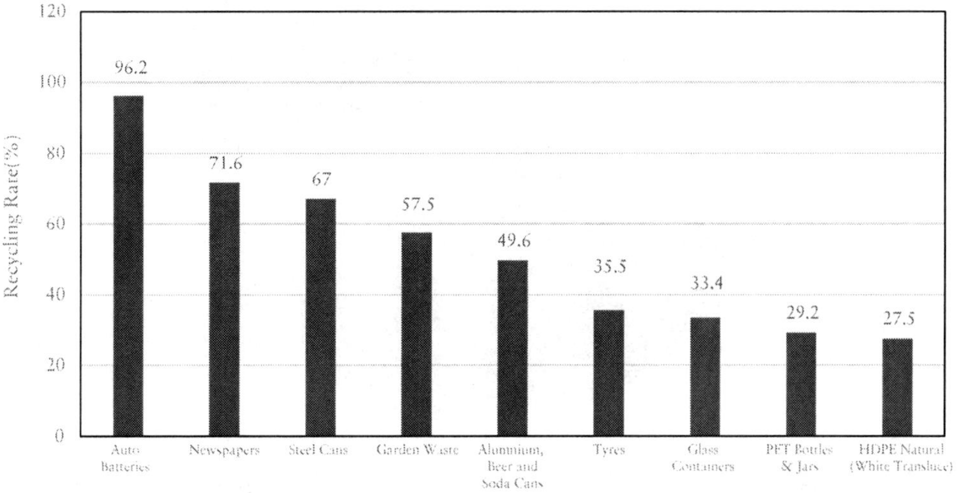

Figure 7.3 Recycling rates of selected products, 2010 [18]

It is important to recognize that the routes travelled by domestic and municipal waste after being discarded are long. On the one hand, recycling rates are stipulated by government. Many companies treat waste material as a liability, while other companies take certain waste and use it as a commodity. Furthermore not all industrial waste and municipal waste separation standards are alike hence extra efforts to attain a homogenous inflow may be required.

Note that the recycling rates in **Figure 7.3** only refer to how consumers separate their waste and not to how much of that waste will actually be recycled into products. **Figure 7.4** shows the types and amounts of municipal solid waste generated in the USA in 2010. Of the total municipal waste stream the 12.4% that consists of plastics is among the most valuable for recycling:

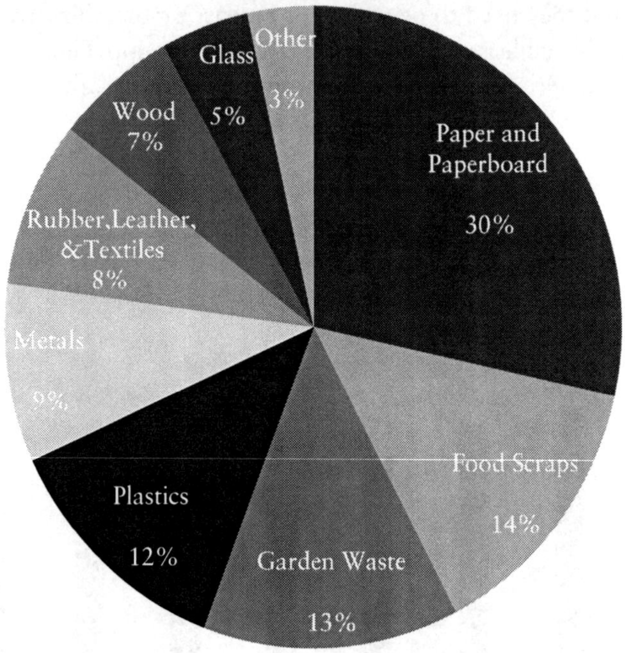

Figure 7.4 Total municipal solid waste generation: 250 million tons (before recycling) in the USA in 2010 [18]

This Figure does not include the large numbers of PET drinks bottles which are often collected separately by private companies. PET bottles are the success story of recycling; they are collected effectively, the PET is re-extruded and sold, and rather than being a liability, every link in the recycling chain can turn a profit. However, with respect to NFRC, polymers like PP and PE are of greater interest. It is still estimated that only eight out of ten PET bottles are being recycled in the USA. As most of the HDPE, LDPE and PP waste comes from packaging made with different additives in different colours, it is hard to obtain one homogenous polymer waste stream [14]. This could limit the use of recycled polyolefins for making NFRC.

PP and PE (polyolefins) are the two major recycled plastics that are used for making NFRC because the processing temperature of natural fibres has to be below 200 °C and there is an abundance of post-consumer polyolefins available [14]. It is reasonable to compare the properties of NFRC prepared from recycled plastics with those made from virgin plastics. If their properties and appearances are comparable, the composite manufacturers have a better reason to convince consumers to use recycled materials. However, in practice, it is not easy to make a side-by-side comparison. The reasons are simple; the source and quality of recycled plastics are very different from place to place. It is very easy to purchase virgin polyolefin with specific physical properties. However, when plastics are recycled, it is almost impossible to separate different grades of polyolefin since their densities are similar to one other and the chemical structures of their monomers are exactly the same. In addition, various plastics additives are commingled during the recycling process and their impact on the recycled polymer remains unknown. Since there are many factors which may affect the physical properties of recycled plastics, it is unfair to compare the properties of recycled plastics with those of the virgin one. Because of these issues, there are many research articles that have worked on using recycled polyolefins to make NFRC, but only a few of them have compared their properties with those of composites prepared from virgin polymers and the conclusions are not clear. Some of the researchers used actual recycled plastics taken from post-consumer products and others simulated recycled plastics by mixing different virgin polymers or by the re-extrusion of virgin plastics or NFRC. Many of the research articles claimed that the properties of composites made from recycled plastics are superior to those made of virgin plastics, but this is not always true. The results of the research articles that compared NFRC made from both virgin and recycled polyolefins are listed in **Table 7.2**, whilst the results of studies which used only recycled polyolefin as the composite matrix are listed in **Table 7.3**.

7.5 Natural Fibre Composites made with Recycled Polystyrene and Acrylonitrile Butadiene Styrene

As previously mentioned, PS only accounts for 7.3% of the total annual plastic usage in Europe, which is far less than polyolefins (59.5%) [14]. The main applications of PS are plastic cups, expandable polystyrene (EPS) foam and extruded PS foam board for insulation applications. Foamed plastics, especially low density foam, cannot be easily recycled. The high specific volume of foam makes it difficult for kerbside collection. Although foam can be easily separated from other plastics, feeding it into an extruder for reprocessing becomes another challenge. Extra steps such as melt compression are needed to increase its density. The specific gravity of EPS foam increases from 0.022 to 0.5 after melt compression [57]. Therefore, although there are many research articles that use PS as the matrix of NFRC, only a few research articles have used recycled PS [57-59]. Since PS is hydrophobic, it is not compatible with WF and coupling agents are needed to improve the strength of the composites. The best candidate is styrene-*graft*-maleic anhydride (SMA) [57].

Table 7.2 References that directly compared the properties of NFRC made from virgin plastics and recycled plastics				
Polymers used	Properties compared	Type of fibres	Comments	References
PP	Mechanical properties and moisture absorption.	Newsprint	Recycled newsprint fibres and nanoclay were compounded with 70 wt% of PP or recycled PP (r-PP). The mechanical properties of r-PP composites were inferior to virgin-PP (v-PP) composites and the rate of water absorption of r-PP composite was higher than v-PP composites. A high loading of PP-*graft*-maleic anhydride (PP-*g*-MA) (10 wt%) was used to disperse the nanoclay. A high content of nanoclay (5 wt%) did not improve the mechanical properties of the composite significantly due to poor dispersion but the flame retardancy improved with the clay loading. Adding nanoclay did not change the diffusion coefficient of the material.	[19, 20]
HDPE	Mechanical properties, dimensional stability, fungal resistanceand appearance durability.	Virgin pine, recycled urea-formaldehyde bonded particle board, and recycled chromated copper arsenate (CCA)-treated red pine wood.	Virgin and recycled HDPE (r-HDPE) were made into wood/plastic composite board by hot pressing. The mechanical properties of samples made from virgin plastics are stronger than those made of r-HDPE. CCA-treated wood showed a much lower level of fungal existence than other samples. The appearance of samples made with r-HDPE did not change after long-term weather exposure. This is one of the advantages of using r-HDPE as the matrix.	[21]
PP, HDPE and PS	Mechanical properties	Wood flour (WF)	Recycled plastics were simulated by mixing virgin or r-PP, HDPE and PS at a ratio of 7:1.5:1.5. The coupling agents used were elastomer-based styrene-(ethylene-*co*-butylene)-styrene grafted with maleic anhydride (SEBS-*g*-MA) or *in situ* grafting with maleic anhydride onto the fibres by reactive extrusion. Samples made of mixed recycled plastics showed inferior mechanical properties to those made of mixed virgin plastics.	[22]

Table 7.2 Continued				
Polymers used	**Properties compared**	**Type of fibres**	**Comments**	**References**
PP + HDPE	Mechanical properties	WF	The mechanical properties of samples made with virgin polyolefin were comparable with those of samples made with recycled ones.	[23]
PP + HDPE	Moisture absorption and diffusion coefficient.	WF	The composites made from recycled plastics showed significantly higher moisture absorption than those made of virgin plastics. The worst results were observed when a ratio of 1:1 r-PP and r-HDPE were mixed and used as the polymer matrix.	[24]
PP	Mechanical properties	WF (sawdust)	r-PP was made by re-extrusion of v-PP in a twin screw extruder. The results showed that passing PP through the extruder more times results in better mechanical properties of wood/plastic composites (WPC).	[25]
HDPE	Mechanical properties	Rice straw fibre	The mechanical properties of r-HDPE were better than those of virgin HDPE (v-HDPE). Thus, the mechanical properties of r-HDPE composites were better than those made of v-HDPE.	[26]
HDPE	Mechanical properties, moisture absorption and dimensional stability.	WF	Virgin and recycled HDPE of similar density and melt flow index were selected and mixed together as the matrix material to compare the physical properties of their WPC. The mechanical properties of r-HDPE were better than those of v-HDPE and the rate of water absorption of composites made of r-HDPE was lower than those made of v-HDPE. The authors claimed that the presence of a mixed type of polymer and chemical impurities in the r-HDPE, which might enhance interfacial bonding.	[27]
PP and HDPE	Diffusion kinetics	WF	The moisture diffusion kinetics of WPC made from v-PP, r-PP, v-HDPE and r-HDPE were determined by long-term moisture absorption experiments. The diffusivity of r-HDPE and r-PP WPC samples were lower than their counterpart, v-HDPE and v-PP WPC samples. Adding coupling agents would significantly reduce the diffusivity.	[28]

HDPE	Dimensional stability	WF and rice husk	Thickness swelling of NFRC made of v-HDPE and r-HDPE was measured at 20, 40 and 60 °C. Composites made with r-HDPE showed less thickness swelling. In addition, composites made with rice husk showed less moisture absorption than those made from WF.	[29]
PP	Mechanical properties	WF and nanoclay	The mechanical properties of samples made with r-PP and v-PP were comparable. Adding nanoclay did not improve the modulus of the composite.	[30]
PP and HDPE	Mechanical properties	Flax fibre mats	The composites were made by overlaying four layers of flax fibre mats between five layers of v-PP or r-HDPE. The polymer was infused into the fibre mats *via* hot pressing. It was found that the fracture toughness of composites made of r-HDPE was comparable to those made of v-PP. The optimised flax content was 20 vol%.	[31]
PP	Mechanical properties	Kenaf and rice husk	In this study, 30 wt% of kenaf or rice husk fibres were compounded with 70 wt% PP by twin screw extrusion and the composites were made by compression moulding. The recycling of composites was simulated by re-extrusion. The strength of both composites decreased after recycling but the modulus of kenaf composites increased after re-extrusion.	[32]
PP	Mechanical properties, thermal stability and moisture absorption.	WF	WPC were made by twin screw extrusion and injection moulding. The recycling of WPC was simulated by re-extrusion of the composites eight times. The WF content was controlled at 40 or 50 wt%. The modulus and strength of samples containing 40 wt% of WF decreased with an increased number of re-extrusion cycles. The mechanical properties of the composites could be predicted by the rule of mixtures. When the fibre loading increased to 50 wt%, the strength and modulus increased until the composites were processed three times and then started to decrease. The impact strength decreased with the number of re-extrusion cycles.	[33, 34]

			The rate of moisture absorption of the composites decreased with the number of recycling times.	
HDPE and PVC	Mechanical properties	Bagasse fibre and WF	Bagasse or WF was compounded with an equal amount of virgin or polymer resin to make NFRC. The virgin resin was HDPE and the recycled resin was either r-HDPE or recycled PVC. The creep behaviour was discussed extensively and several models were compared. The experimental results showed that composites made with virgin plastics possessed superior mechanical properties.	[35]
PP	Mechanical properties	WF and nanoclay	The mechanical properties of v-PP and r-PP-50 wt% WF WPC were studied. Nanoclay was added to reinforce the mechanical properties of the sample. Adding nanoclay did not help to increase the modulus of the composites. When the clay content increased from 3 to 5 wt%, the modulus even decreased.	[30]
HDPE, PP, PS and PVC	Mechanical properties and thermal stability	Cellulose	Virgin HDPE, PP, PS and PVC were mixed at a ratio of 7:1:1:1 to simulate recycled plastics. 12.5 phr cellulose fibres were added to reinforce the mechanical properties of the polymer blends and SEBS-*g*-MA was used as the coupling agent. To compare the properties with actual recycled polymers, a recycled HDPE/PP/PS/PVC blend was also made for comparison. In this study, virgin polymer blends showed better mechanical properties than recycled polymer blends. Adding cellulose fibres could increase the mechanical properties of the polymer blends, but not their thermal stability.	[36]

Table 7.3 NFRC that used recycled polyolefins as the matrix material				
Polymers used	**Properties compared**	**Type of fibres**	**Comments**	**References**
PP and recycled acrylonitrile-butadiene rubber	Mechanical properties, rate of water absorption and dimensional stability.	Rice husk	The modulus and strength of the composites increased with the loading level of rice husk powders but the elongation at break decreased. On the other hand, the thickness of swelling also increased with the fibre loading. The physical properties of the composites could be improved by treating the fibres with silane coupling agents or by adding coupling agents such as PP-*g*-MA.	[37-39]
PP and recycled PET	Mechanical properties	Rice husk	This article proved that recycled PET can replace 40 wt% of the polymers at most. PET did not melt during extrusion but acted as a solid filler in the composite. The modulus of the composite slightly increased with the loading level of PET but the elongation at break decreased. It is interesting to note that the impact strength of the composite increased with the loading level of PET.	[40]
r-HDPE	Mechanical properties and water absorption	Recycled WF from medium density fibre board and particle board.	Recycled WF was compounded with r-HDPE by twin screw extrusion. The wood content was controlled at 60, 70 or 80 wt%. The physical properties of the composite initially increased with the loading level of wood content but started to decrease when the wood content increased to 80 wt%. High content of fibres cannot be bonded well with the polymer matrix.	[41]
Recycled LDPE (r-LDPE) and v-PP	Mechanical properties and dimensional stability.	WF	The author intentionally compounded v-PP with r-LDPE to simulate the recycling of LDPE. WF was added as the reinforcement of the composite and the mechanical properties and rate of moisture absorption of the composites were studied. Two grades of PP-*g*-MA were applied as the coupling agents. The results showed that adding PP-*g*-MA could improve the mechanical	[42]

			properties and reduce the rate of water absorption; it could be applied as the coupling agent in this system.	
r-HDPE	Mechanical properties, water absorption, and dimensional stability.	Rice hull, wood sawdust and sanding flour from medium density fibre board and particle board.	Fibres and r-HDPE were compounded by high speed mixer and then hot pressed into NFRC. The mechanical properties of these composites were similar except that wood sawdust NFRC showed relatively low strength. Rice hull composites showed the lowest rate of water absorption in the short term moisture absorption tests. The thickness swelling could be predicted by modelling.	[43]
r-HDPE	Mechanical properties and water absorption.	WF	The fibres were treated with three surface treatments including alkali treatment, silane treatment and a combination of alkali and silane treatments to improve interfacial adhesion. PP-*g*-MA was also used as the coupling agent. Fibres treated with both alkali and silane showed the highest strength and lowest thickness swelling.	[44, 45]
r-HDPE and r-PP	Mechanical properties	Recycled newspaper	PP-*g*-MA can be used as the coupling agent for recycled newspaper/r-PP or r-HDPE composites. The mechanical properties of the composites were improved by adding PP-*g*-MA.	[46]
r-LDPE	Mechanical properties	WF	r-LDPE can be crosslinked and used as the matrix material of the composite. High degree of crosslinking may result in melt fracture during the extrusion process.	[47]
PP and HDPE	Mechanical properties	WF	Recycled plastics were simulated by blending v-PP and v-HDPE at the ratios of 80:20 or 20:80. Maleic anhydride was grafted onto the WF by reactive extrusion. Samples with high PP content showed superior tensile strength and modulus.	[48]

Table 7.3 Continued				
Polymers used	**Properties compared**	**Type of fibres**	**Comments**	**References**
r-HDPE	Mechanical properties, and biodegradability	Recycled pine wood	Recycled pine wood was compounded with r-HDPE at a ratio of 6:4. The samples were then age accelerated by ultraviolet (UV) exposure and high temperature and tested for biodegradability. Aged samples were more susceptible to termite attack. Termites can destroy the r-HDPE/recycled pine wood interfaces and adversely affect the mechanical properties of the composite. Adding a coupling agent, HDPE-*graft*- maleic anhydride (HDPE-*g*-MA), could slightly retard termite attack.	[49]
r-HDPE	Mechanical properties and water absorption	WF	r-HDPE and WF were compounded by a laboratory scale air mixing process and compression moulded into test specimens. The WF loading levels were 55, 70 or 85 wt%. The coupling agent used was HDPE-*g*-MA. Moisture absorption and mechanical test results showed that the optimum amount of coupling agent was 2 wt%. The mechanical properties results showed that the strength of composites increases with WF loading.	[50]
r-HDPE	Material processing and mechanical properties	WF	r-HDPE/WF composites were prepared by reactive extrusion. Maleic anhydride groups were grafted onto the surface of the fibres and dicumyl peroxide (DCP) was used as the initiator. The optimised DCP content was 0.06-0.09 wt%. High loading levels of DCP did not help to improve the mechanical properties of the composite and caused severe processing difficulties such as high torque and high die pressure.	[51]

HDPE, LDPE and PP	Mechanical properties	Recovered paper fibres and WF.	v-PP was added into v-HDPE or virgin low density polyethylene (v-LDPE) to simulate properties of recycled plastics and used as the matrix material of WPC. Adding v-PP did not reduce the mechanical properties of the composites. The properties of v-LDPE-based WPC even increased with the loading level of PP. Also, HDPE-*g*-MA could be used as a coupling agent in the composite.	[52]
r-LDPE	Mechanical properties and dimensional stability.	Wheat straw	Two coupling agents, waterborne polyacrylate latex, and a blend of polymethylene polyphenylene isocyanate and polyacrylate latex, were synthesised by emulsion polymerisation. After adding the coupling agent, the mechanical properties improved and thickness swelling reduced.	[53]
r-PP	Mechanical properties and foaming ability.	WF	WPC foam containing r-PP and 25 wt% WF was produced by injection moulding. The optimised foaming pressure and temperature were 12.5 MPa and 180 °C, respectively. The impact strength of the WPC tripled with increased cell density.	[54]
r-HDPE	Screw and nail withdrawal resistance.	WF from medium density fibre board and particle board.	r-HDPE was compounded with WF by twin screw extrusion and hot pressed into test specimens. The screw withdrawal resistance decreased when the fibre content increased from 60 to 80 wt%. The nail withdrawal resistance did not change very much with the fibre loading	[55]
Recycled PE (r-PE) and recycled PS (r-PS)	Mechanical properties, dimensional stability, UV resistance.	Kenaf, WF.	NFRC containing 50 wt% of kenaf or WF were compounded with r-PE or r-PS to make the composites. The mechanical properties, moisture absorption and UV resistance of these composites met the standards for medium density particle board and fibre board.	[56]

Acrylonitrile butadiene styrene (ABS) is an engineering plastic and its major applications are for electronic housings, toys and safety helmets. The amount of ABS used for these applications, is much lower than the quantities of ABS used for packaging and the market for recycled ABS is small. In addition, ABS is usually mixed with other plastics and flame retardants to increase its thermal stability. Plastic alloys containing all kinds of additives and impurities are almost impossible to recycle. In the early 2000s, ABS alloys in goods which had reached the end of their service life were regarded as waste materials and the market price was extremely low [60]. However, millions of electronic items are disposed of in the world every year and most of them are computers. In 2006, it was estimated that 60 million computers enter the market every year and about 12 million are thrown out. In the following few years, it was predicted that 680 million computers would become obsolete. The London-based Economist Intelligence Unit, a research and analysis firm, predicted that there would be £4 billion of recycled plastics generated from electronic waste [61]. Generally, plastics used in electronic housings include ABS, polycarbonate (PC) and high impact PS. They are engineering plastics and possess higher mechanical properties and thermal stability than commodity plastics such as polyolefins. If they can be recycled, the end applications and price of recycled ABS or PC could be very different from recycled polyolefins since their physical properties are superior to those of polyolefins.

One of the recycling solutions for these alloys is to recompound them and the compounded plastics are classified based on their mechanical properties. Here, NFRC may provide another solution. Since the stiffness of polyolefins is much lower than that of wood, the modulus of conventional NFRC made with polyolefins is not comparable to that of wood or other natural fibres. ABS is an engineering plastic which possesses stronger mechanical properties than polyolefins. If one could make NFRC using ABS, the mechanical properties of the composites would be improved and the problem of low stiffness may be solved. The use of ABS to make NFRC has been slow due to the relatively high cost of ABS and there are fewer than ten research articles that have used virgin or recycled ABS to make NFRC. As previously mentioned, the price of recycled ABS is considerably lower than its virgin material due to the issues outlined above, particularly the presence of additives and impurities. The most damaging impurities are those that are either not miscible with ABS or not processable at the same temperature. In other words, the presence of solid inclusions in ABS is the cause of the loss of large-strain properties. This problem can be easily overcome by seeking applications where solid fillers or reinforcements are added to the polymer [62, 63]. Using recovered ABS to formulate NFRC is, therefore, ideal from this viewpoint since the added natural fibre particulates have the potential to mask the composition-driven and molecular weight-driven variations in the properties of the base polymer [16].

There is only one reference that discussed using recycled ABS as the matrix material to produce NFRC. The results showed that the mechanical properties of the NFRC made with recycled ABS were comparable to those made of virgin ABS [16]. The authors also tested three different coupling agents: SEBS-*g*-MA, styrene-acrylonitrile-*graft*-maleic anhydride (SAN-*g*-MA) and SMA. Of these, SAN-*g*-MA provided the most pronounced effect since SAN is compatible with the base material of ABS.

7.6 Natural Fibre Composites made with Recycled Nylon and Polyethylene Terephthalate

The major limitation of processing NFRC is temperature because natural fibres burn and darken above 200 °C. Therefore, polymers that have to be processed above 200 °C cannot be applied as the matrix material for NFRC. For example, Muzzy and co-workers used post-consumer nylon from recycled carpet as the matrix and wood fibres were used as the reinforcement. Using nylon meant that the processing temperature had to be raised above 215 °C and the wood fibres were severely degraded and darkened [64]. However, SABIC Innovative Plastics developed a specialty compound which is nylon reinforced with 20% curauá fibre. The mechanical properties of the compound are similar to those of nylon reinforced with 20% glass fibre [65].

Due to the thermal degradation problem, recycled PET cannot be used as the matrix either. However, PET can be converted into unsaturated polyester *via* a glycolysis reaction and used as the matrix material for making unsaturated polyester/kenaf composites [66, 67]. This is related to chemical recycling and it is beyond the scope of this chapter.

In summary, HDPE and PP are the best candidates to be used as the matrix material of NFRC because of their availability and low processing temperatures. However, it has also been shown that it is possible to use recycled PS, ABS and polymer blends as matrices for NFRC.

7.7 Case Studies

At the moment, the way in which biocomposites, recycled polymers and recycled polymer-based biocomposites are bought and sold is much more open than the market in virgin polymers where a few multinational conglomerates dominate the landscape. This is typical for a growing market with many smaller competitors. Most of the companies have fewer than 200 employees and only a few of them are publicly traded companies. A short list of major resin, natural fibre, additives and NFRC manufacturers can be found in reference [68]. According to reference [68], the applications of NFRC are mainly in building and construction. With the development of injection moulding technology, many different shapes of NFRC could be fabricated for a wide variety of applications. Conventional engineers may think of using NFRC to replace glass fibre-reinforced composites for making injection moulded automobile or electronic parts [69]. However, with value-added design concepts, NFRC are not limited to the applications mentioned above. The utilisation of NFRC could be very different and the margin of profit could be significantly increased. Miniwiz, a Taiwan-based company founded in 2007, became known by embedding fashion design elements into recycled materials and now have several profitable products. Previously, recycled materials were generally considered as materials of inferior quality. However, if the product made of recycled material is designed with modern styling and the consumers realise that using recycled materials can help to reduce carbon emissions, the story will be different. Here, we list three innovative examples of NFRC made from recycled plastics which have been developed by by Miniwiz. We hope that these case studies will open new horizons for NFRC.

7.7.1 Case Study 1: Mobile Phone Protective Cases

Recycled PP was compounded with rice husk and injection moulded to produce iPhone cases. Additionally, the packaging material of the case is made up of 100% recycled material. According to the recommended retail price, the price per gram of the NFRC iPhone case is more than 300 times than that of regular WPC deck boards. The iPhone case is shown in **Figure 7.5**.

Figure 7.5 iPhone case made from NFRC [70]

7.7.2 Case Study 2: Wine Carrying Case

Re-wine is a Consumer Electronic Show 2012 Innovations Design and Engineering Awards winning product that provides a wine carrying solution. NFRC made from recycled plastics were injection moulded into reusable wine carrying cases which can be transformed into a desk lamp or can be interlocked and stacked to form a wine bottle rack. The wine carrying case is shown in shown in **Figure 7.6.**

Figure 7.6 Wine Bottle Carrier made from NFRC [70]

7.7.3 Case Study 3: Sunglasses

It is well known that natural fibre will burn when the processing temperature is above 200 °C. Thus, high temperature engineering plastics such as PC were not considered to be suitable as the matrix for NFRC. Nevertheless, in this design scenario, post-consumer PC (from compact discs) was infused with carbonised rice husk fibres to make the frame of the glasses. The carbonised fibres not only served as a decoration but an impact modifier. The lenses of the glasses were also made of recycled PC. The sunglasses are shown in **Figure 7.7**.

Figure 7.7 Sunglasses frame made of NFRC [70]

7.8 Summary

Natural fibres are some of the best candidates to replace glass or carbon fibres in thermoplastic composites. The content of this chapter shows that the development of NFRC is still in its early stages and will enjoy a double digit compound annual growth rate over the next few years. It would be preferable to use recycled thermoplastics in the manufacture of NFRC since it can substantially extend the life cycle of these plastics. However, the quality and supply of recycled plastics are usually unpredictable since recycled plastics are usually mixed with various different additives and the source of recycled plastics can vary. Thus, it would be meaningless to compare the properties of NFRC made by recycled or virgin plastics because the physical properties of their polymer matrices are very different. Currently, NFRC are mainly considered as building and construction materials or as alternatives for glass fibre-reinforced composites, but the applications of NFRC could reach far beyond building and construction. Here, three design scenarios are provided. With innovative design concepts embedded, the value of NFRC can be significantly increased. The idea of integrating innovative design concepts in NFRC could open a new route for NFRC.

Acknowledgements

The authors would like to thank Mr. Arthur Huang, the founder and managing director of Miniwiz Sustainable Energy Ltd for his support and allowing his colleagues to co-author this chapter. The authors would also like to thank Mr. Chia-Chun Hsieh for collecting the references.

References

1. *Natural Fiber Composites Market Trend and Forecast 2011–2016: Trend, Forecast and Opportunity Analysis*, Lucintel, Las Colinas, TX, USA, 2011. *http://www.lucintel.com/reports/Composites/natural_fiber_composites_market_trend_forecast_2011_2016_trends_forecast.aspx*

2. A.K. Bledzki and J. Gassan, *Progress in Polymer Science*, 1999, **24**, 2, 221.

3. D.B. Dittenber and H.V.S. GangaRao, *Composites Part A: Applied Science and Manufacturing*, 2012, **43**, 8, 1419.

4. A.J. Stamm in *Wood and Cellulose Science*, Ronald Press Co., New York, NY, USA, 1964, p.233.

5. A.K. Mohanty, M. Misra and L.T. Drzal in *Natural Fibers, Biopolymers, and Biocomposites*, Taylor and Francis, Boca Raton, FL, USA, 2005, p.833.

6. K. Bledzki, S. Reihmane and J. Gassan, *Polymer-Plastics Technology and Engineering*, 1998, **37**, 4, 451.

7. H. Ku, H. Wang, N. Pattarachaiyakoop and M. Trada, *Composites Part B: Engineering*, 2011, **42**, 4, 856.

8. S. Kalia, B.S. Kaith and I. Kaur, *Polymer Engineering and Science*, 2009, **49**, 7, 1253.

9. K.G. Satyanarayana, G.G.C. Arizaga and F. Wypych, *Progress in Polymer Science*, 2009, **34**, 9, 982.

10. A.K. Bledzki, O. Faruk and V.E. Sperber, *Macromolecular Materials and Engineering*, 2006, **291**, 5, 449.

11. O. Faruk, A.K. Bledzki, H-P. Fink and M. Sain, *Progress in Polymer Science*, 2012, **37**, 11, 1552.

12. A.K. Mohanty, M. Misra and L.T. Drzal, *Journal of Polymers and the Environment*, 2002, **10**, 1-2, 19.

13. S.M. Al-Salem, P. Lettieri and J. Baeyens, *Waste Management*, 2009, **29**, 10, 2625.

14. J. Hopewell, R. Dvorak and E. Kosior, *Philosophical Transactions of The Royal Society B: Biological Sciences*, 2009, **364**, 1526, 2115.

15. A. Azapagic, S. Perdan and R. Clift in *Sustainable Development in Practice: Case Studies for Engineers and Scientists*, John Wiley & Sons, Chichester, West Sussex, UK and Hoboken, NJ, USA, 2004, p.250.

16. S.K. Yeh, S. Agarwal and R.K. Gupta, *Composites Science and Technology*, 2009, **69**, 13, 2225.

17. F. Vilaplana and S. Karlsson, *Macromolecular Materials and Engineering*, 2008, **293**, 4, 274.

18. *http://www.epa.gov/epawaste/nonhaz/municipal/index.htm*

19. M.A. Danesh, H. ZiaeiTabari, R. Hosseinpourpia, N. Nazarnezhad and M. Shamse, *Bioresources*, 2012, **7**, 1, 936.

20. B. Kord, M.A. Danesh, R. Veysi and M. Shams, *Bioresources*, 2011, **6**, 4, 4190.

21. P. Kamdem, H.H. Jiang, W.N. Cui, J. Freed and L.M. Matuana, *Composites Part A: Applied Science and Manufacturing*, 2004, **35**, 3, 347.

22. Y-M. Song, Q-W. Wang, G-P. Han, H-G. Wang and H. Gao, *Journal of Forestry Research*, 2010, **21**, 3, 373.

23. S.K. Najafi, E. Hamidinia and M. Tajvidi, *Journal of Applied Polymer Science*, 2006, **100**, 5, 3641.

24. S.K. Najafi, A. Kiaefar, E. Hamidina and M. Tajvidi, *Journal of Reinforced Plastics and Composites*, 2007, **26**, 3, 341.

25. S.K. Najafi, M. Mostafazadeh-Marznaki, M. Chaharmahali and M. Tajvidi, *Journal of Composite Materials*, 2009, **43**, 22, 2543.

26. F. Yao, Q.L. Wu, Y. Lei and Y.J. Xu, *Industrial Crops and Products*, 2008, **28**, 1, 63.

27. K.B. Adhikary, S. Pang and M.P. Staiger, *Composites Part B: Engineering*, 2008, 39, 5, 807.

28. K.B. Adhikary, S. Pang and M.P. Staiger, *Chemical Engineering Journal*, 2008, **142**, 2, 190.

29. A. Najafi and S. Kazemi Najafi, *Polymer Composites*, 2009, **30**, 11, 1570.

30. H. Khanjanzadeh, T. Tabarsa, A. Shakeri and A. Omidvar, *Wood Material Science and Engineering*, 2011, **6**, 4, 207.

31. A.C.N. Singleton, C.A. Baillie, P.W.R. Beaumont and T. Peijs, *Composites Part B: Engineering*, 2003, **34**, 6, 519.

32. V. Srebrenkoska, G.B. Gaceva, M. Avella, M.E. Errico and G. Gentile, *Polymer International*, 2008, **57**, 11, 1252.

33. M.D.H. Beg and K.L. Pickering, *Composites Part A: Applied Science and Manufacturing*, 2008, **39**, 7, 1091.

34. M.D.H. Beg and K.L. Pickering, *Composites Part A: Applied Science and Manufacturing*, 2008, **39**, 9, 1565.

35. Y. Xu, Q. Wu, Y. Lei and F. Yao, *Bioresource Technology*, 2010, **101**, 9, 3280.

36. S. Ha, H.D. Park and W.J. Cho, *Journal of Applied Polymer Science*, 1999, **74**, 6, 1531.

37. H. Ismail, S. Ragunathan and K. Hussin, *Journal of Vinyl and Additive Technology*, 2011, **17**, 3, 190.

38. S. Ragunathan, H. Ismail and K. Hussin, *Key Engineering Materials*, 2011, **471-472**, 478.

39. H. Ismail, S. Ragunathan and K. Hussin, *Polymer-Plastics Technology and Engineering*, 2010, **49**, 13, 1323.

40. A. Ershad-Langroudi, F. Jafarzadeh-Dogouri, M. Razavi-Nouri and A. Oromiehie, *Journal of Applied Polymer Science*, 2008, **110**, 4, 1979.

41. M. Chaharmahali, J. Mirbagheri, M. Tajvidi, S.K. Najafi and Y. Mirbagheri, *Journal of Reinforced Plastics and Composites*, 2008, **29**, 2, 310.

42. S. Rodríguez-Llamazares, A. Zuñiga, J. Castaño and L.R. Radovic, *Journal of Applied Polymer Science*, 2011, **122**, 3, 1731.

43. A. Najafi and H. Khademi-Eslam, *Bioresources*, 2011, **6**, 3, 2411.

44. Y. Cui, S. Lee, B. Noruziaan, M. Cheung and J. Tao, *Composites Part A: Applied Science and Manufacturing*, 2008, **39**, 4, 655.

45. Y-H. Cui, S. Lee and J. Tao, *Journal of Vinyl and Additive Technology*, 2008, **14**, 4, 211.

46. A. Ashori and A. Nourbakhsh, *Waste Management*, 2009, **29**, 4, 1291.

47. G. Grubbstrom, A. Holmgren and K. Oksman, *Composites Part A: Applied Science and Manufacturing*, 2010, **41**, 5, 678.

48. H. Gao, Y. Xie, R. Ou and Q. Wang, *Composites Part A: Applied Science and Manufacturing*, 2012, **43**, 1, 150.

49. J. López-Naranjo, L. M. Alzate-Gaviria, G. Hernández-Zárate, J. Reyes-Trujeque, C.V. Cupul-Manzano and R.H. Cruz-Estrada, *Journal of Applied Polymer Science*, 2012, DOI: 10.1002/app.38212.

50. A. Nourbakhsh and A. Ashori, *Journal of Composite Materials*, 2009, **43**, 8, 877.

51. Z. Li, H. Gao and Q. Wang, *Journal of Applied Polymer Science*, 2012, **124**, 6, 5247.

52. S.E. Selke and I. Wichman, *Composites Part A: Applied Science and Manufacturing*, 2004, **35**, 3, 321.

53. X. Kuang, R. Kuang, X. Zheng and Z. Wang, *Carbohydrate Polymers*, 2010, **80**, 3, 927.

54. Xie, Y-H. Cui, J. Xu, X-X. Wang, H-H. Zhang and Z-D. Zhang, *Journal of Vinyl and Additive Technology*, 2012, **18**, 2, 105.

55. M. Chaharmahali, M. Tajvidi and S.K. Najafi, *Polymer Composites*, 2008, **29**, 6, 606.

56. T. Sellers, G.D. Miller and M. Katabian, *Forest Products Journal*, 2000, **50**, 5, 24.

57. M. Poletto, J. Dettenborn, M. Zeni and A.J. Zattera, *Waste Management*, 2011, **31**, 4, 779.

58. M. Poletto, M. Zeni and A.J. Zattera, *Journal of Applied Polymer Science*, 2012, **125**, 2, 935.

59. J. Lisperguer, X. Bustos and Y. Saravia, *Journal of Applied Polymer Science*, 2011, **119**, 1, 443.

60. T.A. Osswald and M.W. Dattner in *Society of Plastics Engineers Annual Technical Conference*, Boston, MA, USA, 2005, p.296.

61. M. Griswold, *Plastics News*, 2006, **17**, 52, 7.

62. R. Liang and R.K. Gupta in *Society of Plastics Engineers Annual Technical Conference*, Dallas, TX, USA, 2001, p.2753.

63. R. Liang and R.K. Gupta, *Society of Plastics Engineers Annual Technical Conference*, San Francisco, CA, USA, 2002, p.2948.

64. J. Muzzy, X. Xu and A. Ragauskas in *Society of Plastics Engineers Annual Technical Conference*, Cincinnati, OH, USA, 2007, p.2227.

65. Anonymous, *Additives for Polymers*, 2009, **9**, 5.

66. N. Farahani, I. Ahmad and Z. Mosadeghzad, *Polymer-Plastics Technology and Engineering*, 2012, **51**, 6, 634.

67. I. Ahmad, Z. Mosadeghzad, R. Daik and A. Ramli, *Journal of Applied Polymer Science*, 2008, **109**, 6, 3651.

68. *http://www.wpcinfo.org*

69. B. Hunnicutt, *Plastics Technology*, 2007, **53**, 12, 56.

70. *http://www.miniwiz.com*

8 Joining of Natural Fibre-reinforced Thermoplastic Composites

Inderdeep Singh, Kishore Debnath and Akshay Dvivedi

8.1 Introduction

Glass- or carbon fibre-reinforced plastics are now extensively used in many engineering applications and have replaced conventional materials in various sectors. However, these materials face major problems due to their dependency on petroleum-based resources which are depleting rapidly. Being non-biodegradable in nature, synthetic fibre reinforced plastics are not environmentally friendly. Due to several environmental issues, natural fibre-reinforced biocomposites have attracted the attention of researchers and technologists worldwide. Therefore, the application spectrum of the biocomposites is increasing at a very fast pace. Now, both structural and internal components of cars are incorporating more and more polymeric composite materials due to the advancements in the field of biopolymers and the natural fibre industry. Recently, natural fibre-reinforced polymer composites have been used as automotive parts because of their excellent mechanical properties and light weight characteristics. The most common natural fibres which are used in the automotive, aerospace, textile, paper, chemical and construction industries include jute, coir, sisal, nettle, hemp, kenaf, flax and bamboo. Natural fibre-reinforced composites have been employed as substitutes for synthetic fibre reinforced composites over the past few years. Natural fibres are widely used in various composite industries because the reinforcement cost is very low as compared to synthetic fibre. Moreover, natural fibres are derived from renewable resources and so are environmentally friendly. This resurgence of interest in using natural fibres as reinforcement for polymer-based composites is mainly due to their techno-ecological advantages, such as lower cost, light weight, high strength-to-weight ratio, renewability, acceptable specific properties, lower density, less wear and tear on equipment during processing, lower energy requirements for processing, biodegradability, wide availability and relative non-abrasiveness over traditional reinforcing fibres such as glass, carbon and aramid. The constituents for most natural fibres, except for cotton, are cellulose, hemicelluloses, lignin, waxes, and certain water-soluble compounds, where cellulose, hemicelluloses, and lignin are the major constituents. The properties of the constituents decide the overall properties of the fibre. The percentage composition of each of these components varies for the different fibres. A typical natural fibre contains 60-80% cellulose, 5-20% lignin and up to 20% moisture.

Natural fibre composites as fabricated cannot always be used as the end product. For complex composite structures, the assembly of various components through joining is essential. The most popular joining techniques for natural fibre composites are adhesive and mechanical joining, but both of these processes have several disadvantages. Due to this, an emerging joining technique, namely microwave joining, is now being used to join natural fibre composites as it increases the joint strength effectively and reduces the required joining time compared with conventional joining methods. It has been observed that natural fibre-reinforced thermoplastic composites can be used as potential candidates for both structural and non-structural applications where joining of these materials is very important to complete the structure. Therefore, it is necessary to focus on various aspects of the joining techniques to ensure an excellent quality joint. The following section discusses the three major joining techniques: adhesive bonding, mechanical fastening and microwave joining of natural fibre-reinforced thermoplastic composites.

8.2 Purpose of Joining

Generally, joining is an undesirable secondary operation in any application as the joint forms a weaker region at which a product will often fail. There are many applications for which joining cannot be avoided, e.g., intricate shape products cannot be manufactured in a near-net shape and hence individual components are manufactured and assembled together to create a complex composite product. The major purposes of joining are:

- To transfer load from one component to another component and to allow relative motion between the two components.
- The complete range of features of a complex composite product cannot be obtained during the primary processing of each individual part, but the same features can be satisfactorily obtained by joining several components together.
- The different components of a structure. are subjected to different types of loading which requires components with different properties. The components with different properties which form the whole structure can not be obtained in a single shot through primary manufacturing, therefore, several components are joined together to meet the functional requirements.

- Repair, maintenance and service of the product is much easier when the components can be easily disassembled, which is only possible if the components have been assembled in the first place.

8.3 Adhesive Bonding

Adhesive bonding is a type of bonding in which two substrate materials are joined together by means of an adhesive. The adhesives which are mainly used for thermosetting matrix composites are thermosetting polymers, whereas thermoplastic polymer adhesives are used for thermoplastic matrix composites. The few of the advantages of using a thermoplastic polymer adhesive over a thermosetting polymer adhesive are short processing time (because the curing time of thermoplastic polymers is less than that of thermosetting polymers), high degree of thermal stability, high resistance to chemicals, low moisture absorption and retention of bond strength at elevated temperature. The efficiency of adhesive bonding depends upon various factors, such as surface preparation of the substrates or adherends, type and amount of polymer to be used as an adhesive, fitting of parts to be joined and the curing procedure. Bonding of a thermoplastic polymer is very simple as it is readily bonded by means of heating; sometimes a small amount of adhesive is mixed with solvent to allow easy application. The most common thermoplastic adhesives used for joining thermoplastic matrix composites are polyvinyl acetate, polyvinyl acetals, polyvinyl alcohols, polyamides, acrylics and phenoxies. The adhesives used for bonding polymer matrix composites may be cured at room temperature or at elevated temperature.

Adhesive bonding of natural fibre-reinforced thermoplastic composites includes the following steps: (a) assembly and joint design; (b) selection of an appropriate adhesive; (c) preparation of surfaces; (d) applying an appropriate amount of the selected adhesive; (e) applying pressure and temperature at the interface of the parts which are to be joined together; and (f) curing of the adhesive. Depending upon the bond strength, an adhesive is generally chosen on the basis of the following properties: ability to wet the substrate surface, resistance to heat, resistance to moisture and chemicals, shear, peel strength and toughness of the resulting joint.

8.3.1 Types of Adhesively Bonded Joints

Adhesively bonded joints can have different types of joint configuration, e.g., lap joint, butt joint, strap joint, step joint and scarf joint. These types of joints can also be used in natural fibre-reinforced thermoplastic matrix composites. The various types of joints used for composite materials are shown in **Figure 8.1**. The most common type of joint used in any engineering application is a single lap joint. In a single lap joint, the load is transferred from one substrate to another substrate by means of shear stress in the adhesive. The load transferred from one substrate to another is off-centered which results in the action of normal stress or cleavage stress in the thickness direction of the adhesive. The cleavage stress and normal stress play an important role in defining the bond strength of the single lap joint. The effect of bending in the single lap joint can be largely avoided in the double lap joint. In a double lap joint, both the bending force and the normal stress are eliminated. Therefore, the bond strength of the double lap joint is always greater than that of the single lap joint. For attaining maximum joint strength, step and scarf joints are used for composite structures, but the machining of stepped or scarf ends is a considerably challenging operation.

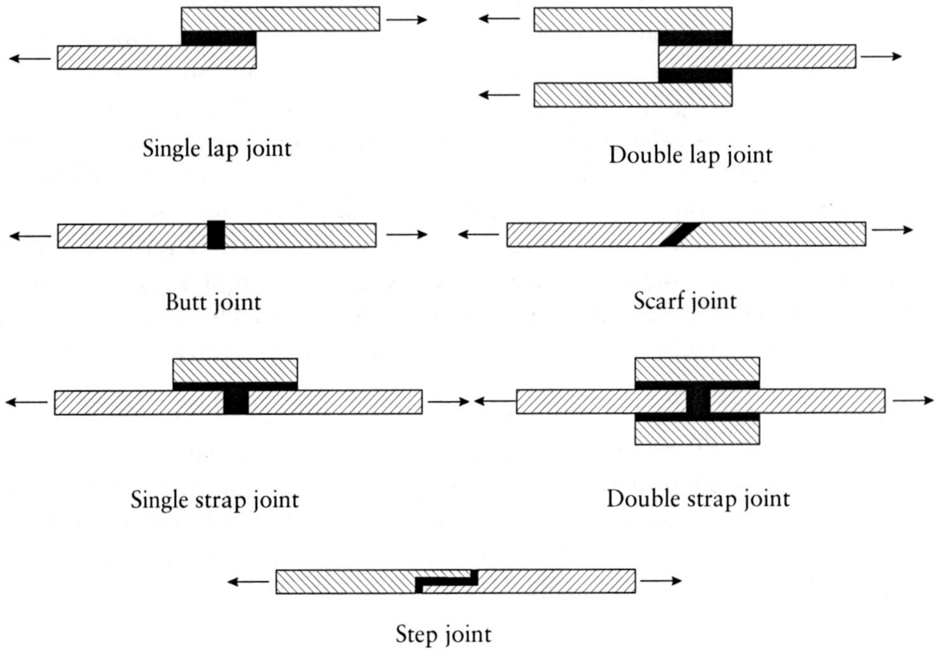

Figure 8.1 Common types of adhesively bonded joints

8.3.2 Types of Adhesives

Adhesives may be broadly classified into three major categories, these are: (a) natural adhesives; (b) inorganic adhesives; and (c) synthetic organic adhesives. Further, the synthetic organic adhesives may be of different types: chemically reactive, pressure sensitive, hot melt, reactive hot melt, evaporative, film and tape, delayed tack and electrically and thermally conductive adhesives. Based on the specific chemistry, adhesive systems may be of three different types: thermosetting adhesives, thermoplastic adhesives and elastomeric adhesives. Adhesives are available in various forms such as pastes, liquids, solutions, emulsions, powder, tape and film. Thermoplastic adhesives and elastomeric adhesives are generally used for non-structural bonding, whereas thermosetting adhesives are primarily used for structural bonding [1]. Depending upon the types of adhesives available on the market, adhesive systems may be two-component mix adhesives, two-component no-mix adhesives and one-component no-mix adhesives [2].

8.3.3 Selection Criteria and Guidelines for Adhesives

Selection of an adhesive is an important design consideration for any adhesively bonded joints. Selection of the correct adhesive for a particular application determines the quality of the joints. The selection of an adhesive depends on various factors and these are listed below:

- Type of substrate material to be joined.
- Curing conditions.
- Required joint strength.
- Impact resistance.
- Resistance to chemical attack, electric current and humidity.
- Service temperature.
- Cost requirements.

8.3.4 Advantages of Adhesive Bonding

Adhesively bonded joints have the following advantages over mechanical joints:

- In an adhesively bonded joint, the load is not concentrated at a single point. Therefore, the stress distribution is uniform throughout the joint area.
- The adherends bonded by an adhesive can sustain more flexural and fatigue stress compared to mechanical joints because there is a possibility of less stress concentration.

- Adhesive bonded joints also provide superior vibration and damping properties.
- The adhesive bonding technique is more suitable for joining irregular or uneven surfaces than mechanical joining.
- This technique can be used to join very thin substrates or adherends.
- Very large surfaces can be joined easily without facing any difficulty.
- An adhesive joint is a gas- or liquid-proof joint which prevents crevice corrosion on the bonding surface.
- The joints prepared by adhesive bonding are less expensive as it eliminates the use of mechanical fasteners and takes less processing time than mechanical joining.

8.3.5 Disadvantages of Adhesive Bonding

The main disadvantages of adhesively bonded joints are:

- Proper surface preparation of the adherends to be joined together is essential and cannot be avoided in adhesive bonding.
- Adhesive bonding is a permanent type of bonding. The disassembly of bonded parts is not easy as it creates damage at the joint surfaces and the surfaces which surround them.
- The adhesives used for adhesive bonding may be expensive and require attention because they need to be stored at a specific temperature, as recommended by the manufacturer.
- Adhesive bonding mostly depends upon the pressure and temperature which directly control the curing time. Sometimes, curing of an adhesive takes a long time and requires special environmental conditions.
- The inspection of joints prepared by adhesive bonding is very difficult. Non-destructive tests are often performed on the joints to observe the presence of any voids and unbonded zones.

8.3.6 Guidelines for Surface Preparation

Surface preparation of the substrate materials is a key task that needs to be performed when joints are made by adhesive bonding. Basically, the surface is pre-treated to increase the wettability of the surface which, in turn, increases the degree of contact between the adhesive and adherends and hence enhances the joint strength. Moreover, the surface has to be free from contaminants such as dust, dirt, foreign particles, grease and oil to avoid substandard joining. Most polymeric materials contain weak boundary layers which creates certain difficulties during the processing of joints. For instance, composite material surfaces are often contaminated with mould release agents and additives. These can be removed mechanically by abrading the surface with sandpaper or by alkaline treatment. For thermoplastic composite materials, special treatments, such as oxidation by flame treatment, plasma and corona treatment, ionised inert gas treatment, or application of primers or adhesion promoters are usually adopted to clean the surface [2]. The three common types of surface preparation methods available for composite materials are mechanical abrasion, chemical treatment and degreasing.

In mechanical abrasion, the substrate surfaces are roughened in order to increase the real surface contact area of the substrates, which results in more interlocking of the adhesive materials into the microvoids of the substrate. The same technique can also be used to remove loose particles and solid contaminants from the surface. Sandpaper, emery cloth and wire brushes are commonly used for the abrasion process.

In chemical treatment, a strong detergent solution and primer are used for surface modification and additional cleaning of the substrate surfaces. Polymeric materials, such as polyolefins and polytetrafluoroethylene require pretreatment of the bonding surfaces. Surface preparation of such plastic materials is often done by applying a primer, but the use of a primer is not as effective as ionisation techniques such as corona discharge. The use of primers is a very simple method that involves spraying or brushing a thin layer of primer onto the substrates. After a short drying time, typically 10 s to 1 min, the adhesive is applied to the joint in the usual way. By using a primer, high bond strengths can be achieved with many polymers and others (most notably fluoropolymers) may be joined with a limited degree of success whereas, without the primer, they are impossible to bond. Primers often contain components which enhance the environmental resistance and thermal stability of the bond [3]. Degreasing involves solvent cleaning, alkaline cleaning and spray cleaning. These techniques are usually used to remove the cutting fluids, greases, waxes and mould release agents. For degreasing of a polymeric composite surface, solvents such as acetone and methyl alcohol are usually used to remove the contaminants.

8.3.7 Design Guidelines for Adhesive Bonding

The general guidelines for designing a good adhesively bonded joint in fibre-reinforced thermoplastic composites are discussed below:

- The most important aspect that needs to be considered when designing an adhesive joint is the amount of load which will be borne by the structure. Tensile loading causes the development of peeling or cleavage stress in the joint area which has to be minimised in order to get the maximum bond strength.
- The joint should be designed in such a way that the maximum load transfer to the joint is in the form of shear or compressive load only.
- The bonding area should be maximised as far as possible because the higher the degree of overlap between an adhesive and the adherends, the higher will be the load carrying capacity.
- The adhesive chosen must be compatible with the substrate materials. Furthermore, an adhesive should be chosen which is suitable for a given application. For example, an adhesive may be required to work at elevated temperature so, for this, the temperature resistance of the adhesive should be high. Other aspects, such as chemical resistance, resistance to environmental conditions, and adequate joint strength are also major factors in the selection of an adhesive.
- The thickness of the adhesive layer should be optimised to achieve maximum bond strength. The thickness of the adhesive layer should not be too low or too high.

8.3.8 Basic Theory of Adhesive Bonding

The complete understanding of adhesive bonding mechanisms involves a number of adhesion theories. The bonding theories explain how adhesion occurs at the interface between the adhesive and adherend. Adhesive bonding of materials involves five basic theories, which are discussed below.

8.3.8.1 Mechanical Theory

The mechanical adhesion theory explains the bonding mechanism when two materials adhere mechanically to each other. According to this theory, adhesion occurs due to the physical interlocking of the adhesive in the surface microvoids of the substrate. Fundamentally, at the microscopic level, all surfaces consist of peaks and valleys. As the adhesive is applied to the substrate surface, it fills the microcavities. When the adhesive becomes hard after curing or solidification it holds the substrates mechanically. The application of pressure plays an important role during bonding as it displaces the trapped air from the interface and helps the adhesive to penetrate into the microcavities. The bond strength of the substrates is highly influenced by the surface roughness of the substrate material. The relative area of interlocking of the adhesive in the surface microvoids of the substrate is increased with increase in the surface roughness. However, surface roughness will only help when the adhesive has good rheological properties.

8.3.8.2 Adsorption Theory

This theory states that the adhesion of two materials takes place due to the developed surface forces usually known as van der Waals forces. The surface forces are considered to be high when the molecular contact between the adhesive and substrate surface is greater. The molecular contact is highly dependent on the wetting of the adherend surface by the adhesive. The wetting further depends on the surface tension of the adhesive. Wetting is more effective when the surface tension of the substrate is higher than that of the adhesive. For this reason a low surface tension polymer (e.g., epoxy, with a surface tension of 0.047 N/m) wets a high surface tension metal (e.g., aluminium, with a surface tension of 0.5 N/m) quite effectively.

8.3.8.3 Electrostatic Theory

According to the electrostatic theory of adhesion, the surfaces of the adhesive and adherends are joined together by means of electrostatic forces. The electrostatic forces are generated due to the transfer of electrons from one surface to the other which form an electrical double layer at the interface between the adhesive and adherends.

8.3.8.4 Diffusion Theory

According to this theory, adhesion occurs when interdiffusion of molecules takes place at the adhesive/adherend interface. This theory is especially appropriate for polymer-based substrates and adhesives. When polymer granules are heated and pressed together, the molecules of the polymer-based substrate and adhesive both diffuse one by one across the interface and hence, adhesion of the particles takes place.

8.3.8.5 Chemical Bonding Theory

This bonding theory involves the formation of chemical bonds such as covalent, ionic or hydrogen and metallic bonds across the joint interface. These types of bonds are achieved by using added coupling agents or with appropriate surface treatments.

8.3.9 Failure Mechanisms in Adhesive Bonding

Adhesive failure and cohesive failure are the two main modes of failure which are commonly observed in adhesively bonded joints. Both the failure mechanisms are represented in **Figure 8.2**. Adhesive failure is a type of failure which occurs at the interface between the adhesive and the substrate. There may be a number of reasons for adhesive failure some of which are inadequate surface preparation, wrong selection of adhesive and high peeling stress. If the failure occurs only in the adhesive or substrate material, then it is termed as cohesive failure. This type of failure often occurs due to a stronger bond strength between the adhesive and substrate than the individual strength of the adhesive or the substrate material. A joint is considered to be a poor joint if the failure occurs adhesively at the interface between the adhesive and substrate, because this confirms that the bond strength is weaker than the individual strength of the adhesive or the substrate material.

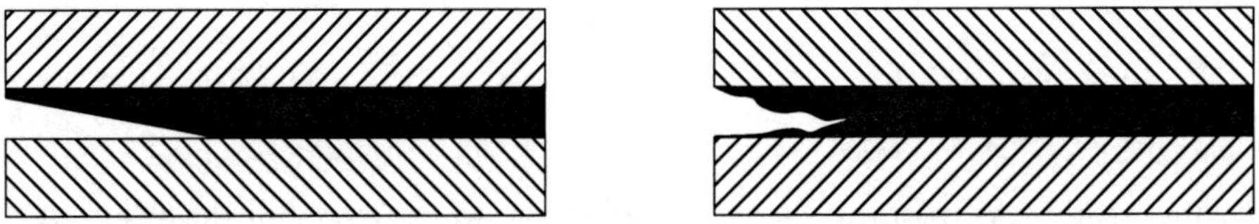

Adhesive failure Cohesive failure

Figure 8.2 Types of failure mode in adhesively bonded joints

8.4 Mechanical Joining

Mechanical joining or fastening of different components by means of fasteners is probably one of the oldest joining techniques used by man. It is the only joining technique which is extensively used for joining metallic components in manufacturing industry. However, the same technique can be used to join natural fibre-reinforced thermoplastic composites. Mechanical joining involves the use of mechanical fasteners for the assembly of two components. The mechanical fasteners which are ordinarily used for composite materials may be of several types, such as rivets, bolts, screws and nails. Based on the type of joint produced, a mechanical joint can be classified as a temporary joint or a permanent joint. Mechanical joining is useful for joining both similar and dissimilar materials. For instance, in the automotive and aerospace industries, most of the plastic and metal parts are joined by mechanical fastening method.

One of the most important aspects that needs to considered in mechanical joining is to estimate the magnitude and direction of the applied load. The applied load carried or transferred by the joint may be static or dynamic. Therefore, the selection of mechanical joints depends on the nature of the applied load. The joints are designed according to the load carrying capacity of the structure. The factors that affect the fastened joints are fatigue load, bending load, vibration and corrosion. One of the major benefits of using a mechanical joint is that there is no need for surface preparation. Moreover, the disassembly of the components for repair, maintenance, and inspection is much easier than with other joining techniques. Most of the mechanical fasteners which are are used for metallic components can also be used to join fibre-reinforced composite materials. Screws and bolts used for joining purposes make only temporary joints, whereas, rivets make a semi-permanent joint. These types of joints can be easily separated without damaging the components. Of all mechanical fasteners, the rivet is one of the most widely used. Joining by means of rivets gives a semi-permanent joint; the joined components can be detached only by destroying one of the two rivet heads. Threaded fasteners can also be used to join thermoplastic composites, but it will not provide sufficient joint strength. This is because the threads created in the plastic parts are not strong enough to carry the load. Hence, most of the time, metallic inserts such as rivets and bolts are preferred for joining thermoplastic composite materials.

Though mechanical joints have a number of disadvantages, this joining method cannot be avoided, especially in situations where it is necessary to transmit high mechanical loads from one component to the other. The design of mechanical joints for fibre-reinforced composites is more challenging than for metals. This is because there are a number of factors that can affect the joint strength. These factors are: (a) different weave pattern of fibre; (b) various fibre orientations; (c) a virtually unlimited number of possible combinations of composite materials; and (d) complex distribution of stress and strain in the joint area.

8.4.1 Types of Mechanical Joining

A wide variety of mechanical joints is available for the fastening of composite materials. These are riveted joint, bolted joint, screw joint, knuckle joint, cotter joint, and pin joint. The three most common types of mechanical joints that can be used for joining components made up of natural fibre-reinforced thermoplastic composites are discussed below.

8.4.1.1 Riveted Joint

The fasteners that are usually used in the riveted joint are termed rivets. A typical rivet is an unthreaded mechanical fastener. A common type of riveted joint is illustrated in **Figure 8.3.** In general, rivets are used to join thin metallic plates, but they can also be used to join plastics and thermoplastic composite materials. The riveted joint may be of a permanent or semi-permanent type. The fastening of plates by means of rivets is a very simple technique compared with other types of mechanical fastening. A typical rivet consists of a cylindrical shaft with a head at one end, which is also known as the factory head, and the other end opposite to the factory head is called the buck-tail. The shaft or shank of the rivet is passed through a hole and the buck-tail is deformed or upset, which expands it to about 1.5 times the shank diameter. The expansion of the buck-tail holds the mating plates together. There are various types of rivets available to meet specific requirements such as strength and stiffness, accessibility and cost. The three basic types of rivets are solid, hollow and blind. Several types of rivet, with different types of head, are available. These are: round head, cone head, countersunk head, pan head and flat head rivets.

The major advantage of using a riveted joint is that it provides adequate strength with a minimum sized fastener. The main disadvantage of riveted joints is that the disassembly of the mating parts is troublesome. Moreover, a pilot hole is required for the assembly of different parts. A riveted joint may fail because of three major failure modes namely tearing of the plate, shearing of the rivet and crushing of the rivet. The efficiency of a riveted joint can be found out by calculating the ratio of the strength of the riveted joint to the strength of the solid or unriveted plate.

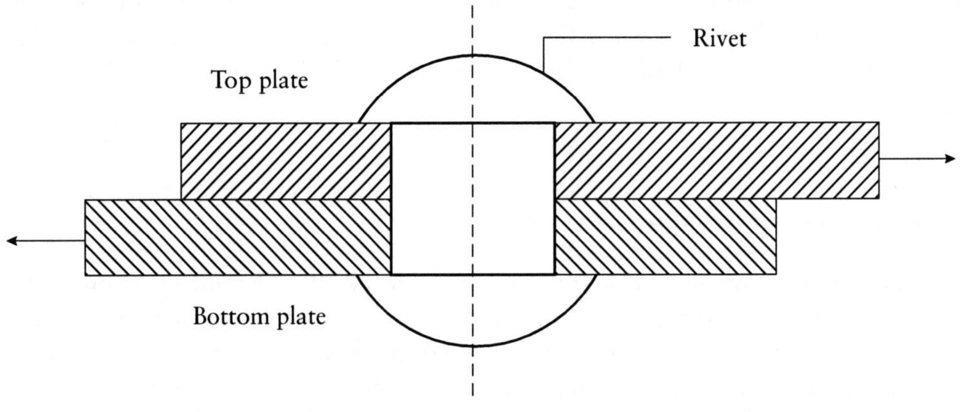

Figure 8.3 Schematic illustration of a riveted joint

8.4.1.2 Bolted Joint

Bolted joints are created by using nuts, bolts and washers. **Figure 8.4** represents a bolted joint, where a nut with a threaded hole and a threaded bolt are used to fasten thermoplastic composite parts. A bolted joint is usually made by drilling a a hole through the parts to be joined and then a bolt is placed in the hole and a nut is screwed onto the threaded bolt. The strength of a bolted joint is affected by various factors, such as clamping force, quality of the hole, fibre orientation, and lay-up sequence. The major types of bolted joints can be categorised as single lap joint, double lap joint and butt joint. The head of the fastener plays an important role, because the stress distribution is dependent on the head geometry. For instance, most of the time round heads are preferred to conical heads, because round heads uniformly distribute the developed stress, whereas the use of a conical head results in undesirable tensile stress [4]. This undesirable stress can be controlled by using a flat washer under the nut and bolt head. Various modes of failures may cause the failure of a material fastened by bolted joint. The three main modes of failure in bolted joints are shearing, crushing and tensile failure of the mating parts.

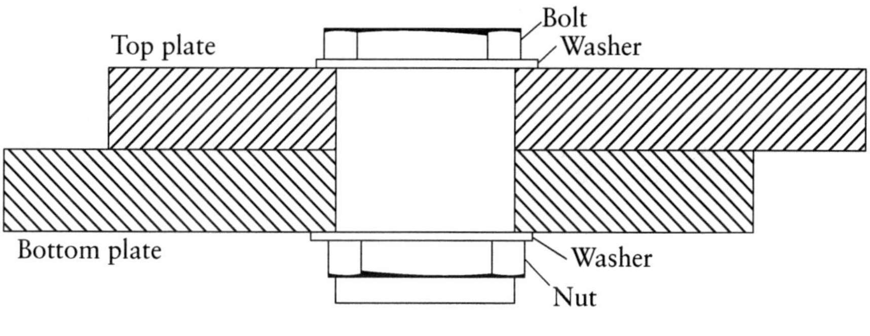

Figure 8.4 Schematic illustration of a bolted joint

8.4.1.3 Screw Joint

Screw joints are the most economical and effective joints for plastics and thermoplastic composite materials. The screws used for fastening purpose are of two types: machine screws and tapping or self-tapping screws. **Figure 8.5** schematically illustrates a typical screw joint. Screws are similar to bolts; the only difference is that their length and diameter are always smaller than those of the bolts. The engineering materials that are used to manufacture bolts can also be used to make screws. Screws are available with a variety of heads, such as, round head, square head, hexagonal head, flat head and oval head. One of the main advantages of using screws is that the mating parts can be disassembled very easily and quickly. For this reason, screw joints are probably one of the most widely used mechanical joints. A self-tapping screw can be used to fasten composite material, a self-tapping screw can be installed but this type of screw builds a high stress level. Moreover, the use of self-tapping screws is not safe for this application, because when self-tapping screws are used there is a chance of formation of new threads in the substrates to be joined together. In the case of thermoplastic materials the chance of thread formation is greater because these materials are soft and deform on application of small amounts of load.

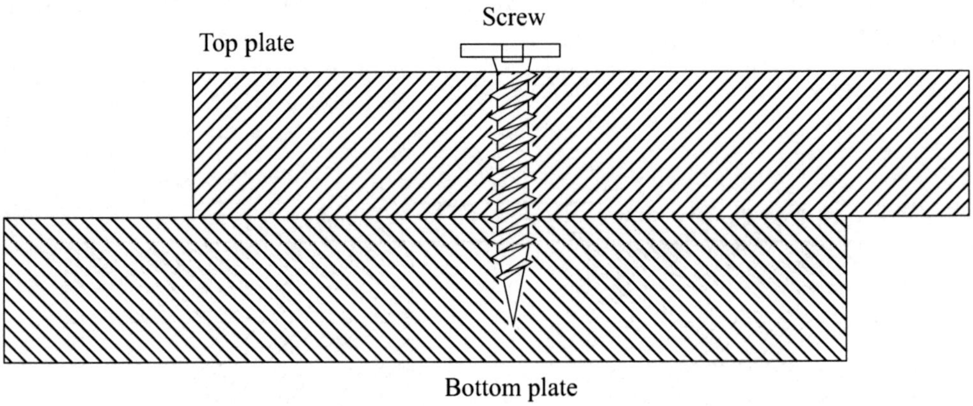

Figure 8.5 Schematic illustration of a screw joint

8.4.2 Advantages of Mechanical Joining

Mechanical joining has several advantages over the other joining techniques. The benefits of mechanical joining that make it one of the widely used joining techniques are summarised below:

- Mechanical joining can be used to join metallic, non-metallic, and a combination of both metallic and non-metallic materials.

- Mechanical joining requires no (or very little) surface preparation of the substrate materials.
- The substrate materials are not heated during mechanical joining processes, therefore the thermal stress and embrittlement of the substrates due to a rise in temperature are completely avoided.
- Mechanical joining facilitates maintenance, repair, service, inspection and quality control.
- It allows for easy assembly and disassembly of components without damaging the parent materials.
- Mechanical joining does not make any changes in the microstructure or composition of the parent materials.

8.4.3 Disadvantages of Mechanical Joining

In spite of all the advantages of mechanical joining, there are some disadvantages that cannot be ignored. These are listed below:

- The induced stress concentration is one of the major demerits of mechanical joining. Stress concentration around the fastener caused by the presence of holes affects the performance of the structure.
- The drilling of thermoplastic composite materials to prepare holes causes various types of surface damage around the hole, such as splintering, delamination and fraying.
- The mechanical fasteners used in mechanical joints add some mass to the structure, which is a weight penalty compared with the adhesive and microwave joining processes.
- This process creates discontinuities in the fibres around the drilled hole which adversely affects the performance of the joint.
- The joints made by this process are often not sealed which allows the intrusion of moisture and other fluids and thus permit leakage through the hole generated for mechanical fastenings such as rivets and bolts, and this can result in corrosion.
- The mechanical joining process is a labour intensive process especially when high performance structural components are joined together.

8.4.4 Design Guidelines for Mechanical Joining

The appropriate selection of mechanical fasteners for a given application or situation improves the joint strength and therefore enables the joint to carry or transfer the load for a longer period of time. The following aspects needs to be considered when designing a mechanical joint:

- Type of loading to which the components of a structure are subjected.
- Size of pilot holes and the distance between adjacent holes.
- Selection of appropriate materials for the fastener, because the compatibility of the fastener material with the parent material to be fastened is important.
- Number of fasteners to be used, as this is directly proportional to the cost and the weight of the structure. An increased number of fasteners will increase the cost and also causes additional weight for the structure.
- Location of holes is also an important consideration as it directly affects the chances of failure. When the holes are positioned close to the corners or edges, there is a chance of of the materials tearing when they are subjected to high external loads.

8.5 Microwave Joining

As discussed in the previous sections some of the well-established joining techniques available for metallic substrates are not suitable for thermoplastic composite materials. The joining of fibre-reinforced composite components by means of adhesive bonding or mechanical fastening is troublesome because there are many challenges associated with these types of joints, such as high stress concentration, excessive delamination and extensive weight penalty. Therefore, an advanced technique named microwave joining has been established to eliminate many of these problems. This promising joining technique is also suitable for joining natural fibre-reinforced thermoplastic composites. The main benefit of using microwave energy for joining is that it does not change the microstructure or properties of the parent materials. The process basically involves the use of high frequency electromagnetic radiation to heat a susceptor material located at the joint interface. The generated heat melts the thermoplastic materials and these then join upon cooling. A susceptor is a high dielectric constant material which can absorb electromagnetic energy and then convert it to heat energy. The dielectric constant of a material determines its ability to absorb microwave radiation. A susceptor may be of a ceramic or metallic material. The important process parameters that that can affect the joint strength are exposure time, power level, applied pressure and percentage of susceptor to be used. Microwave joining demonstrates several advantages over traditional joining methods, such as:

- Rapid processing.
- Energy saving.
- No polluting gases.
- Ability to join complex geometries.
- Volumetric heating and uniform temperature distribution.
- High strength and homogenous microstructure of the joint.

8.5.1 Microwave Joining of Natural Fibre-reinforced Thermoplastic Composites

Microwave joining of natural fibre-reinforced thermoplastic composites is an innovative joining technique which is getting worldwide attention from researchers and technologists. Microwave joining of thermoplastic composites necessitates the application of pressure at the junction by which parts or objects are joined together. The pressure can be applied throughout the process or just during the solidification stage. The applied pressure has to be optimised in order to attain maximum heating and to minimise the amount of material squeezed out of the joint [5]. Bajpai and co-workers [6] experimentally investigated the load carrying capacity of joints bonded through microwave radiation. Different combinations of natural fibres and matrix, such as nettle/polypropylene (PP), *Grewia optiva*/PP, nettle/poly(lactic acid) (PLA) and *Grewia optiva*/PLA were used in the experiments. The joint strength was determined in each case using standard procedures. The results showed that microwave joining offers higher joint strength than adhesive bonding. A multimode applicator with a fixed frequency of 2.45 GHz and a maximum power output of 900 W was employed for the experiments. Microwave power and time are the two main controlling parameters in this applicator for microwave joining. In order to obtain good bond strength, PLA-based composite specimens were exposed for 200 s, whereas PP-based composites were exposed for 250 s. To accelerate the heating process, a susceptor material was used. Charcoal, which quickly couples with microwave energy and produces rapid heating effects, was used as the susceptor material. An initial holding pressure was applied to the joint to ensure effective bonding by using a Teflon (polytetrafluoroethylene) tape. The top surfaces of the materials to be joined were heated through conventional modes of heat transfer from the microwave-heated charcoal. The temperature rise was eventually sufficient to cause melting and fusion of the interfaces. The results (see **Table 8.1**) showed that microwave joined *Grewia optiva* fibre-reinforced PLA had the greatest load carrying capacity, whereas nettle fibre-reinforced PP showed the lowest joint strength. The joining of thermoplastic-based materials with microwave heating is a viable and sustainable alternative as it is a green processing technology. Microwave joining of fibre-reinforced thermoplastic composites involves five different steps. These are:

- The use of susceptor material.
- Clamping of composite parts to be joined together.
- Heat generation due to absorption of microwave energy by the susceptor.
- Heat conduction and melting in the joint areas.
- Cooling and solidification of the joint.

8.6 Comparisons of Joint Strength under Tensile Loading

8.6.1 Joint Strength of Microwave Bonded Joint

Failure testing of microwave joined composite specimens was carried out under tensile loading. Lap joint failure loads for different composite specimens are shown in **Table 8.1**. From **Table 8.1**, it is clear that, for microwave bonded joints, the greatest load was taken by *Grewia optiva*/PLA, whereas the nettle/PP adherend showed the lowest joint strength. In each case, the load taken by the joint was less than that of respective parent material. In the case of *Grewia optiva*/PLA and nettle/PLA composites, the joints did not fail but the adherends failed as shown for nettle/PLA composites in **Figure 8.6**. This indicates that, after microwave processing, the adherends became weaker. The reason behind this may be unwanted heating of the specimen near the joint interface as, in some specimens, a hot spot region developed near the joint interface. In microwave joining, hot spots were observed in the specimen due to the wave interference in the microwave system. This unwanted heating can be avoided by changing the location of samples in the cavity of the testing equipment and redesigning the fixture properly to hold the adherends [6].

Table 8.1 Load carried by the joint formed with microwave heating and adhesive bonding with epoxy adhesive

Sample number	Fibre/matrix	Failure load (N)		
		Parent material	Microwave bonded joint	Adhesively bonded joint
1	Nettle/PLA	2250	1207	163
2	*Grewia optiva*/PLA	3500	2200	173
3	Nettle/PP	1205	623	160
4	*Grewia optiva*/PP	1447	833	173

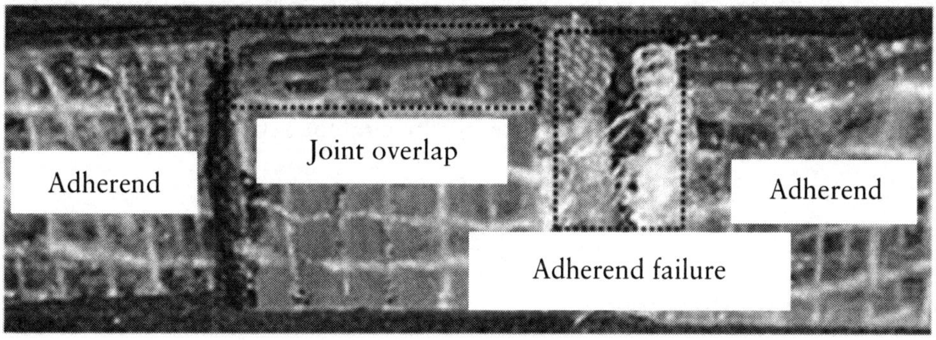

Figure 8.6 Adherend failure under tensile loading for nettle/PLA composites

8.6.2 Joint Strength of Adhesively Bonded Joint

As shown in **Table 8.1**, the load taken by adhesively bonded joints of natural fibre-based composite specimens was substantially less than the joint formed by microwave heating. The reason behind this may be the poor chemical bonding between the matrix materials and the epoxy adhesive used. In the case of nettle/PLA and *Grewia optiva*/PLA adherends, the joint failure load was 7.2% and 4.9% of the parent material failure load, respectively. The joint failure load for nettle/PP and *Grewia optiva*/PP adherends was 13.2% and 11.9% of the parent material failure load, respectively. In the present case, adhesive failure at the joint interface was observed in nettle/PLA composites as shown in **Figure 8.7**. **Table 8.1** shows the failure load under tensile loading for each type of adhesively bonded composite adherends [6]. Surface preparation of specimens and selection of the most suitable adhesive may improve the joint strength of an adhesively bonded composite specimen.

Figure 8.7 Adhesive failure under tensile loading for nettle/PLA composites

8.7 Concluding Remarks

Natural fibre-based composite materials have reached a decisive stage where they can act as effective alternatives to synthetic fibre-based composite materials. The biodegradable nature of the natural fibre-reinforced composites makes them potential candidates for structural and non-structural applications. The secondary processing such as joining of these low weight low cost materials is a practical necessity because components need to be assembled together to obtain a complete structure. However, joints are potential sources of weakness, failure and additional weight. The present chapter highlights the important techniques which have been employed to join parts made of natural fibre-reinforced thermoplastic composites. Adhesive bonding and mechanical fastening are established joining techniques for polymer composites, which have their own merits and demerits. There is an urgent need to focus on both the existing and new joining techniques in order to make good quality natural fibre-based products. The joining of composite parts by means of microwave energy is an emerging technique which overcomes the limitations of conventional joining techniques. Microwave joining of polymer composites is still at an early stage of development, therefore a better understanding of this joining technique will definitely result in a much broader range of applications in the future.

References

1. S. Kalpakjian and S.R. Schmid in *Manufacturing Processes for Engineering Materials*, 5th Edition, Prentice Hall, Upper Saddle River, NJ, USA, 2007, Chapter 12.

2. S.K. Mazumder in *Composites Manufacturing: Materials, Product, and Process Engineering*, CRC Press, New York, NY, USA, 2001, Chapter 9.

3. G. Savage, *Engineering Failure Analysis*, 2007, **14**, 321.

4. R.A. Malloy in *Plastic Part Design for Injection Molding: An Introduction*, Hanser Gardner Publications, Cincinnati, OH, USA, 1994, Chapter 6.

5. *Handbook of Plastics Joining - A Practical Guide,* Plastics Design Library, Norwich, NY, USA, 1997, Chapter 8.

6. P.K. Bajpai, I. Singh and J. Madaan. *Materials and Design*, 2012, **35**, 596.

9 Natural Plant Fibre Biocomposites for Structural Vehicle Components

James Njuguna and Elias Nassiopoulos

9.1 Introduction

All natural fibres can be subdivided into three categories based on their origins, coming from plants, animals or minerals [1]. Plant fibres consist mainly of cellulose whilst animal fibres are composed of proteins. Among them, plant fibres are the most appealing for the construction of fibre-reinforced plastics as their properties, combined with their very low density, make them ideal to be used as reinforcements. Plant fibres can be divided into bast (stem), leaf, seed, wood fibres and other grass fibres [2]. **Figure 9.1** below presents the classification of natural fibres according to their origin, together with examples of the most popular plant fibres currently used in the design of composite materials.

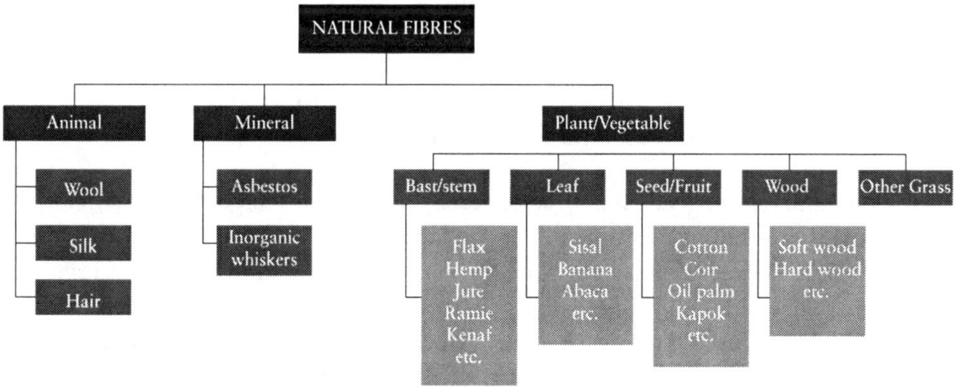

Figure 9.1 Classification of natural fibres

Bast fibres such as hemp, jute, kenaf and flax are fibrous bundles found in the inner bark of the plant stem. Leaf fibres, known also as hard fibres, run along the length of the leaves. Sisal is the most common, whilst others include banana, abaca and pineapple fibres. The most commonly used fruit fibres in composite applications are coir and oil palm [1-3].

Bast fibres have the highest mechanical properties and this makes them the first choice as reinforcement for polymers in the manufacturing of natural fibre composites. However, the most suitable type of fibre depends on the specific application, the properties and the performance needed.

9.1.1 Structure of Plant Fibres

Plant fibres are technically referred to as 'elementary fibres' [4-6]. Depending on the category and nature of the plant, the elementary fibres are located and extracted differently. However they are always grouped, forming a 'technical fibre' (a technical fibre consists of about 10 to 40 elementary fibres), held together mainly by a lignin and hemicellulose matrix. The technical fibres have an average length of up to a metre with a typical diameter between 50 to 100 μm and, in many cases, also form larger groups which are most often referred to as 'bundles' [2]. The characteristic schematic representation of the flax stem in **Figure 9.2** illustrates well the physical form of the fibres as described above. In composites, elementary fibres, technical fibres or even bundles can be used, depending on the specific requirements and the preferred production methods since the fibre extraction method and final properties dictate the application limitations.

The elementary fibres (plant fibres), which are responsible for the overall mechanical properties of the final product, are basically single plant cells, with a typical length of 10 to 50 mm and a diameter between 10-50 μm. Due to their complex structure that resembles microscopic tubes made out of layers of different constituents, plant fibres are often considered as natural composites.

Each elementary fibre (cell) has four different layers (as depicted schematically in **Figure 9.3**) surrounding the lumen [4]. The lumen is an open channel in the centre of the fibre. The outer layer is called the 'primary' cell wall and is usually a very thin wall, whilst the inside cell wall is called the 'secondary' cell wall and consists of three layers (S1, S2, and S3). Of these, S2 is much thicker than the other two, contains large quantities of cellulose and is the major contributor to the overall properties. From the outside to the inside the hemicellulose content remains constant, in contrast with the cellulose content of each layer which increases steadily. Consequently the lignin content decreases from the outer to the innermost S3 layer.

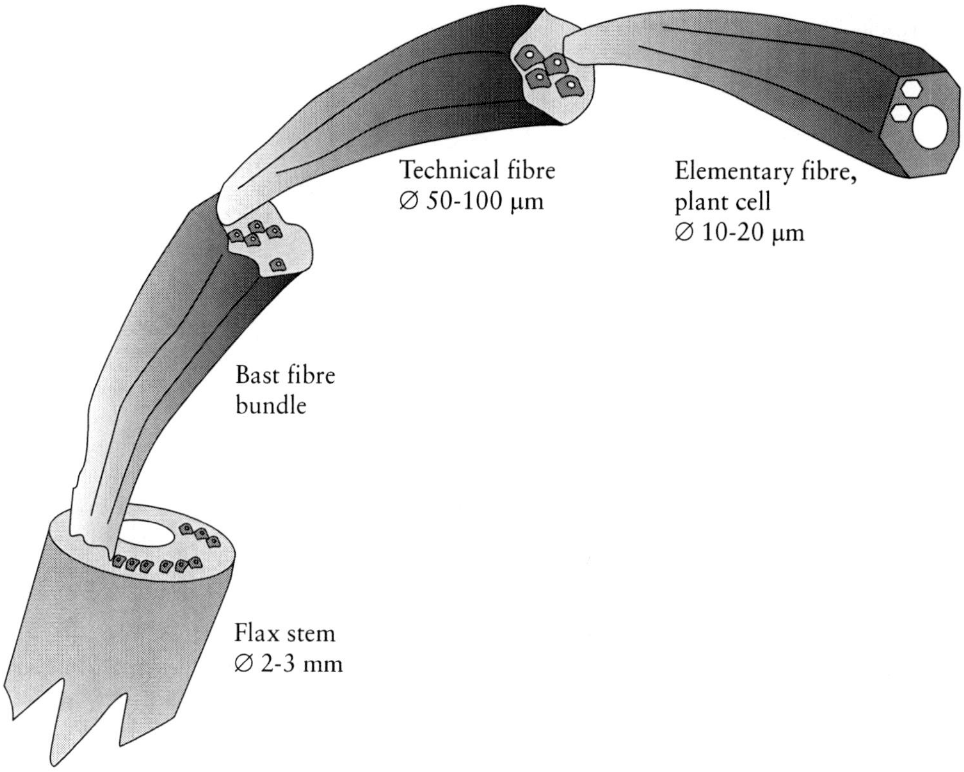

Figure 9.2 Schematic representation of flax fibre, from stem to elementary fibres. Ø is the diameter of each component. Reproduced with permission from H. Bos, D.O. Van and O. Peters, *Journal of Materials Science*, 2002, **37**, 1683. ©2002, Springer [6]

Although the primary wall is porous, and is composed of disorderly arranged crystalline cellulose, the secondary layers are formed from orientated cellulose microfibrils 'glued' together in a complex hemicellulose and lignin matrix (sometimes pectin can also be found in the matrix) [2]. In each layer the microfibrils are fairly parallel to each other and follow a helical arrangement around the cell (fibre axis). They consist of semicrystalline cellulose with alternate crystalline and amorphous regions and their diameter has a typical value from 10 to 30 nm. The angle of the helical arrangement to the normal fibre axis in every layer has a huge impact on the mechanical properties of the fibres and is known as the microfibrillar angle (**Figure 9.3**). The structure, microfibrillar angle, cell dimensions, defects, and the chemical composition of fibres are the most important variables that determine the overall properties of the fibres.

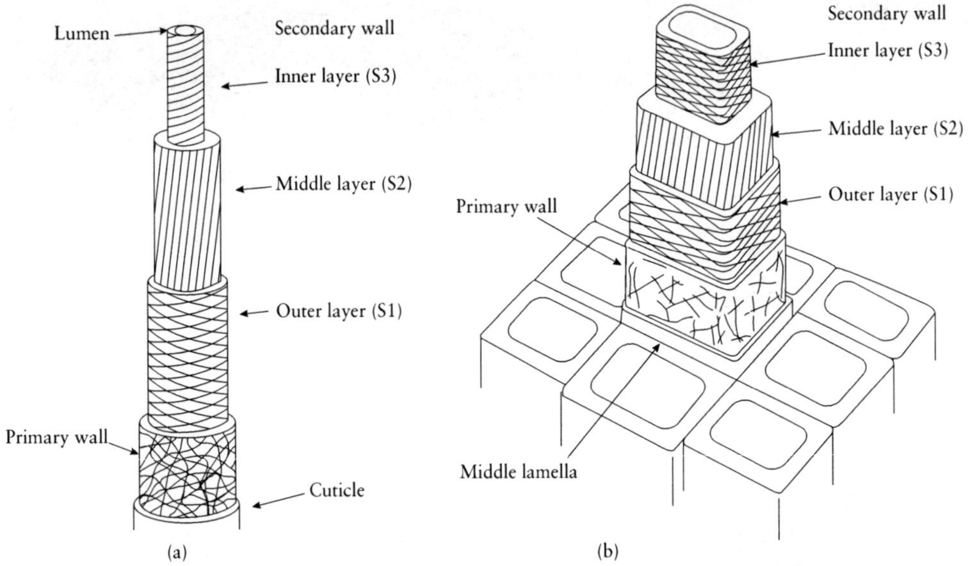

Figure 9.3 Schematic structure of plant fibre architecture. (a) Cotton fibre; (b) a delignified spruce wood. Reproduced with permission from J. Biagiotti, D. Puglia and J.M. Kenny, *Journal of Natural Fibers*, 2004, **1**, 2, 37. ©2004 Taylor and Francis [4]

9.1.2 Chemical Composition

The chemical composition as well as the structure of plant fibres is fairly complicated, with the main polymers involved in the composition of plant fibres being cellulose, hemicellulose and lignin. Other extractives present in plant fibres include pectin and waxes [2, 4, 7].

The major component of most plant fibres is cellulose (α-cellulose). Cellulose is a linear macromolecule consisting of D-anhydroglucopyranose (often abbreviated as anhydroglucose) repeating units ($C_6H_{11}O_5$) joined by β-1,4-glycosidic linkages. The Haworth's projection formula of cellulose is given in **Figure 9.4**.

The molecular structure of cellulose is responsible for the chemical and physical properties of the fibres. In the fully extended molecule, sequential chain units are orientated by their mean planes at an angle of 180° to each other. Thus, the repeating unit in cellulose is the anhydrocellulobiose unit and the number of repeating units per molecule is half the degree of polymerisation. This, in native cellulose, can be as high as 14,000 units.

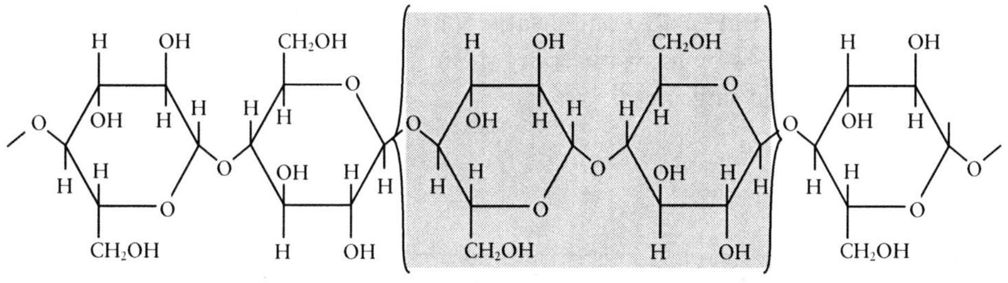

Figure 9.4 Chemical structure of cellulose

Different plants have different types of cellulose, depending on the length of the chain and the cell geometry. The adjacent chains are linked due to strong hydrogen bonds between the three hydroxyl groups in each unit. Cellulose has a semicrystalline structure with regions of high order (crystalline)and regions of low order (amorphous). Several different crystalline structures of cellulose are known, corresponding to the location of hydrogen bonds between and within the strands (chains). The crystal structure of naturally occurring cellulose (largely obtained from plants and wood) is known as cellulose I, in which all the cellulose strands are parallel to each other and no intersheet hydrogen bonding occurs [4]. In nature cellulose I (naturally occurring) is metastable and can be transformed irreversibly to the more stable cellulose II.

On average, natural fibres contain 60-80 wt% cellulose, 10-20% hemicelluloses, 5-20 wt% combined lignin and pectin) and up to 20 wt% moisture. The chemical composition of some of the most common plant fibres is shown in **Table 9.1** [2, 4, 5, 8].

Hemicellulose is made up of polysaccharides composed of a combination of 5-carbon and 6-carbon ring sugars (e.g., glucose, mannose and galactose) [2]. The chains are much shorter than in cellulose (degree of polymerisation approximately 50 to 300) and branched, containing pendant side groups giving rise to its non-crystalline nature (amorphous). Hemicellulose is very hydrophilic, soluble in alkali and easily hydrolysed in acids [1].

Lignin is thought to be a complex, three-dimensional copolymer of aliphatic and aromatic constituents with a very high molecular weight [4]. The monomers and their proportions depend on the source in nature. Its chemistry has not yet been precisely established as there is still no method of isolating lignin in its native state, however hydroxyl, methoxyl and carbonyl groups have been identified. Lignin is a thermoplastic polymer with melting temperature of around 170 °C; it is structurally amorphous and hydrophobic in nature [2]. Its mechanical properties are lower than those of cellulose whilst lignin is not hydrolysed by acids, but soluble in hot alkali, it is readily oxidised and easily condensable with phenol. Lignin has the advantage that it can resist attacks by most microorganisms, and anaerobic processes tend not to attack the aromatic rings at all.

Table 9.1 Chemical composition of examples of natural plant fibres

Name	Cellulose (wt%)	Hemicelluloses (wt%)	Lignin (wt%)	Pectin (wt%)
Flax	60-81	14-20.6	2-3	1.8-2.3
Sisal	43-88	10-14	5-14	0.8-2
Hemp	70-78	17.9-22.4	3.7-5.7	0.9
Ramie	68.6-76.2	13.1-16.7	0.6-1	1.9-2
Jute	51-72	12-20.4	5-20	0.2
Oil palm	65	-	29	-
Abaca	56-64	21	12-13	0.8-1
Banana	60-65	6-19	5-10	3-12
Coconut (coir)	43	0.3	45	4
Stinging nettle	86	10	-	-
Cotton	82.7-92	2-5.7	0.5-1	5.7
Bamboo	26-43	30	21-31	-
Kenaf	36-57	21-21.5	8-18	2-5
Wood	45-50	23	27	-

- Data not available

Pectins are a family of complex structural heteropolysaccharides which consist essentially of polygalacturonic acid, whilst waxes are different types of alcohols which are insoluble in water as well as in several acids.

9.1.3 Mechanical Properties

The overall properties of the fibres depend on the individual properties of each of their components, their physical and chemical properties as well as their morphology. The structure, microfibrillar angle, cell dimensions and defects, and the composition are among the most important variables. Increasing the cellulose content will in general increase the strength and modulus, whilst the orientation of the microfibrils will also control the stiffness. The more parallel the microfibrils are in respect to the fibre axis, the higher the resulting rigidity of the fibre [2, 5]. Hemicellulose is responsible for the biodegradation, moisture absorption and thermal degradation of the fibres whilst, on the other hand, lignin (like pectin) is thermally stable but is responsible for ultraviolet degradation of the fibres. Moisture absorption and the hydrophilic nature of plant fibres results in a reduction of the overall mechanical properties. The length of the fibres is another property that has been proven to affect their strength, depending on the degree of homogeneity or amount of defects along the fibre length [10].

The main mechanical properties found in the literature for the most important natural fibres are collected in **Table 9.2** and compared with those of glass fibres [5, 6, 9-11].

Table 9.2 Mechanical properties of examples of natural plant fibres [5, 6, 9-11]

Name	Diameter (μm)	Density (g/cm^3)	Young's Modulus (GPa)	Tensile strength (MPa)	Elongation at break (%)
E-glass	5-25	2.55	70-73	2400	-
Flax	40-620	1.45-1.5	27-80	343-1500	2.7-3.2
Hemp	16-50	1.4-1.5	30-90	350-1100	1.3-4.7
Sisal	100-300	1.3-1.5	9-28	300-855	2-5
Jute	30-140	1.3-1.5	20-50	187-773	1.4-3.1
Bamboo	-	0.8-1.4	30-89	391-1000	2
Ramie	40-60	1.5	44-128	400-938	3.6-3.8
Abaca	17-21	1.5	72	980	10-12
Banana	50-280	1.3-1.35	7.7-32	529-914	3-10
Coconut (coir)	100-450	1.2-1.5	3-6	106-270	15-47
Kenaf	40-90	1.22-1.4	25-50	295-930	1.7-6.9
Cotton	16-50	1.4-1.5	5.5-12.6	287-597	2-10

- Data not available

Adapted from A.K.Bledzki and J. Gassan, *Progress in Polymer Science*, 1999, **24**, 2, 221 [5]; H. Bos, D.O. Van and O. Peters, *Journal of Materials Science*, 2002, **37**, 1683 [6]; H.L. Bos, J. Müssig and M.J.A. van den Oever, *Composites Part A: Applied Science and Manufacturing*, 2006, 37, 10, 1591 [9]; H. Bos in *The Potential of Flax Fibres as Reinforcement for Composite Materials*. Technische Universiteit Eindhoven, Eindhoven, Netherlands, 2004 [10]; and H. Ku, H. Wang, N. Pattarachaiyakoop and M. Trada, *Composites Part B: Engineering*, 2011, **42**, 4, 856 [11].

The scattering of the values for all properties is underlined as one of the major disadvantages of natural fibres, and can be explained by differences in the fibre structure, composition and morphology due to different overall environmental conditions during the plant growth (e.g., area, climate and age of the plant when harvested).On average the performance of natural fibres in terms of stiffness and strength cannot compete with those of glass fibres (**Figure 9.5**). However their specific properties, especially stiffness, can be comparable or even superior as the density of natural fibres is half that of glass as shown on (**Figure 9.5**) whilst **Figure 9.6** demonstrates the comparison of tensile strength properties.

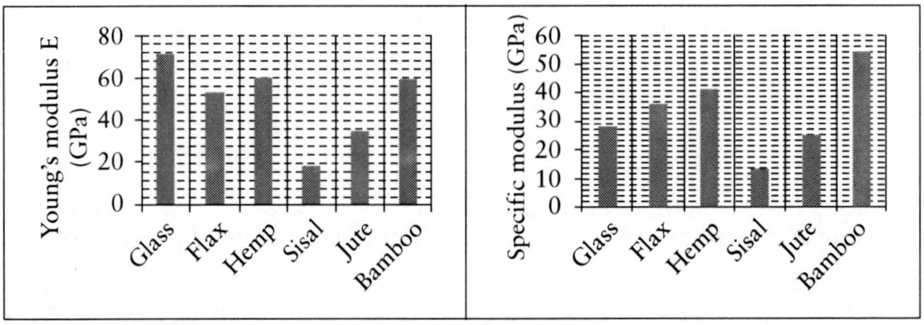

Figure 9.5 Young's modulus and specific modulus of most commonly used natural fibres

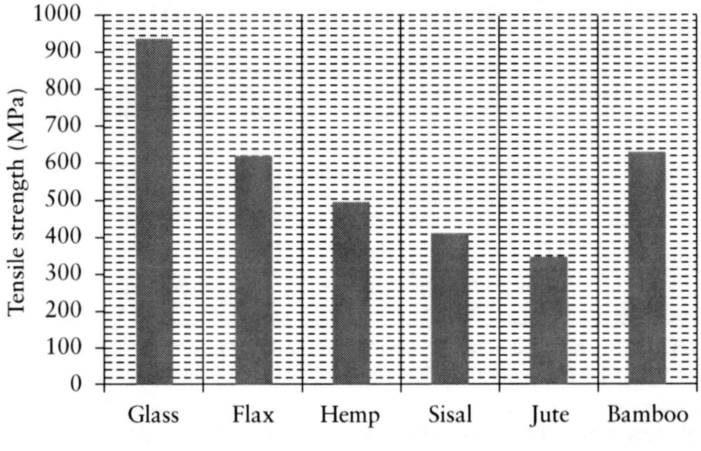

Figure 9.6 Tensile strength of most commonly used natural fibres

9.1.4 Fibre Modifications

All natural fibres (although to different extents) are hydrophilic in nature. This is attributed mainly to the lignocellulose into their structure, which contains strongly polarised hydroxyl groups [1]. These fibres, therefore, are inherently incompatible with many already established composite manufacturing resins, which are hydrophobic. Only a few thermosets such as the phenol-formaldehyde resins and related polymers are less hydrophilic and thus less problematic.

This discrepancy leads often to the formation of an ineffective interface between the fibres and the matrix. The major limitations of using these plant fibres as reinforcements in such matrices include poor interfacial adhesion between polarhydrophilic fibres and a non-polar hydrophobic matrix, and difficulties in mixing due to poor wetting of the fibres with the matrix. The role of the matrix in a fibre-reinforced composite is to transfer the load to the stiff fibres through shear stresses at the interface. This process requires a good bond between the polymeric matrix and the fibres [12]. Poor adhesion at the interface means that the full capabilities of the composite cannot be exploited and leaves it vulnerable to environmental attacks that may weaken it, thus reducing its life span. Insufficient adhesion between the polymer and the fibres results in poor mechanical properties of the natural fibre-reinforced polymer composites.

Pretreatment of the fibres can clean the fibre surface, chemically modify it, stop the moisture absorption process and increase the surface roughness [5, 7]. Physical treatments like cold plasma treatment or corona treatment, chemical treatments such as graft copolymerisation, esterification and isocyanates, or the use of coupling agents can improve the composite's mechanical properties and protect it from environmental attack.

9.1.5 Selected Natural Fibres

9.1.5.1 Flax

Flax is an erect annual plant growing up to 1.2 m tall, with slender stems and green leaves 20 to 40 mm long and 3 mm broad. Flax (with the Latin name *Linumusitatissimum*) is one of the most popular and commonly used natural fibres, but also one of the oldest textile fibres known. The first documented applications date back to 5000 BC when it was used by the Egyptians to wrap their mummies [10]. Since then, flax has been successfully applied to a vast number of applications and products from clothing and packaging to the naval and automotive industries.

Presently, two types of flax are grown: fibre flax and seed flax. Flax is harvested for fibre production after approximately 100 days of growth or two weeks after the seed capsules form when the base of the plant will begin to turn yellow [2, 10]. If the plant is still green, the seed will not be useful, and the fibres will be underdeveloped. The fibres degrade once the plant is brown. Flax fibre is suitable for the production of thin strong fibres. Flax grows in moderate climates and is presently cultivated in large parts of Europe, Canada, the USA and Russia, among other regions. Approximately 5 million hectares of flax are grown worldwide. In the European Union 122,000 tonnes of flax fibre were produced in 2007 [13], making it the world's biggest producer, followed by China with about 25,000 tonnes. The demand for flax fibres in plastic composites is growing by more than 50% annually in Europe and this trend has now started in North America. By far the largest users are automotive part manufacturers who are being pressured to make cheaper and lighter weight vehicles with lower fuel consumption, and to use more environmentally friendly materials in their construction. However at present, flax fibres are able to substitute only for the lower glass grades (lower price fibreglass), mainly due to the fibre inconsistency.

Flax belongs to the category of bast fibres and is considered to be one of the strongest and easily available. Its strength varies between 350 and 1,500 MPa whilst its modulus can be as high as 80 GPa [1, 4-6, 9, 10, 14]. These properties, combined with their very low density, make flax fibres a very competitive contestant against glass. Flax fibres are also generally cheaper, lighter in weight and impart more springiness than fibreglass. In addition they take less energy to manufacture and are easier to degrade or burn than fibreglass.

Bos and co-workers [9, 10] conducted a range of studies on the stiffness and strength of flax technical and elementary fibres. The tensile properties of flax fibres depend on the clamping length due to the structure of the technical flax fibres, which also explains the scatter in flax fibre tensile strengths reported in the literature. They [9, 10] also concluded that single elementary flax fibres have considerably higher strengths than technical fibres as the fibre strength strongly depends on the decortication process. The strength distribution of flax fibres was also studied by Andersons and co-workers [15] using a modified Weibull distribution.

Liu and co-workers [16] fabricated flax-epoxy composites using different types of woven fibre reinforcement and volume fractions, and showed that the presence of flax improved the stiffness by up to four times that of the unreinforced epoxy. They also concluded that toughness is dominated by the fibre volume fraction, rather than the architecture of the weave. Another study, on the reinforcement of polypropylene (PP) with flax nonwovens, was conducted by John and Anandjiwala [17]. Tensile strength and modulus were found to increase with the increase in fibre loading, whilst flexural strength and modulus also registered an increase. Zein modification of the nonwovens was found to enhance the mechanical properties of the composites due to better interfacial bonding. However these authors also observed a decrease in the impact strength, explained by the presence of fibre breakage in the chemically modified composites. Thermoplastic composites based on flax fibres and a PP matrix was manufactured by Garkhail and co-workers [18]. The materials' stiffness, strength and impact performance were then studied with respect to fibre length, fibre volume fraction and interface modification through the use of a maleic anhydride grafted PP grade. It was concluded that composites based on a PP matrix and flax fibres can compete with E-glass-based materials in stiffness-critical structures, whereas for strength- and impact-critical applications these materials still need to be optimised further. Baiardo and co-workers [19] also studied flax fibre composites with a biopolyester matrix (Bionolle). The study showed again the need for surface modification to improve the adhesion between fibres and matrix. The use of acetylation on the fibre surface remarkably increased the strength of the composite. Many more studies have been carried out on flax fibre-reinforced composites, their properties, their environmental impact and the effect of different treatments and modifications [1].

9.1.5.2 Hemp

Hemp is the soft, durable fibre that is cultivated from plants of the *Cannabis* species (*Cannibis sativa*) and has long been used for industrial purposes including paper, textiles, biodegradable plastics, construction, automotive applications, health food and fuel with modest commercial success [13].

For centuries, hemp was used to make rope, canvas and paper [20]. In 2800 BC the first ropes were produced in China using hemp fibres. In 100 BC China also produced the first paper in the world from hemp fibres. From the Middle Ages until the end of the sailing ship period, hemp was an important crop in many European countries. Long hemp fibres can be spun and woven to make a crisp, linen-like fabric used in clothing, home furnishing textiles and floor coverings. Nowadays, in Europe, hemp fibres are used mainly in the special paper industry; thanks to its lower lignin content, it can be pulped using fewer chemicals than wood. Hemp fibres are also used to reinforce moulded thermoplastics in the automobile industry. The short core fibres go into insulation products, fibreboard and erosion control mats, whilst the fibrous core cuttings can be blended with lime to make strong, lightweight concrete.

China, Canada and Europe are the main hemp cultivation areas in the world [13]. In 2009 the total cultivation area in the European Union was around 15,000 hectares producing around 24,000 tonnes of hemp fibres. The main European hemp producers are France, the UK, Germany, The Netherlands and Poland. The competitive prices of the different qualities against other domestic and exotic fibres, together with the high mechanical properties, make hemp ideal for technical applications.

Although hemp has the advantage of being a very fast growing plant without the need for fertilisers or pesticides, the production and extraction of the primary bast fibres has traditionally been a very labour-intensive process [1]. However, recently, an alternative fibre separation process has been developed using technologies such as ultrasound and steam explosion, which are much less labour-intensive. Once separated, the bast fibres are ready for spinning and weaving into textiles, or for pulping into high quality pulp.

Long, strong and durable, hemp fibres contain about 70% cellulose and low levels of lignin (less than 10%). The fibre diameter ranges from 16 to 50 microns [2]. Further, hemp has among the highest overall mechanical properties with the stiffness of the individual fibres achieving values of 90 GPa and a strength of 1100 MPa. Combined with its very low density of between 1.4 and 1.5 g/cm^3 the specific stiffness of hemp is higher than that of glass fibre, and thus has, for a long time, been very appealing for use in composite materials, and especially to the automotive industry [5, 7, 8].

In their work on natural fibre composites, Wambua and co-workers [12] investigated the mechanical properties of sisal, hemp, coir, kenaf and jute-reinforced PP composites. Of all the fibre composites tested, coir-reinforced PP composites registered the lowest mechanical properties whereas hemp composites showed the highest.

In an interesting study, Lee and co-workers [21] compared a series of hemp-reinforced vinylester and polyester resins, made with the sheet moulding compound process, with the equivalent glass-reinforced materials for selected automotive applications. The study indicated that hemp composites can meet the requirements and, taking into account the very low density, hemp-reinforced materials can be ideal for the replacement of glass composites used in the automotive industry.

Beckermann and Pickering [22] investigated the properties of hemp fibre-reinforced PP composites after alkali treatment. It was found that an injection moulded hemp fibre-reinforced PP composite consisting of 40 wt% sodium hydroxide/sodium sulfite treated fibre, 4 wt% maleic anhydride-modified PP and 56 wt% PP had the highest tensile strength (50.5 MPa) and Young's modulus (5.31 GPa) of all the composites studied. Another study on hemp fibre modification was conducted by Pracella and co-workers [23] who studied PP composites containing hemp fibers modified *via* different methods. All the modified composites showed improved fibre dispersion in the matrix and higher interfacial adhesion with respect to the unmodified PP/hemp. Higher stiffness, strength and thermal stability were also observed.

Yuanjiang and Isaac [24] conducted a series of experiments to characterise the impact strength and residual tensile and fatigue properties following impact tests on nonwoven hemp fiber mat-reinforced polyester. For comparison they used nonwoven glass-reinforced polyester material. The most significant difference between the two materials was the substantially poorer impact performance and the inferior residual properties of the hemp composite following impact damage.

9.2 Resins

Growing environmental concern has made synthetic composites a target of criticism due to their lack of degradability [25]. The manufacturing, use and removal of traditional composites have been studied for many years proving that the disposal of composites after their intended life span is, even when it is possible, very expensive whilst the recycling, as well as reuse, of these composites is not easy since they are made of two or more dissimilar materials. Most traditional thermoplastics and thermosets are considered to be non-biodegradable and there has been a lot of interest in research committed to the design of biodegradable plastics. The concept of environmentally conscious materials is being rapidly accepted by countries all over the world and biodegradable polymers are now considered to be the environmental waste management option.

9.2.1 Selected Natural Resins

9.2.1.1 Bio-epoxy

Epoxy resins are a large family of resins that represent some of the high performance resins available in the market and are produced when a thermosetting epoxide polymer cures (polymerises and crosslinks) when mixed with a catalysing agent or 'hardener'. Epoxies generally outperform most other resin types in terms of mechanical properties including impact resistance.

The chemistry of epoxies and the range of commercially available variations allow the cured polymers to be produced with a very broad range of properties [26]. In general, epoxies are known for their excellent adhesion, chemical and heat resistance, their high mechanical properties and very good electrical insulating properties. However almost all of these properties can be modified and adjusted.

A range of bio-epoxy resins can be extracted from natural sources such as epoxidised vegetables oils (including pine oil waste and soya oil), or waste streams of other industrial processes, such as wood pulp and biofuel production [27]. Important bio-epoxies are glycerol and sorbitol, two abundant and inexpensive bio-based aliphatic polyols, which can be derived from triglyceride vegetable oil and corn starch, respectively. Glycerol-based epoxy resins such as glycerol polyglycidyl ether and polyglycerol polyglycidyl ether are industrially available, inexpensive epoxy resins and have been used in textile and paper processing agents and reactive diluents. Takada and co-workers [28] investigated the use of glycerol-epoxy bioresins for the production of composites and reported that, due to their high mechanical performance and thermal properties, glycerol-epoxy resins could be a potential substitution for petroleum-based epoxy resin composites. The same conclusions were drawn by Shibata and Nakai [29] who experimented with glycerol and sorbitol polyglycidyl ether cured with a tannin-based agent that showed comparatively good and balanced mechanical and thermal properties.

Vegetable oil-based epoxy resins are another vast category with the most utilised today being linseed and soybean epoxidised vegetable oils. Aboobucker Sithique and co-workers [30] studied the tensile and flexural properties of biopolymers based on epoxidised soybean oil at various concentrations modified with different types of bismaleimides. They concluded that although the initial soy-based resin had very poor properties, the bismaleimide modifications resulted in a pronounced improvement of the mechanical performance of the resin which could be used for automotive and other structural applications. Gerbase and co-workers [31] studied the mechanical and thermal behaviour of epoxidised soybean oil modified with various cyclic acid anhydrides using dynamic mechanical analysis and thermogravimetry. All the samples had thermoset characteristics with properties dependant on the type of anhydride, the ratio of epoxy to anhydride and the epoxy group content. Ratna [32] reached the same conclusions experimenting with soyabean oil modified epoxy resins where, although increasing the content of soyabean oil decreased the tensile and flexural modulus, the resins showed higher elongation together with an improved impact performance. Miyagawa and co-workers [33] manufactured and tested nanocomposites from biobased epoxidised soyabean oil epoxy and organoclay reinforced with carbon fibres to investigate their mechanical and thermal performance. Their efforts resulted in novel advanced materials which had high modulus and strength properties. Ahmetli and co-workers [34] used fatty acid from sunflower wastes to modify a commercial synthetic epoxy. All modifications improved the mechanical performance, with higher strength and elongation, whilst a significant improvement in the hardness of the resin was also observed.

9.2.1.2 Polylactides

Another biopolymer used today is poly(lactic acid) or polylactide (PLA). PLA is an aliphatic thermoplastic polyester derived from lactic acid found in renewable resources, such as corn, wheat, barley, cassava, and sugar cane [35]. Lactic acid is then polymerised to PLA, either by gradual polycondensation or by ring-opening polymerisation. Due to the chiral nature of lactic acid, several distinct forms of PLA exist: poly-L-lactide (PLLA) is the product resulting from the polymerisation of L, L-lactide (also known as L-lactide), while poly-D-lactide is produced from D-lactide. Polylactic acids are hydrophobic in nature and are among the few polymers in which the stereochemical structure can easily be modified by polymerising a controlled mixture of the L- or D-isomers to yield high molecular weight amorphous or crystalline polymers that can be used for food contact applications [35].

The commercially attractive features of PLA include its production from renewable resources as well as its good mechanical properties [36]. PLA has good tensile strength, low extension and a high modulus (approximately 4.8 GPa) and hence has been considered to be an ideal biomaterial for load bearing applications. PLA-based polymers are completely degradable and after use can be recycled or disposed of by incineration or by landfilling. PLA polymers also have a relatively high hardness. In order for PLA to be processed on large-scale production lines such as injection moulding, compression moulding and extrusion, the polymer must possess adequate thermal stability to prevent degradation and maintain its molecular weight and properties. PLA has a glass transition and melt temperature of about 55 °C and 175 °C respectively whilst it undergoes thermal degradation at temperatures above 200 °C.

Although PLA has been frequently used for packaging, it is also suitable for use as a matrix in which fibres can be embedded to form composites. For instance Tokoro and co-workers [37] mixed three types of bamboo fibres into a PLA matrix to improve its impact strength and heat resistance. Their studies resulted in good quality composite materials, with the presence of the bamboo fibres greatly increasing the impact performance and thermal properties. Ochi [38] investigated kenaf/PLA composites with different fibre loadings. The tensile and bending strength, as well as Young's modulus, increased linearly up to a fibre content of 50 wt%. He also concluded that unidirectional biodegradable composites fabricated using an emulsion type PLA resin and kenaf fibres at a fibre content of 70 wt% had high tensile and flexural strengths of 223 MPa and 254 MPa, respectively. Shibata and co-workers [39] prepared composites with lyocell fabric and PLA by compression moulding. The tensile modulus and strength of lyocell/PLA composites improved with increasing fibre content. The impact strength was considerably higher than that of pure PLA. Pan and co-workers [40] also produced kenaf/PLA composites by injection moulding with fibre contents ranging between 0 and 30 wt%. At 30 wt% a tensile strength improvement of 30% was observed.

9.3 Mechanical Performance of Natural Fibre Composites

In a composite the properties of the fibres are combined with those of the matrix, which is responsible for transferring the external loads to the stiffer fibres through shear stresses at the interface, as well as keeping the fibres together in a specific structural form. Thus the properties of the composite are a combination of the properties of the ingredients and their prediction and estimation is not trivial.

9.3.1 Stiffness and Strength

The mechanical properties of natural fibres are much lower than those of glass fibres. However their specific properties, especially stiffness, are comparable to the stated values of glass fibres. Moreover, natural fibres are about 50% lighter than glass, and in general, cheaper. It is widely acknowledged that natural fibre composites combine good mechanical properties with a low specific mass and offer an alternative material to glass fibre-reinforced plastics in some technical applications. For example, Bledzki and Gassan [5] observed that the characteristic values of natural fibres are comparable to those of glass fibres. Experimental data giving the tensile strength, flexural strength, modulus, impact force and compressive force are available in the literature for different types of natural-fibre composites [5, 12].

The ultimate strength of any composite depends on several factors, the most important of which are the properties of the components and the volume fraction. Wambua and co-workers [12] studied the importance and effect of the volume fraction on the tensile strength of natural fibre composites. They reported that an increase in the fibre weight fraction produces an increase in the tensile strength. Testing different fibre reinforcements they also found that hemp/PP composites with a 30% volume fraction displayed a tensile strength of 52 MPa, higher than equivalent glass-reinforced composites with the same volume fraction. Further, hemp/PP and kenaf/PP composites registered a high tensile modulus of 6.8 GPa compared to 6.2 GPa for the equivalent glass composites. The increase of the modulus and the tensile strength with increase of the volume or weight fraction was also shown by Bos [9, 10] on flax/PP composites with maleic anhydride grafted PP for improved adhesion.

Studies and results of tensile tests on flax fibre-reinforced PP composites were conducted by Garkhail and co-workers [18] and concluded that fibre length affects the strength and modulus of the composites for short fibre lengths whilst after a specific value for the length, the two parameters are constant. The stiffness of a flax/PP composite was shown to be comparable to an E-glass-based composite, especially with regard to the specific properties due to the very low density of flax. However the results also showed a relatively low tensile strength for flax/PP composites compared to the E-glass-based composite.

Nishino [3] studied the mechanical properties of kenaf/PLLA composites. He concluded that the modulus of the composites increases with the increase of the volume fraction, but only up to a certain level. When this threshold is achieved, further increase of the fibre fraction leads to a dramatic reduction of the composite properties.

In addition, water content has a dramatic effect on the properties of natural-fibre composites. Espert and co-workers [41] showed this effect on cellulose/PP composites by submerging samples in distilled water at different temperatures. The samples were removed from the water at certain times and the water absorption was measured. The results of tensile tests showed a significant effect of the water content on the Young's modulus of the samples, and an even bigger effect on the tensile strength. The studies also concluded that the effect of the water on the properties was highly influenced by the fibre content, the matrix and, mainly, the temperature. Thwe and co-workers [42] investigated the same effect on bamboo fibre composites and the results showed that both the tensile strength and modulus decreased after ageing in water at 25 and 75 °C for prolonged periods. The extent of strength and stiffness loss depended on the ageing time and temperature. They also concluded that tensile strength and stiffess were enhanced by the inclusion of a coupling agent, MAPP, in the matrix material as a result of improved interfacial bonding.

9.3.2 Impact Performance

There are only a few studies known about the impact behaviour of natural fibre-reinforced composites. The impact performance of several natural fibre composites was compared and reviewed by Wambua and co-workers [12]. Using kenaf-, coir-, sisal-, hemp-, and jute-reinforced PP the study concluded that natural fibre composites display low impact strengths compared to glass composites, whereas their specific impact strength can be comparable with those of glass mat composites. Among the materials studied, sisal and hemp showed the highest impact strength.

Pavithran and co-workers [43] determined the fracture energies for sisal, pineapple, banana and coconut fibre-polyester composites in a Charpy impact test. They concluded that increased fibre toughness results in increased fracture energy, and found that fibres with higher fibril angles have higher fracture toughness than those with a small spiral angle.

Fibre content and fibre length also contribute to the impact performance of the composite. Tobias [44] examined this effect with banana fibre composites and concluded that short fibre lengths have higher impact strength and this also increases for higher fibre contents. However, different results were obtained by Garkhail and co-workers [18] on flax/PP composites. Their results showed that (as in glass fibre composites) the impact strength increases with increasing fibre length until a plateau level is reached. After that level the impact performance drops, depending on the pretreatment of the fibres and the adhesion of the fibre/matrix interface.

Müller [45] investigated the effect of several material parameters on the impact strength of compression moulded components from hemp, flax and kenaf/PP composites. The studies showed a strong influence of the thermal processing conditions during the moulding. He concluded that, for every material studied, there is an optimum temperature that results in a peak of the impact strength. Higher and lower processing temperatures resulted in lower mechanical property values that could be explained by a thermal decomposition of the fibres. Composites containing finer fibres showed better impact performance. The effect of temperature and water on the impact properties of natural fibre/thermoplastics were reviewed by de Bruijn [46], and showed no significant effect on the impact properties of the composites. However the results showed that the impact strength was 20 to 25% that of glass-reinforced thermoplastics.

It has been reported that coupling agents have a significant effect on the impact strength. When the composites have no coupling agent a part of the energy is lost at the interface by, for example, debonding and friction effects. Maleic anhydride grafted polypropylene (MAPP) jute composites showed higher impact strength than untreated samples prepared under the same processing conditions.

9.3.3 Fatigue Behaviour

Few investigations have been carried out on the cyclic loading of natural fibre composites. Gassan [47] investigated the fatigue behaviour of flax and jute/epoxy resin composites. Fibre type, textile architecture, interphase properties, fibre properties and fibre content were found to affect the fatigue behaviour strongly. It was also found that natural fibre-reinforced plastics with higher fibre strength and modulus, stronger fibre-matrix adhesion or higher fibre fractions required higher critical loads for damage initiation and higher failure loads. In addition, damage propagation rates were reduced. Furthermore, unidirectional composites were less sensitive to fatigue-induced damage than woven reinforced ones.

Savastano and co-workers [48] presented the results of experimental studies of crack resistance curve behaviour and fatigue crack growth in cementitious matrices reinforced with natural fibres such as sisal, banana and bleached eucalyptus pulp.

Fatigue crack growth was observed to occur in three stages: an initial decelerated growth, a steady-state growth, and a final catastrophic crack growth. In the case of the composites reinforced with sisal and banana fibres, most of the fatigue life was spent in the second stage of steady-state crack growth. The results showed that fatigue crack growth in the composites occurred *via* matrix cracking, crack deflection around fibres, and crack bridging by uncracked fibres and ligaments, whilst fibre pull-out was also observed.

The fatigue performance of sisal/epoxy composites was also studied by Towo and Andsell [49, 50] who looked into the effect of surface modification on the fatigue performance of the composites. The results show that a sodium hydroxide surface treatment has a significant effect on the tensile modulus and strength of the material but the fatigue life is not highly influenced, especially at low stress levels. Their conclusion states that the behaviour of sisal fibre composites is similar to that of conventional synthetic fibre composites and static and fatigue strengths are high enough for many commercial applications. Towo and Ansell also studied the fatigue properties of flax/polyester composites with alkali-treated and untreated fibres. In this case they observed that surface modification improved the fatigue life of the components and they also showed that the polyester matrix samples had lower fatigue life span than the epoxy samples [49, 50].

A comparison between hemp- and flax-reinforced polyester composites with the focus on the fatigue behaviour was conducted by Yuanjian and co-workers [24]. A steeper gradient of the S-N curve for the hemp fibre composite was indicative of a higher rate of reduction in fatigue strength. (S-N curve is a graph of the magnitude of a cyclic stress [S] against the logarithmic scale of cycles to failure [N].) However the fatigue performance levels of this hemp mat composite were comparable to, and slightly greater than, those of the glass fibre composite which was also included in the study.

9.4 Current Applications of Natural Fibre-reinforced Composites: Overview

Natural fibre composites have been used in a vast number of applications, which have shown constant growth over the past decade. Lines, ropes, textiles, clothes, baskets and paper are some of the traditional applications [7, 10]. However biocomposites have been used also in the design and manufacturing of sensing flanges, grinding discs, fan rotors and seats for vehicles. Their use has also been widely used in the packaging industry (e.g., packages for cosmetics), but also for speaker boxes, chairs and tableware [45]. Other examples include steam cleaners, toilet seats, ironing boards and a range of other domestic applications. Natural fibre composites have made their appearance in sports equipment (e.g., golf). A catamaran (made of flax fibres) and body structures for caravans and motorhomes are two more examples.

The list of automotive applications is long; door panels, seat backs, dashboards and package trays, head restraints and seat back linings are just some examples. Mercedes used jute-based composites for the door panels of the E-class, banana fibre-reinforced composites for the A-class and different other biocomposites for the S-class [51]. BMW followed the same example using a considerable amount of green materials and green composites. The brand reports the use of approximately 10,000 tonnes of natural fibres in 2004 [52].

Daimler-Chrysler was another example with up to 50 components in its European vehicles some among which were the innovative abaca fibre exterior underfloor impact protection for passenger cars and front bumpers made of flax fibres. General Motors used a mixture of kenaf and flax to produce package shelves and door panel inserts for the Saturn L300 and the European version of the Opel Vectra [52]. Audi's A2 car carries a door trim made of mixed flax and sisal fibres. Audi used natural fibre composites in other vehicles of its range (A3, A4, A6, A8, TT Roadster) for seat backs, side and back door panels, boot linings and other components [10] The East German Trabant car had body frames made of cotton fibre-reinforced polyester. The Toyota Lexus HS250h hybrid vehicle is yet another car containing plant fibre-reinforced composites. Others include Ford, Renault (Clio), Peugeot (406), Fiat (Punto, Brava, Marea and others), Saab, Seat, Volkswagen (Golf A4, Passat, Bora) and Volvo (C70, V70). Lotus incorporated natural fibre composites in the Eco Elise bodywork, using hemp fibres, while its interior trimmings are made of sisal composites [51]. Goodyear, found that corn-infused tyres have lower rolling resistance providing lower fuel consumption. These tyres are now used on several European vehicles. The route is followed by many Tier 1 and Tier 2 suppliers.

In 2003, the Araco Corporation in Japan, presented the Grasshopper, a fully electric vehicle. Its body was made completely out of plant-based composites, mainly kenaf [3]. The efforts of Fiat on a more environmental friendly vehicle, produced the Fiat Ecobasic prototype that saves weight and the need for painting through the application of natural fibre composites on the external panels. The BioConcept Car project by Four Motors GmbH successfully built an endurance racing car made with natural fibre-reinforced plastics. In 2006 the racing car competed successfully in a 6 h race for the first time [53]. More recently, Rinspeed prepared a new concept car to be presented at the 2011 Geneva Motor Show. The Rinspeed Bamboo, as its name reveals, has major components of the interior made of bamboo fibre-reinforced composites.

The European Union has been very supportive of the use of natural fibre composites in the automotive industry. The ECOFINA project, part of the FP5 Growth Project, investigated synthetic polymers reinforced with natural fibres for use in automotive parts [54]. The project included the standardisation of a processing technology to produce natural fibre products of consistent quality. Several other European Union projects such as (currently) ECOSHELL [55] encourage the production of natural fibres and their use in structural applications.

9.5 Conclusions

This chapter has addressed the current state-of-the-art in natural fibre biocomposites with special focus on their potential to be used in load bearing applications. The mechanical properties of both the constituents and the composites, manufacturing techniques and methods of improving both of these have been discussed. The work has also addressed the modelling of natural fibre composites and presented several methods and existing models. Finally, a wide range of current applications has been presented.

The good mechanical properties combined with the very low density, the availability and cost, the degradability and recyclability, and finally the ecofriendly production, makes natural fibre-reinforced biocomposites ideal for use in the automotive industry. A number of processes now exist to improve the quality of the fibres, as well as the adhesion between the fibres and the matrix. Well-established manufacturing techniques can now ensure very good final products with a relatively high production rate and low cost. A number of the applications mentioned show the suitability of these materials to be used in structural applications, but, a number of technical considerations have so far limited their use on a larger scale. The variable properties of the fibres, the interfacial adhesion, the reliability issues associated with the different properties in different environments, are just some of these issues. A gap in the knowledge is also underlined in this paper concerning the impact behaviour and fatigue properties of natural fibre biocomposites. Modelling is also still in its infancy, with the available models being able to provide good predictions of modulus and strength of specific composites and geometries, but failing to accurately represent these materials under general loading conditions.

Acknowledgements

The authors are grateful for the financial support by the European Commission and the FP7 ECOSHELL Project (Project Reference No.265838).

References

1. J. Njuguna, P. Wambua, K. Pielichowski and K. Kayvantash in *Cellulose Fibres: Bio- and Nano-Polymer Composites*, 1st Edition, Eds., S. Kalia, B.S. Kaith and I. Kaur, Springer-Verlag, Berlin, Heildelberg, Germany, 2011, Chapter 23.

2. *Natural Fibres, Biopolymers, and Biocomposites*, Eds., A.K. Mohanty, M. Misra and L.T. Drzal, CRC Press, Boca Raton, FL, USA, 2005.

3. T. Nishino in *Green Composites: Polymer Composites and the Environment*, Ed., C. Baillie, Woodhead Publishing, Cambridge, UK, 2004, Chapter 4.

4. J. Biagiotti, D. Puglia and J.M. Kenny, *Journal of Natural Fibers*, 2004, **1**, 2, 37.

5. A.K.Bledzki and J. Gassan, *Progress in Polymer Science*, 1999, **24**, 2, 221.

6. H. Bos, D.O. Van and O. Peters, *Journal of Materials Science*, 2002, **37**, 1683.

7. J. Summerscales, N.P.J. Dissanayake, A.S. Virk and W. Hall, *Composites Part A: Applied Science and Manufacturing*, 2010, **41**, 10, 1329.

8. A.K. Bledzki, O. Faruk and V.E. Sperber, *Macromolecular Materials and Engineering*, 2006, **291**, 5, 449.

9. H.L. Bos, J. Müssig and M.J.A. van den Oever, *Composites Part A: Applied Science and Manufacturing*, 2006, **37**, 10, 1591.

10. H. Bos in *The Potential of Flax Fibres as Reinforcement for Composite Materials*. Technische Universiteit Eindhoven, Eindhoven, Netherlands, 2004.

11. H. Ku, H. Wang, N. Pattarachaiyakoop and M. Trada, *Composites Part B: Engineering*, 2011, **42**, 4, 856.

12. P. Wambua, J. Ivens and I. Verpoest, *Composites Science and Technology*, 2003, **63**, 9, 1259.

13. International Year of Natural Fibres, 2009. *http://naturalfibres2009.org/en/index.html*

14. A.K. Mohanty, M. Misra and G. Hinrichsen, *Macromolecular Materials and Engineering*, 2000, **276-277**, 1, 1.

15. J. Andersons, E. Sparniņš, R. Joffe and L. Wallström, *Composites Science and Technology*, 2005, **65**, 3-4, 693.

16. Q. Liu and M. Hughes, *Composites Part A: Applied Science and Manufacturing*, 2008, **39**, 10, 1644.

17. M.J. John and R.D. Anandjiwala, *Composites Part A: Applied Science and Manufacturing*, 2009, **40**, 4, 442.

18. S.K. Garkhail, R.W.H. Heinjenrath and T. Peijs, *Applied Composite Materials*, 2000, **7**, 351.

19. M. Baiardo, E. Zini and M. Scandola, *Composites Part A: Applied Science and Manufacturing*, 2004, **35**, 6, 703.

20. European Industrial Hemp Association. *http://www.eiha.org*

21. E.C. Lee, C.M. Flanigan, K.A. Williams, D.F. Mielewski and D.Q. Houston in *Proceedings of 5th Annual SPE Automotive Composites Conference and Exhibition*, 12-14th September ,Troy, MI, USA, 2005, p.1.

22. G.W. Beckermann and K.L. Pickering, *Composites Part A: Applied Science and Manufacturing*, 2009, **40**, 2, 210.

23. M. Pracella, D. Chionna, I. Anguillesi, Z. Kulinski and E. Piorkowska, *Composites Science and Technology*, 2006, **66**, 13, 2218.

24. T. Yuanjian and D.H. Isaac, *Composites Science and Technology*, 2007, **67**, 15-16, 3300.

25. R.M. Maryudi, A.H.N. Yunus and M.H. Abidin, *Journal of Applied Sciences*, 2009, **9**, 17, 3156.

26. B. Ellis in *Chemistry and Technology of Epoxy Resins*, Blackie Academic and Professional, London, UK, 1993.

27. S.G. Tan and W.S. Chow, *Polymer-Plastics Technology and Engineering*, 2010, **49**, 15, 1581.

28. Y. Takada, K. Shinbo, Y. Someya and M. Shibata, *Journal of Applied Polymer Science*, 2009, **113**, 1, 479.

29. M. Shibata and K. Nakai, *Journal of Polymer Science Part B: Polymer Physics Edition*, 2010, **48**, 4, 425.

30. M. Aboobucker Sithique, S. Ramesh and M. Alagar, *International Journal of Polymeric Materials and Polymeric Biomaterials*, 2008, **57**, 5, 480.

31. A.E. Gerbase, C.L. Petzhold and A.P.O. Costa, *Journal of the American Oil Chemists' Society*, 2002, **79**, 797.

32. D. Ratna, *Polymer International*, 2001, **50**, 2, 179.

33. H. Miyagawa, R.J. Jurek, A.K. Mohanty, M. Misra and L.T. Drzal, *Composites Part A: Applied Science and Manufacturing*, 2006, **37**, 1, 54.

34. G. Ahmetli, H. Deveci, U. Soydal, S.P. Gurler and A. Altun, *Journal of Applied Polymer Science*, 2012, **125**, 38.

35. D. Garlotta, *Journal of Polymers and the Environment*, 2001, **9**, 2, 63.

36. S.S. Ray and J. Ramontja in *Biodegradable Polymer Blends and Composites from Renewable Resources*, Ed., L. Yu, John Wiley and Sons, Inc., Hoboken, NJ, USA, 2009, p.389.

37. R. Tokoro, D. Vu, K. Okubo, T. Tanaka, T. Fujii and T. Fujiura, *Journal of Materials*, 2008, **43**, 2, 775.

38. S. Ochi, *Mechanics of Materials*, 2008, **40**, 4-5, 446.

39. M. Shibata, S. Oyamada, S. Kobayashi and D. Yaginuma, *Journal of Applied Polymer Science*, 2004, **92**, 6, 3857.

40. P. Pan, B. Zhu, W. Kai, S. Serizawa, M. Iji and Y. Inoue, *Journal of Applied Polymer Science*, 2007, **105**, 3, 1511.

41. A. Espert, F. Vilaplana and S. Karlsson, *Composites Part A: Applied Science and Manufacturing*, 2004, **35**, 11, 1267.

42. M.M. Thwe and K. Liao, *Composites Science and Technology*, 2003, **63**, 3-4, 375.

43. C. Pavithran, P.S. Mukherjee and M. Brahmakumar, *Journal of Reinforced Plastics and Composites*, 1991, **10**, 91.

44. B.C. Tobias in *Advanced Composites'93: International Conference on Advanced Composite Materials*, Wollongong, Australia, 1993, p.623.

45. H.D. Müller, *International Nonwovens Journal*, 2004, **13**, 4, 31.

46. J.C.M. de Bruijn, *Applied Composite Materials*, 2000, 7, 415.

47. J. Gassan, *Composites Part A: Applied Science and Manufacturing*, 2002, **33**, 3, 369.

48. H. Savastano, Jr., S.F. Santos, M. Radonjic and W.O. Soboyejo, *Cement and Concrete Composites*, 2009, **31**, 4, 232.

49. A.N. Towo and M.P. Ansell, *Composites Science and Technology*, 2008, **68**, 3-4, 915.

50. A.N. Towo and M.P. Ansell, *Composites Science and Technology*, 2008, **68**, 3-4, 925.

51. M. Karus, S. Ortmann, C. Gahle and C. Pendarovski in *Use of Natural Fibres in Composites for the German Automotive Production from 1999 till 2005*, Nova-Institute, Hurth, Germany, 2006.

52. K. van Rijswijk, W.D. Brouwer and A. Beukers in *Application of Natural Fibre Composites*, Structures and Materials Laboratory, Faculty of Aerospace Engineering Delft University of Technology, Delft, The Netherlands, 2001.

53. Composites Europe, 2011. *http://blog.composites-europe.com/tag/natural-fibre-reinforced-plastics/* Accessed 22/08/2013

54. J. Holbery and D. Houston, *Journal of the Minerals, Metals and Materials Society*, 2006, **58**, 11, 80.

55. J. Fan, E. Nassiopoulos, J. Brighton, A. De Larminat and J. Njuguna in *Proceedings of the Society of Plastics Engineers (SPE) EUROTEC 2011 Conference*, Barcelona, Spain, 2011, Code 88784, 5 pp.

56. J. Zhu, H. Abhyankar, E. Nassiopoulos and J. Njuguna, *IOP Conference Series: Materials Science and Engineering*, 2012, **40**, 1, 012030.

10 Biocomposites based on Cellulose Material (Poplar Seed Floss) and High Density Polyethylene: Accelerated Weathering Behaviour

Iuliana Spiridon, Carmen-Alice Teacă, Raluca Nicoleta Darie, Ruxanda Bodîrlău and Ana Maria Resmerită

10.1 Introduction

A growing interest in relation to the production of biodegradable materials has occurred as a result of the environmental concerns worldwide about the diminishing landfill capacity for plastic waste disposal and its slow degradation in the environment [1]. In the last decade, there has been increasing research interest in the value of biobased products recovered from residual lignocellulosic materials. A biorefinery approach involves multistep processes in which the first step typically involves pretreating the precursor (cellulose, lignin or hemicelluloses) containing lignocellulosic biomass to make it more amenable for further processing. The use of lignocellulosic biomass in composite materials can be one such potential processing option. Biocomposite materials are normally produced by mixing the lignocellulosic component with a polymer, or by adding it as a filler within a polymer matrix, and pressing or moulding under high pressure and temperature. Most polymers, especially thermoplastics, have a nonpolar (hydrophobic) character, so are not compatible with polar (hydrophilic) wood fibres and, therefore, poor adhesion between polymer and wood in composites can result. Thermoplastics which are used in biocomposites must melt or soften at or below the degradation point of the lignocellulosic component, normally 200 °C to 220 °C [2]. These thermoplastics include polypropylene (PP), polystyrene, vinyls, and low and high density polyethylenes.

Several considerations must be kept in mind when processing lignocellulosics with thermoplastics. Materials must either be predried, or vented equipment must be used to remove moisture. There are two main strategies for processing thermoplastics in lignocellulosic composites [3]. In the first one, the lignocellulosic component serves as a reinforcing agent or filler in a continuous thermoplastic matrix (composites with high thermoplastic content). In the second one, the thermoplastic serves as a binder to the majority lignocellulosic component.

Compounding is the feeding and dispersing of the lignocellulosic component in a molten thermoplastic to produce a homogeneous material. Compounding may be accomplished using either batch mixers (internal and thermokinetic mixers) or continuous mixers (extruders and kneaders).

Wood/plastic composites (WPC) are primarily produced with high filler loading. Rheological characterisation affords knowledge about both the fundamental flow behaviour of the wood-filled plastics and practical methods for evaluating the flow performance for process development. Appropriate rheology approaches must be used to answer the questions encountered by wood/plastic composites producers, such as the dependence on wood species, particle size, and wood content, as well as the properties of the polymer matrix resin and the additives [4-10].

There are a number of external factors that affect the mechanical and physical properties of the WPC. Ultraviolet (UV) light, moisture, and heat are significant factors that lead to changes in these properties [11-13]. There are many publications related to the photodegradation of WPC that have demonstrated that the changes in colour and mechanical properties occur as a result of exposure to UV light and the action of moisture [11, 12, 14-16].

All the main wood polymer components (cellulose, hemicelluloses, lignin) suffer from photodegradation [17]. Because the location of most wood chromophores is in the lignin component, lignin accounts for 80-95% of light absorption of wood, and so it contributes to the discolouration of wood which is exposed to light [18]. Wood exposed to UV radiation darkens as the exposure time increases [17-20].

The photodegradation of polyolefins is a radical-based oxidative process. The catalyst residues, hydroperoxide groups, and carbonyl groups introduced during polymer manufacturing, processing, and storage absorb UV light above 290 nm and initiate photochemical reactions. Photodegradation causes changes in the monomer unit (oxidation), the chain (crosslinking or chain scission), the morphology, and on the macroscopic scale (colour change, increasing roughness, cracks) [21]. Density gradients, due to oxidation gradients, cause the stresses which, combined with chain scission, could initiate and propagate cracks. Cracks on the surface result in light diffusion and also lead to the loss of mechanical properties [22].

Photodegradation of WPC involves several factors including the wood content, coupling agents, manufacturing methods, and weathering conditions. The influence of accelerated artificial weathering on the colour changes and tensile properties of polyvinyl chloride/wood flour composites was investigated [23]. Composites containing 15 wt% wood darkened more than those containing 30 wt% and 45 wt% wood. Weathered high density polyethylene (HDPE)/wood flour composites containing 50 wt% wood flour lightened more than the counterparts with 25% wood flour [24]. The oxidation rate of composites increased with wood content and decreased with the presence of compatibiliser due to better dispersion of the wood in the polymer matrix for HDPE-based and PP-based WPC [16].

Recent investigations studied the effects of natural and accelerated artificial weathering on the appearance and chemical changes of HDPE-based and PP-based WPC [11, 14, 15, 23, 24-30]. The lightness, total colour change and wood loss increased with exposure time. PP-based WPC experienced quicker photodegradation in terms of lightness (a property of colour -tone- which is changing through wood discolouration) and wood loss. Delignification and oxidation accounted for the increased lightness of weathered composites. WPC underwent two competing redox reactions upon UV exposure: the formation of paraquinone chromophoric structures generated by the oxidation of lignin, resulting in yellowing, as well as the reduction of the paraquinone structures to hydroquinones, leading to photobleaching [11].

Cellulose fibres originate from renewable, inexpensive and readily available resources and may add degradability to plastics. Considering the growing interest in ecofriendly materials, cellulose fibres are convenient for composites with different applications. Formulators of cellulose composites are always looking for additional types of fibres in order to improve performance or reduce cost. Individual seed hairs may be converted into chemical derivatives or can be used in their physical form (usually fibrous) and herein are referred to as 'seed hair fibres', as a component of fibrous structures. Seed hairs may have a wide range of morphology and chemical properties. The seed hair fibres have a high length-to-diameter ratio.

In this context, poplar seed floss (PSF), a cellulose-enriched material resulting from the widely available biomass waste from *Populus* wood species, can represent a potentially valuable environmentally friendly resource. On average, the poplar tree produces 35 l of fresh fruits that yield from 280,000 to 14,850,000 seeds (0.9 kg) depending on species, location and type of tree [31]. Presently, the fibrous poplar seeds are treated as waste or, at best, used as a cheap fertiliser. The microtube morphological structure and enhanced hydrophobicity indicate that poplar seed fibres may serve as a valuable industrial raw material for the production of oil absorbent [32]. Poplar seed hair fibres are hollow hydrophobic microtubes with an external diameter between 3 and 12 mm, an average length of 4 ± 1 mm and average tube wall thickness of 400 ± 100 nm. The solid skeleton of the hollow fibres consists of lignocellulosic material coated with a hydrophobic waxy coating.

In this chapter, we describe the preparation of composites containing poplar seed floss (PSF) as the filler and HDPE as the thermoplastic polymer matrix. HDPE is widely used in composites because of its low cost, good mechanical properties and processability. The behaviour of these composites was then studied under accelerated weathering conditions. Changes in the structure and morphology of the composites were investigated through attenuated total reflectance (ATR)-Fourier transform infrared (FTIR) spectroscopy, scanning electron microscopy (SEM), mechanical and thermal analysis. The rheological behaviour of the composites in the melt state was also investigated.

10.2 Experimental

10.2.1 Materials

HDPE in the form of homopolymer pellets with density of 0.95 g/cm^3 and melt flow index of 0.9 g/10 min at 190 °C was used to obtain composites with 5, 10 and 15 wt% of PSF. PSF was used without any preliminary treatment (apart from boiling with distilled water for residue removal). SEM photomicrographs of PSF samples are presented in **Figure 10.1**.

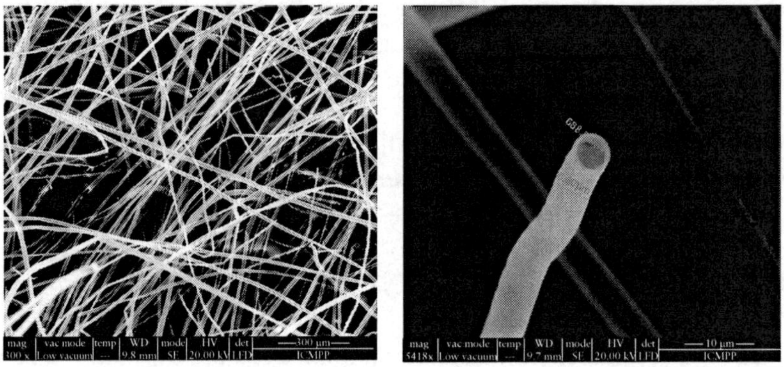

Figure 10.1 SEM images recorded for PSF samples (left - PSF sample at 300x magnification and resolution 300 μm; right - PSF sample at 5418x magnification and resolution 10μm)

The chemical composition of PSF was determined according to the Technical Association of the Pulp and Paper Industry (TAPPI) methods, and the results were:

64 wt% holocellulose (around 43 wt% cellulose, and 21 wt% hemicelluloses (mainly xylan), respectively), 19.4 wt% lignin, and around 7 wt% extractives (mainly wax).

10.2.2 Compounding and Processing of the High Density Polyethylene/ Poplar Seed Floss Composites

Before preparing the blend samples, the HDPE pellets and PSF were dried in a vacuum oven for 6 h at 80 °C in order to reduce the moisture levels. PSF was then mixed with HDPE pellets. The composite materials were obtained by melt blending at 170 °C for 10 min at a rotor speed of 50 rpm using a fully automated Brabender Lab-Station (Brabender, Germany), equipped with a type 30 EHT internal mixer. The polymer blends were formulated with 5, 10 and 15 wt% PSF filler, respectively. Specimens for the mechanical characterisation were prepared by compression moulding using a Carver 4394 Press (Carver Inc., USA). The compression moulding was carried out at 175 °C with a prepressing step of 3 min at 50 atm and a pressing step of 2 min at 150 atm. The code names for composite samples are written as: HDPE/PSF5 (sample containing 5 wt% PSF), HDPE/PSF10 (sample containing 10 wt% PSF), and HDPE/PSF15 (sample containing 15 wt% PSF). Samples after 600 h UV exposure were coded as: HDPE-600, HDPE/PSF5-600, HDPE/PSF10-600, and HDPE/PSF15-600, respectively.

10.2.3 Accelerated Weathering Tests

All composites were exposed to artificial light from a mercury lamp (wavelength range $200<\lambda<700$ nm, incident light intensity 39 mW/cm^2) at a temperature of 40 °C and 65% humidity in a commercial climatic chamber (CH 250, Angelantoni Industrie, Italy) for accelerated weathering tests. The samples were removed from the chamber after 600 h of exposure and were further characterised by FTIR, SEM and mechanical properties analysis. Non-weathered composite samples were used for reference.

10.2.4 Methods for Characterisation of the Composites

The composites were characterised by investigating their mechanical properties (impact and tensile tests), ATR-FTIR spectroscopy, scanning electron microscopy (SEM) and simultaneous thermal analysis (thermogravimetric analysis [TGA]), and differential scanning calorimetry (DSC).

10.2.4.1 Mechanical Testing

All mechanical tests were performed at 50 ± 5% relative humidity (RH) and 23 °C. The composite specimens were conditioned under the same conditions for 24 h before testing. The tensile strength and the Young's modulus were determined according to the European Standard-International Organization for Standardization (EN ISO) 527. An Instron 5 kN test machine operated at a crosshead speed of 30 mm/min was used for testing the specimens (testing area: 80 × 10 × 4 mm; total length 150 mm). The Charpy impact strength of the composites was tested according to EN ISO 179. A CEAST testing machine with a pendulum of 15 J energy was used to measure the un-notched specimens (testing area 80 × 10 × 4 mm). Ten specimens were tested for each composite sample in order to obtain an average value.

10.2.4.2 Fourier-Transform Infrared Spectroscopy

FTIR spectra of PSF and HDPE were recorded on solid samples in potassium bromide (KBr) pellets by using a FTIR Digilab, Scimitar Series Spectrometer (USA) with a resolution of 4 cm^{-1} and the scanning range varied from 400 to 4000 cm^{-1}. The concentration of the sample was 5 mg/500 mg KBr. Accelerated ageing processes in HDPE/PSF composites were monitored by ATR-FTIR spectroscopy using a Bruker Vertex 70 FTIR spectrometer equipped with an ATR device (zinc selenide crystal) with a 45° angle of incidence. Sixty four scans were acquired with a spectral resolution of 2 cm^{-1}. Some indices for composite samples were calculated using **Equations 10.1** and **10.2**, as follows [15, 26]:

$$\text{Carbonyl index} = (I_{1716}/I_{2916}) \times 100 \qquad (10.1)$$

$$\text{Vinyl index} = (I_{908}/I_{2916}) \times 100 \qquad (10.2)$$

where I denotes the peak intensity.

The peak intensity was normalised to the peak at 2916 cm^{-1} which corresponds to alkane C-H stretching vibrations of methylene ($-CH_2-$) groups. This peak was chosen as a reference because it changed the least during weathering.

10.2.4.3 Contact Angle Measurements

The sample films were kept for 48 h at 50% (RH) before being tested in static conditions on an AdveX Instrument. A 2.5 μL droplet of solvent was applied to the composite sample film surface. Water, formamide and diiodomethane were employed as solvents with different polarity. The evolution of the droplet shape was recorded every second by a video camera and image analysis software was used to determine the contact angle values. The contact angle value reported was the average of ten measurements. Surface free energy values were also determined.

10.2.4.4 Water Uptake Measurements

To determine the water absorption [33], the composite specimens used were in the form of thin rectangular strips (10 × 10 × 0.2 mm). These films were thin enough so that the molecular diffusion was considered to be one-dimensional, and were vacuum dried at 90 °C overnight. After weighing, they were conditioned at 25 °C in a desiccator containing sodium sulfate in order to ensure a RH of 95%. They were then removed at specific intervals and gently blotted with tissue paper to remove the excess of water on the surface, and the water uptake was calculated with **Equation 10.3**, as follows:

$$\text{Water uptake (\%)} = [(W_t - W_0)/W_0] \times 100 \tag{10.3}$$

where:

W_t and W_0 represent the weight at time *t*, and before exposure to 95% RH, respectively. The determinations were performed in triplicate.

10.2.4.5 Scanning Electron Microscopy

The fractured surfaces of composites resulting from tensile tests were analysed using an environmental scanning electron microscope (FEI Quanta 200 ESEM). Air dried samples were fixed onto aluminium stubs using carbon adhesive discs and their surfaces were observed with a low vacuum secondary electron detector using the accelerating voltage of 25.0 kV. The samples were analysed at room temperature and at an internal pressure of 0.50 torr.

10.2.4.6 Differential Scanning Calorimetry Analysis

DSC analysis was performed with a Mettler Toledo DSC-12E calorimeter with a heating/cooling rate of 10 °C/min under nitrogen atmosphere. The thermograms were performed under constant nitrogen flow (20 mL/min) with a heating rate of 10 °C/min using a Mettler Toledo TGA/SDTA 851E balance. The heating scans were performed on 1.5 to 3 mg of sample in the temperature range 25-800 °C. The degree of crystallinity of the HDPE samples was obtained by dividing the melting enthalpy of the sample by 277 J/g [34], which is the estimated melting enthalpy of a pure HDPE (100 % crystalline). The crystallinity of the HDPE/PSF composites was estimated as a function of both the fraction of HDPE in the composite and the melting enthalpy.

10.2.4.7 Thermogravimetric Analysis

Thermal analysis was performed with a Netzsch STA 449 F1 thermal analyser (Germany). Samples (approximately 5 mg) were placed in aluminium oxide crucibles and heated under nitrogen atmosphere from 40 up to 600 °C at a heating rate of 10 °C/min. TGA curves recorded with a precision of ±0.5 °C were analysed with Netzsch Proteus analysis software.

10.2.4.8 Dynamic Rheological Measurements

Oscillatory melt rheology tests were performed on an Anton Paar rheometer equipped with a CTD450 convection temperature device, in plate-plate geometry, under oscillatory mode, at a testing temperature of 145 °C. The frequency dependence of the complex viscosity and dynamic moduli were recorded in the linear viscoelastic region at a deformation of 5%. The viscoelastic parameters were followed as a function of the oscillation frequency, ω, at a constant temperature of 145 °C. From the crossover frequency (ω_i), the relaxation time (θ) was calculated as $\theta = 1/\omega_i$.

10.3 Investigation of the Properties of High Density Polyethylene/ Poplar Seed Floss Composites

10.3.1 Fourier-Transform Infrared Spectroscopy

The FTIR spectrum of PSF (**Figure 10.2**) exhibits absorption bands at 2922 cm^{-1} and 2852 cm^{-1} corresponding to the stretching of aliphatic CH_2 and CH_3 groups and indicating the presence of plant wax [35-37]. This consists of *n*-alkanes and a smaller portion of alcohols, fatty acids, aldehydes, ketones and *n*-alkyl esters [38-41]. Absorption bands appear in the region between 2000 and 1150 cm^{-1} (C=O stretching vibrations) which indicate the presence of aliphatic aldehydes, esters, ketones and lignin [38, 42]. Bands at 1595 cm^{-1}, 1505 cm^{-1}, 1461 cm^{-1} and 1424 cm^{-1} correspond to the stretching modes of C-O and C=C bonds in lignin [43, 44]. The band at 1033 cm^{-1} belongs to the stretching of C-O bonds in cellulose, hemicellulose and lignin [45]. The band at 770 cm^{-1} can be associated with vibrational modes of the paraben structure in lignin [46]. Peaks at 1741 and 1048 cm^{-1} correspond to C=O and C-O stretching in cellulose and hemicelluloses of wood [47-50].

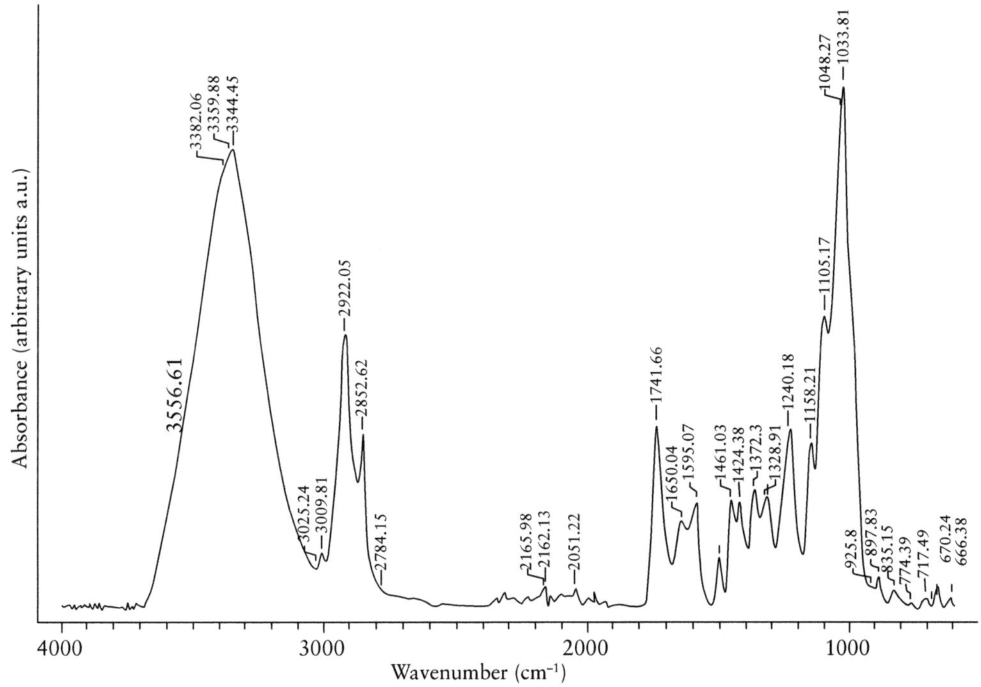

Figure 10.2 FTIR spectrum recorded for PSF

Table 10.1 lists the wavenumber values for HDPE along with the assignments of corresponding functional groups [23, 25-27].

Table 10.1Assignment of absorption bands of the FTIR spectrum for HDPE	
Wavenumber (cm^{-1})	**Assignment group**
2915	CH_2 stretching of CH_2, CH_2 (asymmetrical)
2847	CH_2 stretching (symmetrical)
1713	C=O stretching (acid carbonyl)
1472	CH_2 deformation (scissor vibration), crystalline
1462	CH_2 deformation (scissor vibration), amorphous
1176	C-O stretching (ester)
730	CH_2 rocking vibration, crystalline
718	CH_2 rocking vibration, amorphous

Adapted from L.M. Matuana and D.P. Kamdem, *Polymer Engineering and Science*, 2002, **42**, 8, 1657 [23]; L.M. Matuana, D.P. Kamdem and J. Zhang, *Journal of Applied Polymer Science*, 2001, **80**, 11, 1943 [25]; N.M. Stark and L.M. Matuana, *Polymer Degradation and Stability*, 2007, **92**, 10, 1883 [26]; and N.M. Stark *Journal of Applied Polymer Science*, 2006, **100**, 4, 3131 [27].

The FTIR spectrum for HDPE is presented in **Figure 10.3**. The backbone molecule had a strong peak of (-C-H) at about 2915 and 1470 cm^{-1}, respectively and a moderate peak of (-CH_2-) at around 718 cm^{-1}. Methyl groups (CH_3) also presented a significant peak at around 2848 cm^{-1}.

Polyolefins perform poorly in long-term exposure to temperature, humidity and UV radiation. Carbonyl groups are formed and this is accompanied by the occcurrence of vinyl groups. Carbonyl groups are postulated to be the main light absorbing species responsible for the photochemical-induced degradation reactions in UV-exposed polymers [51]. The degradation reactions proceed from carbonyl group precursors according to Norrish type I and II reactions [51, 52]. If the degradation of the carbonyl groups occurs according to the Norrish I reaction, the resultant free radicals can attack the polyolefin, which may lead to the end of the degradation process *via* crosslinking or chain scission [52]. If the degradation proceeds according to the Norrish II reaction, carbonyl groups and terminal vinyl groups are produced and chain scission occurs. In addition, the carbonyl group formed is capable of further degradation. During the course of accelerated polyethylene photodegradation the two mechanisms, chain scission and crosslinking, compete [51-53]. Chain scission lowers the molecular weight, whereas crosslinking raises the molecular weight by increasing the bonding between polymer chains. The formation of carbonyl and vinyl groups can be studied as a direct indication of main chain scission [53, 54]. The weathering of polyethylene results in an increase in crystallinity [55], which also indicates that chain scission has occurred.

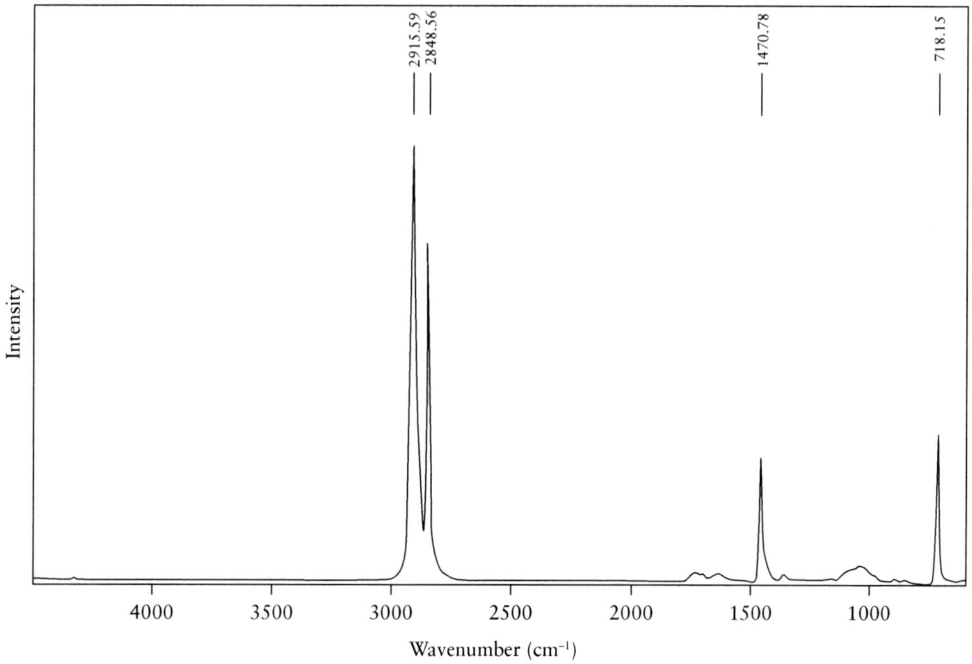

Figure 10.3 FTIR spectrum recorded for HDPE

ATR-FTIR spectroscopy was used to study the chemical changes (carbonyl group, and vinyl group formation) on the surface of HDPE/PSF composites before and after the accelerated weathering process. The main changes occurred in the carbonyl (1600–1800 cm^{-1}) and the vinyl (908 cm^{-1}) regions as well as at the ether bonds (1000–1200 cm^{-1}) range. The intensity at 1716 cm^{-1} and 908 cm^{-1}, which represents carbonyl (C=O) stretching (mainly in cellulose and hemicellulose) [28, 56], and CH_2 out-of-plane deformation of the vinyl groups formed as a result of photodegradation, respectively, was measured.

An increase in the calculated carbonyl index denotes an increase in polymer chain scission. The carbonyl index was higher for HDPE/PSF composites as compared to HDPE, probably due to lignin from PSF (**Figure 10.4**).

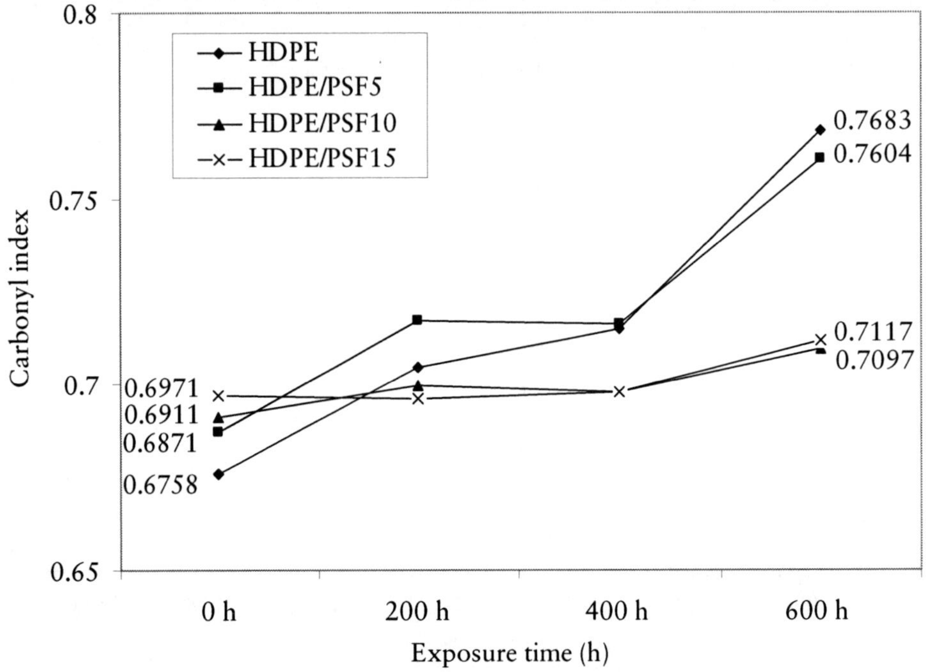

Figure 10.4 Influence of exposure time upon carbonyl index for HDPE and HDPE/PSF composites

The HDPE/PSF composites experienced surface oxidation. This can be clearly seen by the increased carbonyl index over time (**Figure 10.4**). Photodegradation of lignin leads to the formation of carbonyl groups and thus to an increase in intensity of the carbonyl bands. Several authors have correlated the increased integrated area or peak height of this band and the carbonyl index with the oxidative photodegradation of polymer [57, 58], wood [17-19], and WPC [14, 15, 30] due to the presence of oxygenated products.

The calculated vinyl group index can also be used to indicate polymer chain scission (**Figure 10.5**). ATR-FTIR spectroscopy evidenced the formation of both vinyl groups (probably *via* Norrish II reactions, resulting in chain scission) and carbonyl groups.

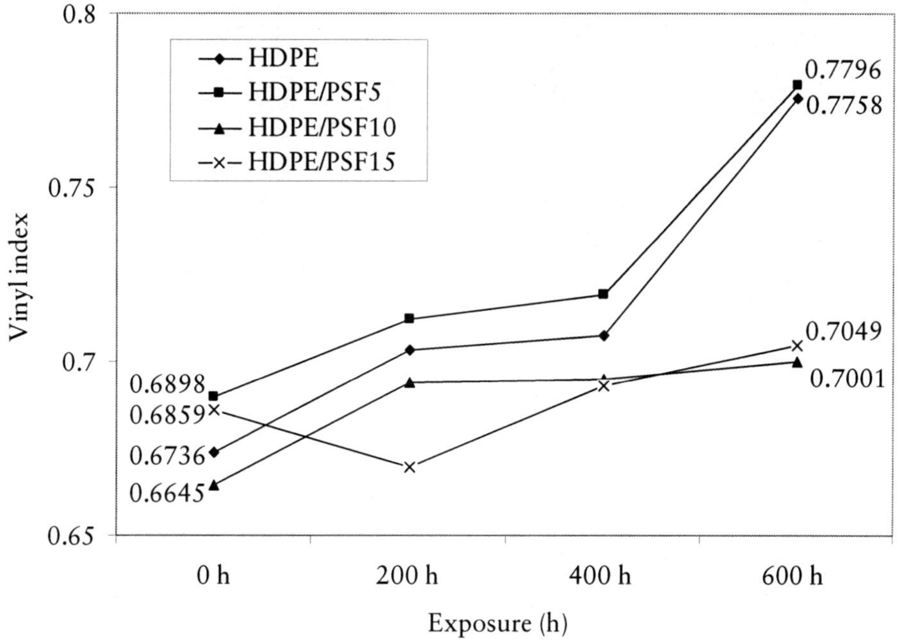

Figure 10.5 Influence of exposure time upon vinyl index for HDPE and HDPE/PSF composites

10.3.2 Investigation of Surface Properties

10.3.2.1 Scanning Electron Microscopy Analysis

Micrographs of the non-weathered and weathered HDPE and HDPE/PSF composites are shown in **Figure 10.6.**

In neat HDPE no significant changes in morphology were observed after exposure to the accelerated weathering conditions. The exposure to controlled weathering conditions caused visible changes in the composites, the most important being the increased surface roughness and occurrence of small cracks. Both photodegraded HDPE and swollen PSF can cause cracks on the surface of composites, due to poor adhesion between components under accelerated weathering conditions.

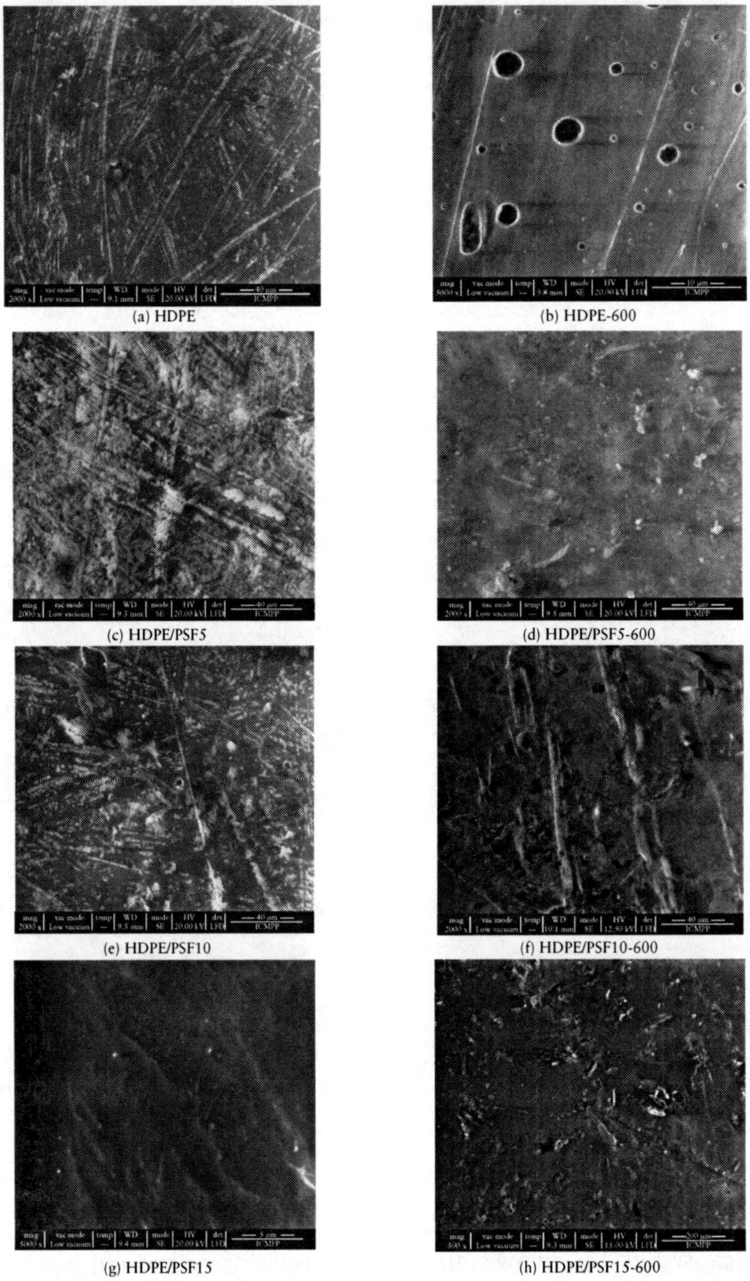

Figure 10.6 SEM micrographs recorded for neat HDPE and HDPE/PSF composites before and after accelerated weathering process (exposure time 600 h) (a) HDPE; (b) HDPE-600; (c) HDPE/PSF5; (d) HDPE/PSF5-600; (e) HDPE/PSF10; (f) HDPE/PSF10-600; (g) HDPE/PSF15; and (h) HDPE/PSF15-600

Before weathering, the surfaces of both the neat HDPE and those of the HDPE/PSF composites were smooth. SEM micrographs shown in **Figures 10.6e and 10.6g** evidenced that high PSF loading resulted in the formation of voids in the composites, due to PSF agglomeration. After exposure to weathering, some cracks occurred and PSF particles were visible on the surface, the cracks being more severe with increasing UV weathering exposure time.

10.3.2.2 Contact Angle Measurements

The contact angle of a water drop is an indication of the hydrophobic properties of a composite surface. All contact angle measurements were performed at ambient conditions of 22 ± 1 °C and 50 ± 2% RH. Experimental data are presented in **Table 10.2.**

In order to characterise the wetting properties, four liquids were used: *n*-heptane as a reference, whilst water, formamide and diiodomethane were the wetting liquids. All liquids (purity 99%) were provided by Sigma-Aldrich Chemie GmbH.

The addition of PSF into neat HDPE influenced the wettability. The hydrophilic character of the film surface increased due to the presence of polar PSF. Thus, with increasing fibre content, the contact angle reduced gradually to about 3° for the composite containing 15 wt% PSF, indicating a slow enhancement in hydrophilic character and wettability of the film.

Table 10.2 The contact angles and surface free energy values for materials under study

Sample	Total surface free energy (mJ/m^2)	Contact angles in different liquids (°)		
		Water	Formamide	Diiodomethane
HDPE	40.33	83.43±0.44	56.53±0.48	42.62±0.43
HDPE/PSF5	39.63	81.8±0.63	59.87±0.77	41.47±0.16
HDPE/PSF10	39.37	80.27±0.59	64.6±0.47	41.25±0.28
HDPE/PSF15	38.98	80.1±0.71	61.29±0.78	41.39±0.45
HDPE-600	46.68	66.61±0.67	63.5±0.74	40.98±1.02
HDPE/PSF5-600	47.97	50.0±1.29	36.87±0.5	39.41±0.32
HDPE/PSF10-600	47.08	54.9±0.81	41.7±0.71	27.57±0.76
HDPE/PSF15-600	48.34	55.88±0.86	54.05±0.14	34.39±1.04

When using diiodomethane, the values of the contact angle were approximately similar for all samples. After weathering, as result of the combined action of humidity, temperature and UV radiation, the contact angle of HDPE decreased from 83.43 to 66.61°. The weathering process led to enhanced hydrophilic character of the composites as compared to the non-weathered samples. After the weathering process, the total surface energy increased for HDPE and the HDPE/PSF composites, which can reduce the stability of the surface properties by increasing surface roughness and appearance of cracks.

10.3.2.3 Water Uptake Measurements

Figures 10.7 and **10.8** show the values for the water absorption recorded before and after accelerated weathering of neat HDPE and HDPE/PSF composites, which vary depending upon the PSF loading and water immersion time. As shown in **Figure 10.7**, the water absorption of the HDPE-PSF composites increased with increasing PSF loading, probably due to the specific characteristics of PSF (it becomes hydrophilic).

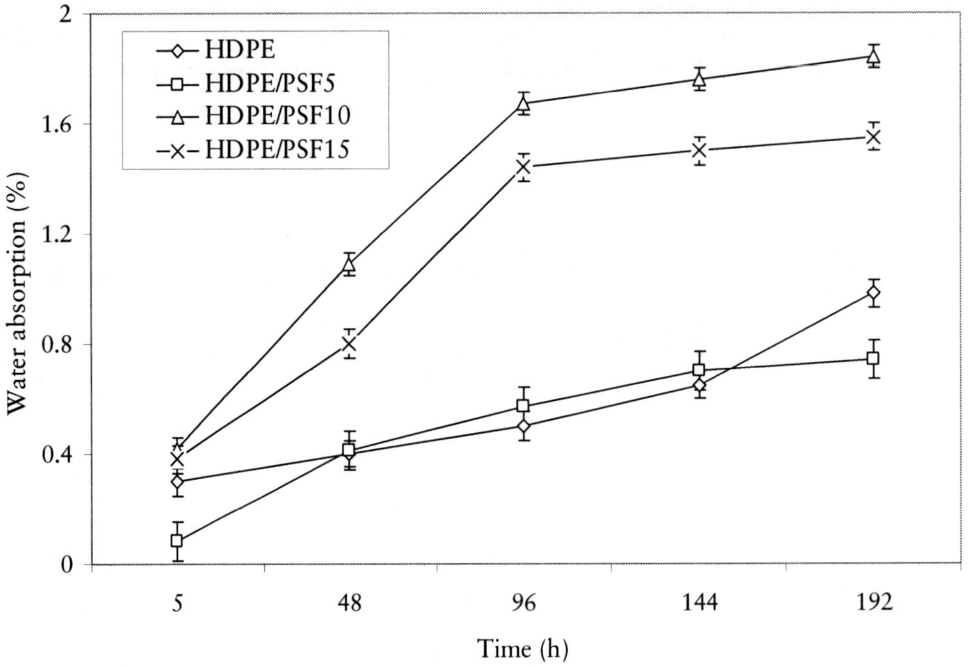

Figure 10.7 Water absorption values recorded for non-weathered neat HDPE and HDPE/PSF composites

The water uptake-time curves exhibit two well-separated zones. At short times, below 96 h, the kinetics of absorption are very fast, whereas at longer times, after 96 h, the kinetics of absorption are slow and this leads to a plateau, corresponding to the equilibrium of water uptake. The water uptake after accelerated weathering (**Figure 10.8**) was significantly higher than that noticed for the initial composites (**Figure 10.7**).

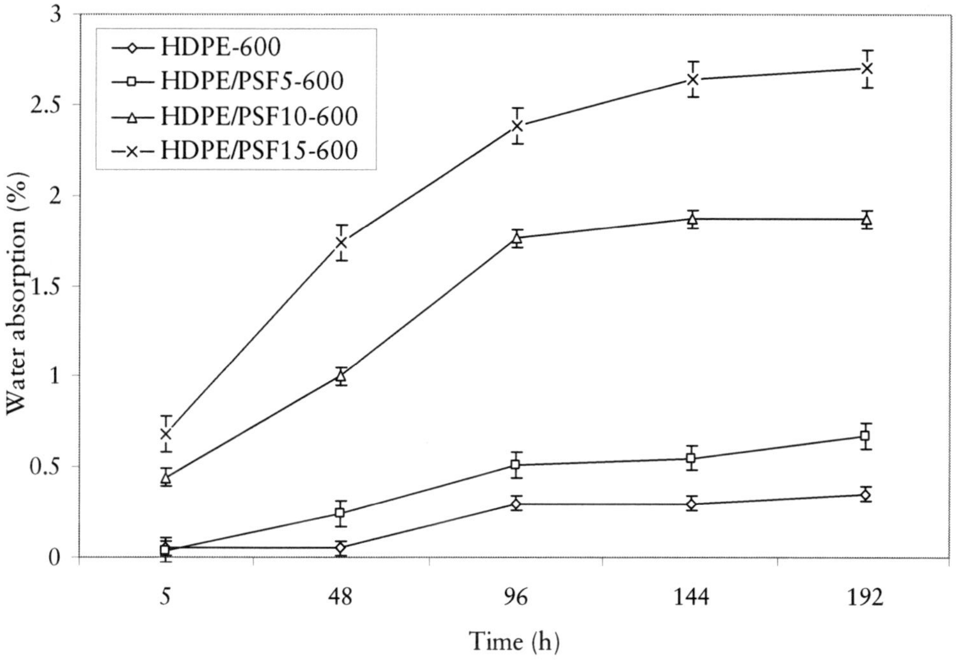

Figure 10.8 Water absorption values recorded for weathered neat HDPE and HDPE/PSF composites containing 5 wt%, 10 wt% and 15 wt% PSF, after 600 h UV exposure

At immersion time 96 h, the water absorption for HDPE is higher (0.5%) than that of HDPE-600 (0.3%), probably due to some re-organisation processes in the HDPE backbone after weathering. The opposite effect occurs for HDPE/PSF composites, when an increase in water absorption is noticed during weathering with increasing PSF loading.

10.3.3 Investigation of Thermal Properties

10.3.3.1 Thermogravimetric Analysis

The thermal behaviour of the HDPE and HDPE/PSF composites was investigated in order to evaluate their thermal stability and degradation profiles. **Figure 10.9** shows thermograms recorded for neat HDPE and HDPE/PSF composites, before and after accelerated weathering.

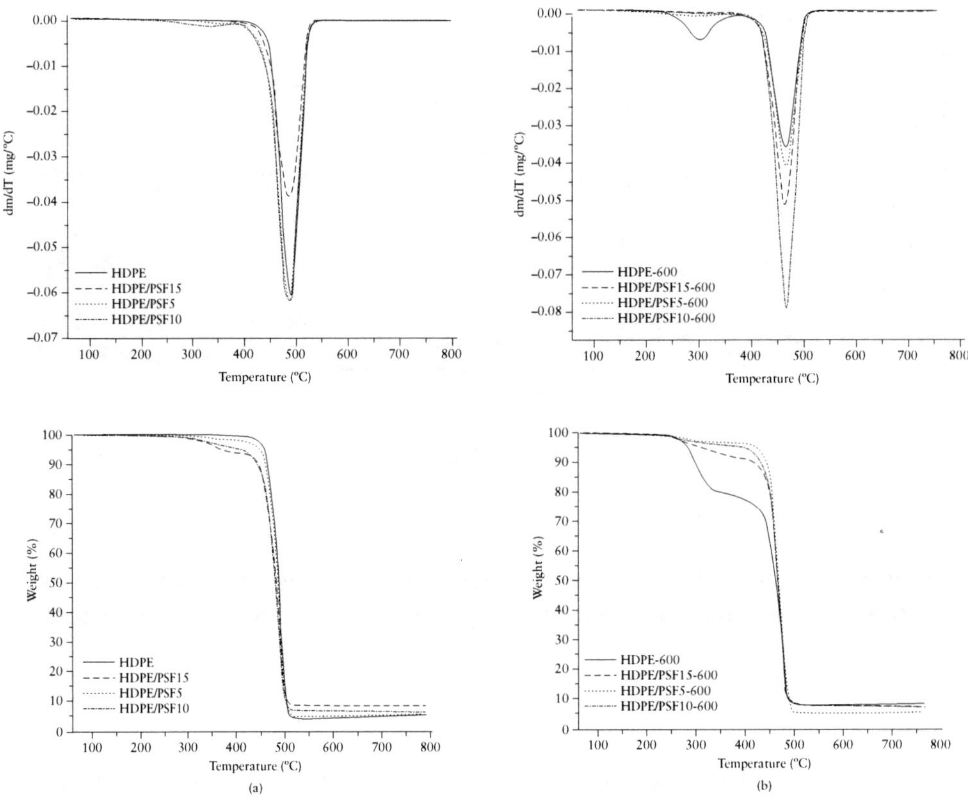

Figure 10.9 TGA-DTG curves recorded for HDPE and HDPE/PSF composites: (a) non weathered and (b) weathered (exposure time 600 h) - dm/dT means change in mass with temperature

Neat HDPE presented a single thermal decomposition process (**Table 10.3**) in the 450-500 °C temperature range, a weight loss of 98.95% being noticed.

After weathering, HDPE exhibited two degradation processes as follows: 1) in the 288-334 °C temperature range (22.02% weight loss), and 2) in the 463-500 °C temperature range (73.22% weight loss).

Table 10.3 Thermal parameters for HDPE and HDPE/PSF composites, before and after weathering

Sample	Step	T_{onset} (°C)	T_{final} (°C)	T_{peak} (°C)	W (%)	Residue (%)
HDPE	I	460	499	485	98.95	1.05
HDPE/PSF5	I	322	367	352	2.24	2.05
	II	463	500	485	95.71	
HDPE/PSF10	I	290	371	351	4.37	5
	II	449	497	481	90.63	
HDPE/PSF15	I	296	366	351	5.96	5.21
	II	453	499	482	88.83	
HDPE-600	I	288	334	315	22.02	4.76
	II	463	499	482	73.22	
HDPE/PSF5-600	I	455	500	484	95.51	4.49
HDPE/PSF10-600	I	457	497	481	94.79	5.21
HDPE/PSF15-600	I	279	366	326	7.28	4.97
	II	456	498	482	87.75	

T_{onset}: The start temperature of the degradation process.
T_{final}: The end temperature of the degradation process.
T_{peak}: Temperature of the maximum rate of the degradation process; residue (char formed at 900 °C).
W: Weight loss on each step of the degradation process.

The total weight loss recorded for HDPE as a result of weathering was around 95.24%, lower than that recorded for HDPE before weathering, and this was attributed to some structural re-arrangements of the HDPE chains that had been disrupted by the degradation process.

After weathering, the HDPE/PSF composites showed two thermal degradation processes as follows: (I) in the 290-370 °C temperature range, attributed to the degradation of hemicelluloses and cellulose from PSF, and (II) in the 450-500 °C temperature range, attributed to the degradation of HDPE, as well as cellulose and lignin from PSF. The total weight loss values decreased with increasing PSF amount in the composites from 97.7% (for HDPE/PSF5) down to 95.0% (for HDPE/PSF10), and to 94.8% (for HDPE/PSF15). The HDPE/PSF composites were less thermally sTable when compared to neat HDPE. Moreover, the thermal stability decreased after samples were subjected to UV radiation.

10.3.3.2 Differential Scanning Calorimetry Analysis

DSC measurements (**Figure 10.10**) were used for estimation of the melting (T_m) and crystallisation (T_c) temperatures. Data related to the fusion enthalpy (ΔH_m) and crystallisation enthalpy (ΔH_c) are presented in **Table 10.4**.

Table 10.4 DSC parameters recorded for HDPE and HDPE/PSF composites, before and after weathering

Sample	T_m (°C)	ΔH_m (J/g)	T_c (°C)	ΔH_c (J/g)	Crystallinity index (%)
HDPE	122	108	111	114	38.9
HDPE-600	120	133	113	123	48.0
HDPE/PSF5	124	120	113	102	45.6
HDPE/PSF5-600	122	127	113	126	48.2
HDPE/PSF10	125	119	114	106	47.7
HDPE/PSF10-600	123	134	115	131	53.8
HDPE/PSF15	126	114	115	108	48.4
HDPE/PSF15-600	124	128	114	122	54.4

In the absence of PSF, there was no hindrance to the self-nucleation of HDPE. Thus, a single well-defined exothermic peak was observed at 113 °C. The HDPE exhibited a maximum melting point at 122 °C. When PSF was added, the melting point and the crystallisation temperature both increased slightly. The weathering treatment shifted the melting temperature to lower values and changed the shape of the DSC curve, as a result of changes in the sizes and distribution of the crystallites, and due to branching and/or crosslinking reactions which took place after irradiation, causing a disruption in the polymer sequences. An increase of ΔH_m for both HDPE and composites was registered.

After accelerated weathering, the degraded chains were characterised by higher mobility. This phenomenon could suggest an increase in crystallinity which influenced both HDPE and PSF. As result, some changes in morphology and mechanical properties were observed.

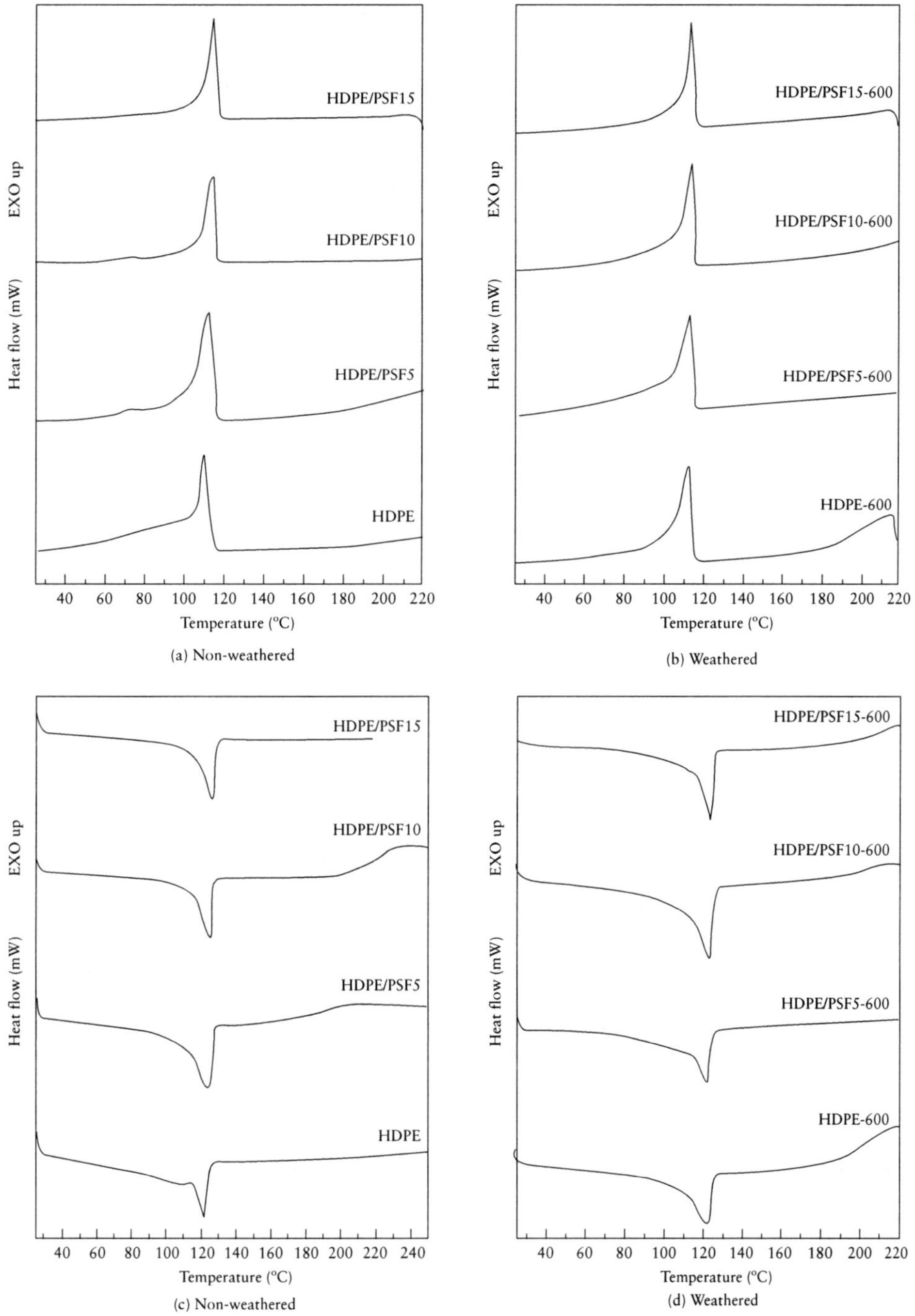

Figure 10.10 DSC curves recorded for HDPE and HDPE/PSF composites: first heating (a and b), and second heating (c and d)

10.3.4 Investigation of Mechanical Properties

The mechanical behaviour of composites is strongly dependent on the distribution of the filler particles within the polymer matrix. Filler particle agglomerates act as defects or stress concentration sites for crack nucleation. The mechanical properties of neat HDPE and HDPE/PSF composites are summarised in **Figures 10.11-10.13**. As compared with HDPE, the HDPE/PSF composite samples registered slightly diminished tensile strength values (see **Figure 10.11**). The tensile test results showed that PSF acts as a filler and not as a reinforcing agent. The decrease in tensile strength at higher fibre loading reflects a poor adhesion between the PSF and HDPE. The same change tendency was registered during the weathering process.

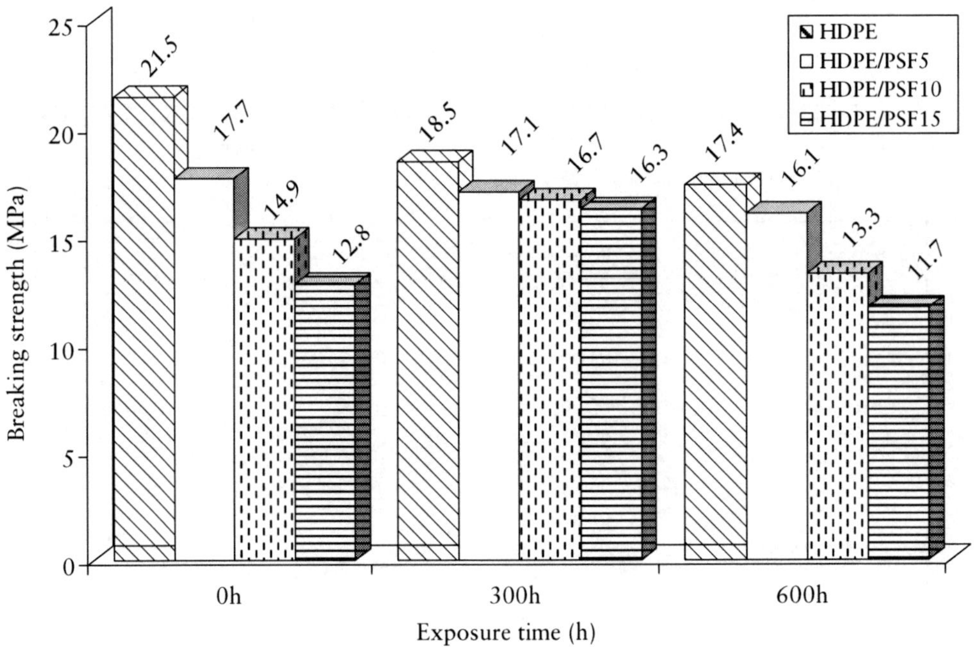

Figure 10.11 Breaking strength values recorded for neat HDPE and HDPE/PSF composites

HDPE/PSF composites showed a significant decrease in Young's modulus compared with that of neat HDPE. This property is slightly enhanced, upon adding the PSF filler, after the weathering process due to the different stiffness of the filler and HDPE.

During weathering, an increase in crystallinity may be observed, composite samples becoming more fragile, and less easy to deform.

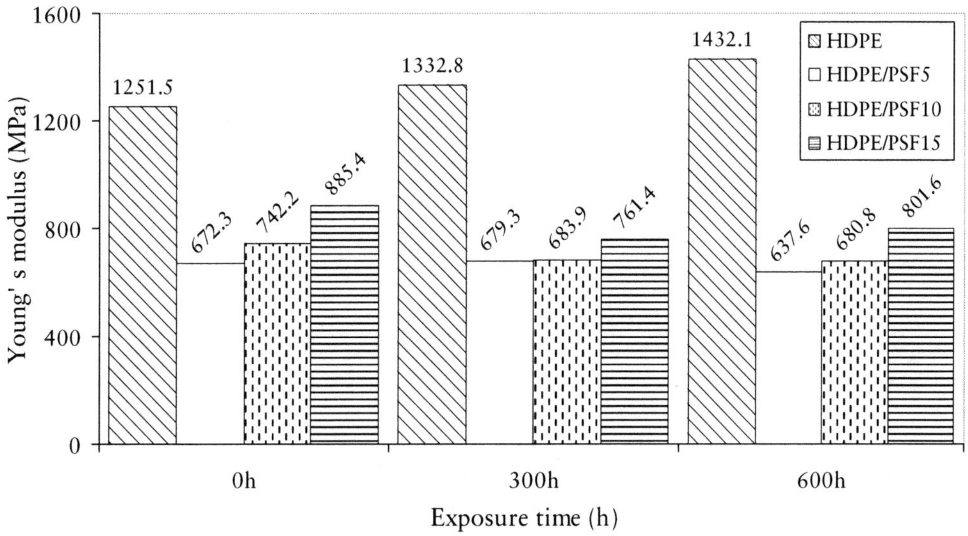

Figure 10.12 Young's modulus values recorded for neat HDPE and HDPE/PSF composites

In WPC the polymer matrix absorbs the impact energy and it serves as a link with the fibre; the matrix acts as the load distributor while the fibre is the carrier of it [59]. The addition of PSF in HDPE resulted in a decrease of impact strength values, probably due to the weak interfacial regions and stress concentrations.

A slight increase in impact strength with PSF addition was recorded mainly for the HDPE/PSF5 sample. This could be explained by the fact that PSF may interfere with the crack formation by acting as a stress transferring medium. This trend was retained to some extent under accelerated weathering conditions (see the HDPE/PSF15 sample, for example, in **Figure 10.13**).

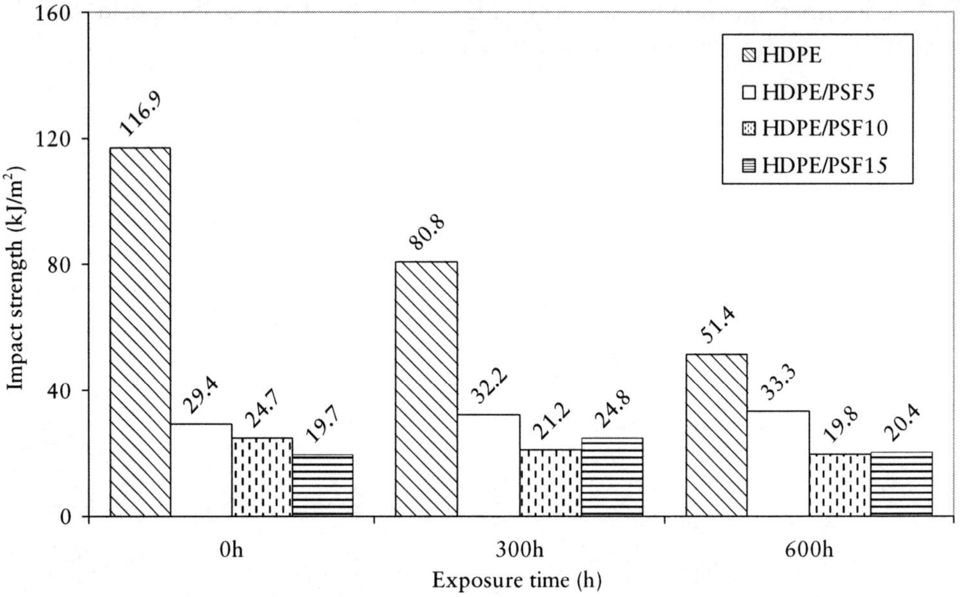

Figure 10.13 Impact strength values recorded for neat HDPE and HDPE/PSF composites

10.3.5 Investigation of Melt Rheology

The presence of fillers in viscoelastic polymers is generally considered to reduce melt elasticity and hence influence dependent phenomena such as die swelling [60]. Oscillatory rheology is a valuable tool for understanding the structural and dynamic properties of soft materials [61].

Li and Wolcott have used capillary rheometry to study steady-state flow of wood/HDPE composites [62]. Low density polyethylene (LDPE)/oak flour composites exhibited decreased melt viscosity compared to the matrix, due to the fact that lignin acts as a plasticiser and determines an improvement of flow [63]. The molten viscoelastic behaviour of HDPE/sisal composites under steady and dynamic mode was studied [64], and the melt viscosity of the composites increased with the addition of fibres and compatibiliser. The complex viscosity and dynamic shear storage modulus of the HDPE/chitosan blend crosslinked with vinyl triethoxysilane were higher than those for non-crosslinked samples [65].

In a previous work, we found that the addition of PSF within a thermoplastic matrix (LDPE) can yield composites with increased stiffness. The low values of mechanical characteristics confirmed a fragile behaviour of the composite due to a poor adhesion between non-polar LDPE and PSF [66].

In the present study, the rheological behaviour of HDPE/PSF composites was investigated in the melt state, the viscoelastic parameters being followed as a function of the oscillation frequency,ω, at a constant temperature of 145 °C. From the cross-over frequency (ω_i), the relaxation time (θ) was calculated as $\theta = 1/\omega_i$, the results obtained are presented in **Figure 10.14** and **Table 10.5**.

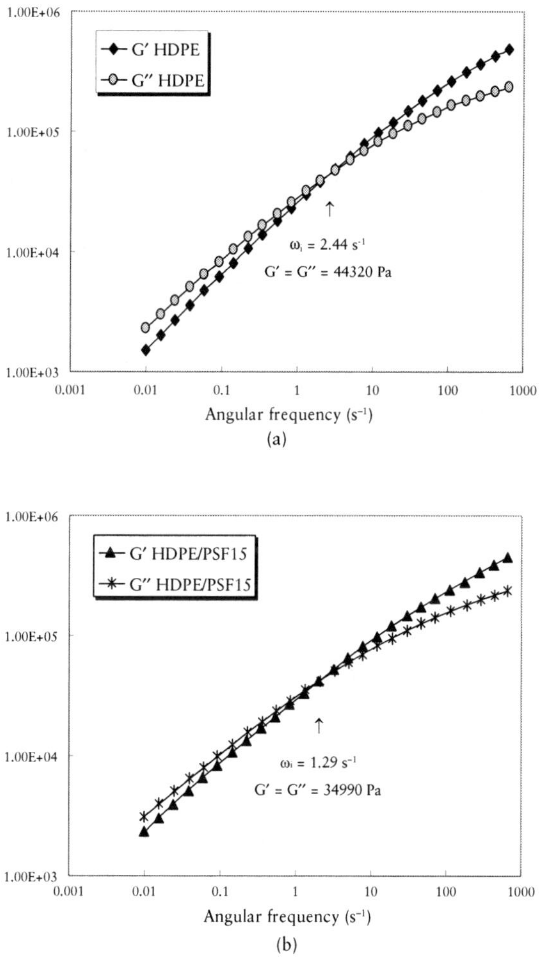

Figure 10.14 Correlation between viscous and elastic behaviour of: (a) HDPE and (b) HDPE/PSF15

There is a critical frequency at which the two responses (viscous and elastic) are equal. As a general observation, it can be seen from **Figure 10.14**, that both the sample composition and the weathering process influenced the values of G′ (storage modulus), G″ (loss modulus) and complex viscosity (η^*) for the composites under study. All the samples studied behaved as viscoelastic materials, a final rubbery state being defined as the cross-over point of G′ and G″ (when G′= G″) where the viscoelastic behaviour changes from a dominant viscous, liquid-like (G″> G′) to a dominant elastic solid-like behaviour (G′> G″) as a function of the oscillation frequency (ω). At low ω values, up to the cross-over point (ω_i), most of the materials had a predominantly viscous behaviour, in which G″ > G′, whereas above ω_i, this changed to an elastic one (G′ > G″). This trend was maintained upon UV exposure (not shown here).

Table 10.5 Crossover frequency (ω_i) and dynamic modulus at ω_i values for HDPE and HDPE/PSF systems before and after weathering

Sample	Crossover frequency (s^{-1})/ relaxation time θ (s)		G′=G″ (Pa)	
	No UV	After 600 h UV exposure	No UV	After 600 h UV exposure
HDPE	2.436/0.410	0.680/1.471	42320	23250
HDPE/PSF5	1.949/0.513	1.440/0.694	44400	31790
HDPE/PSF10	1.425/0.702	1.034/0.967	43820	30710
HDPE/PSF15	1.288/0.776	0.609/1.642	34990	23970

It was observed that with the addition of PSF, the melt viscosity of the composites increased (**Figure 10.15a**). Further, when the dynamic viscoelastic behaviour was measured, an increase in the storage modulus (*G′*) was revealed, indicating higher stiffness in the composites as compared to HDPE (**Figure 10.15b**). At low ω the complex melt viscosity increased with increasing PSF level, while at high rates of angular frequency, the values almost converged. The same evolution is observed for the loss modulus (G″) values of composites which increase with increasing frequency as compared to HDPE (**Figure 10.15c**).

Figure 10.16 plotted the evolution of complex viscosity and storage modulus for neat HDPE before and after weathering as a function of angular frequency. A difference in the values of complex viscosity at lower frequencies can be noticed, this being supported by the G′ data. It is known that G′ is more sensitive to weathering conditions than viscosity and this is evident in the G′ data at low ω which showed an increase with exposure time. Hence, for the neat HDPE sample with zero chain branches in the molecule the degradation is slow and dominated by crosslinking.

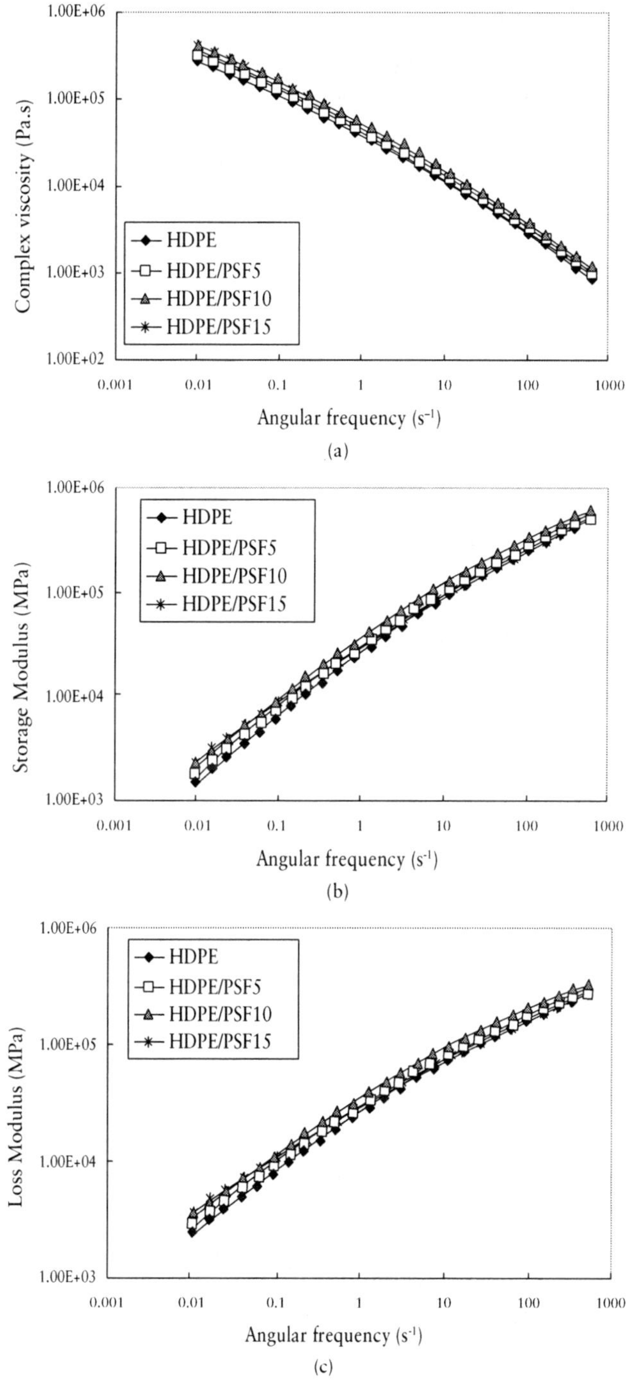

Figure 10.15 Variation of dynamic rheological parameters as a function of angular frequency for neat HDPE and HDPE/PSF composites

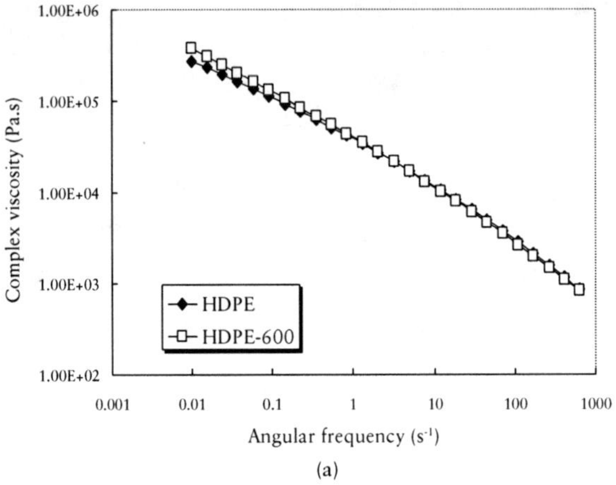

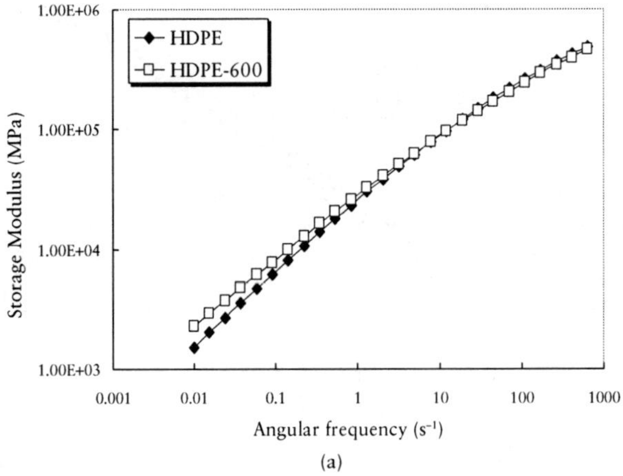

Figure 10.16 Influence of weathering upon dynamic viscoelastic parameters for neat HDPE: (a) complex viscosity and (b) storage modulus

When comparing the dynamic rheological results obtained for HDPE/PSF composites exposed to weathering conditions with those of the neat HDPE, one can observe a different trend (see **Figure 10.17**). After weathering, the decrease in complex viscosity is noticeable for the composites containing 10 wt% PSF as plotted in **Figure 10.17a**, indicating a chain scission. The results show that the composites containing PSF are more susceptible to UV degradation.

It should be noted that, for the composites, increased values of θ were registered when increasing the PSF amount incorporated compared with neat HDPE. Accelerated weathering influenced the viscoelastic behaviour of the samples, evidenced also by changes in relaxation time.

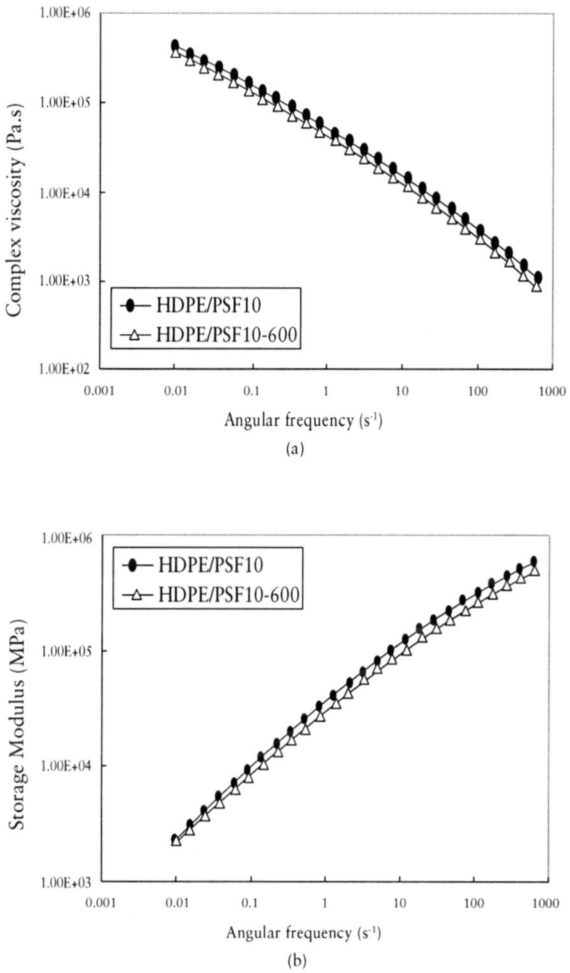

Figure 10.17 Influence of UV exposure over some dynamic viscoelastic parameters for HDPE/PSF10: (a) complex viscosity and (b) storage modulus

Data plotted in **Figure 10.18** show a significant increase in the value of θ, and a decreased cross-over frequency (ω_i) (as shown in **Table 10.5**), for HDPE after weathering. As for the HDPE/PSF composites, the increasing trend for θ values is maintained after weathering, the highest difference in relaxation times being observed for the sample with 15 wt% PSF, which is more affected by weathering.

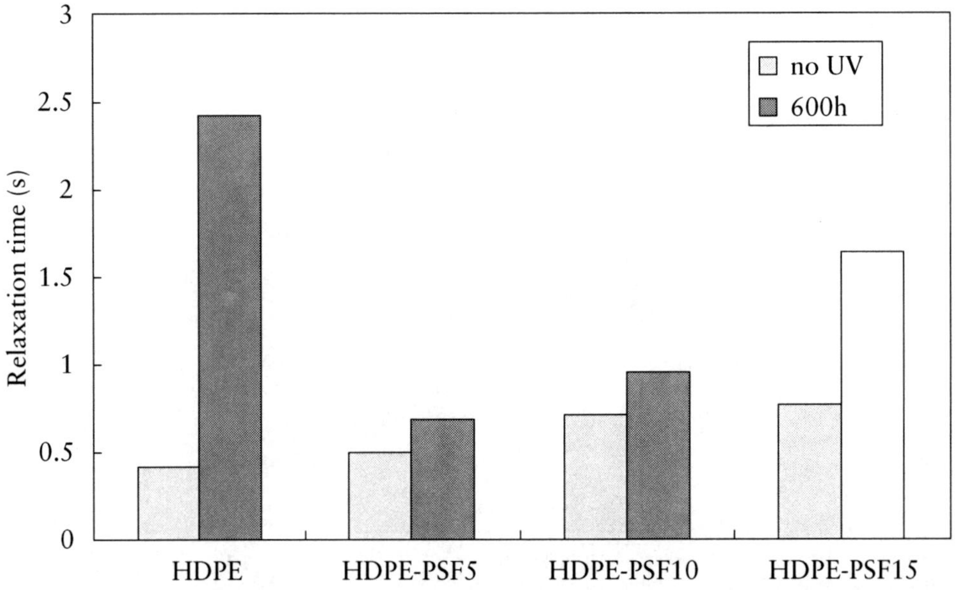

Figure 10.18 Influence of composition and weathering on relaxation time for neat HDPE and HDPE/PSF composites (UV exposure time 600 h)

The non-Newtonian, pseudoplastic behaviour is represented in a simple way by the power law model: $\eta_a = k \cdot \gamma^{n-1}$, where η_a is apparent viscosity, k is the consistency index, γ is the shear rate, and n is the flow behaviour index. For Newtonian fluids n = 1, and for pseudoplastic fluids, n =< 1, with smaller values indicating greater shear sensitivity. At very high shear rates, the polymer chains are, in theory, fully aligned in the direction of flow. Thus, the viscosity cannot decrease further and is constant in the upper Newtonian plateau [67].

Figure 10.19 shows that, at high oscillation frequencies, no significant differences could be observed between the dynamic viscosities under normal conditions for HDPE and HDPE/PSF15.

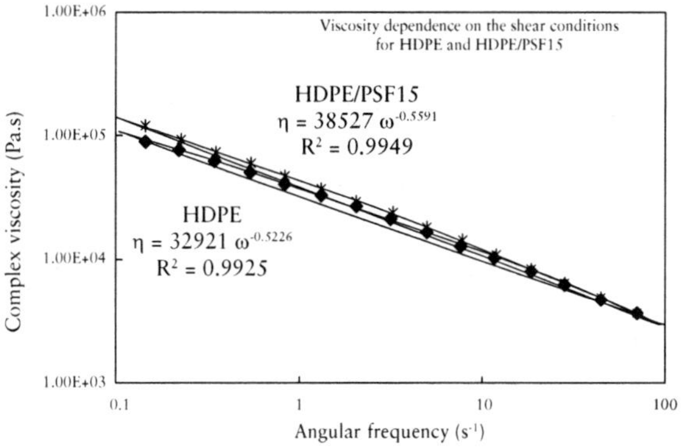

Figure 10.19 Variation of complex viscosity with angular frequency for neat HDPE and HDPE/PSF15 composite

The complex viscosity varies as $\omega^{-0.52}$ for HDPE and as $\omega^{-0.56}$ for composites containing 15 wt% PSF (**Figure 10.19**), the presence of PSF inducing a slight change in the viscosity dependence on the shear conditions. The correlation coefficient R^2 is approximately 1, for both HDPE and HDPE/PSF15 samples, indicating that the regression line perfectly fits the experimental data.

10.4 Conclusions

The mechanical properties and the results obtained by FTIR and SEM clearly demonstrate the existence of interactions between the HDPE and PSF. The non-weathered composites present a very weak physicomechanical adhesion (known as interdiffusion) that allows the formation of a composite structure *via* diffusion of HDPE and PSF.

All these results clearly indicate that the macroscopic properties of the composites depend on the PSF content. The tensile strength, as well the other mechanical properties, depend mainly on the PSF content.

The increase in crystallinity for HDPE/PSF composites resulting from the weathering process may be attributed to chain scission. The dynamic rheology results obtained showed that chain scission is dominant in HDPE/PSF composites, while crosslinking is dominant for the neat HDPE upon weathering. The composites with an increased amount of PSF are more susceptible to UV degradation.

References

1. L. Yu, K. Dean and L. Li, *Progress in Polymer Science*, 2006, **31**, 576.

2. J.A. Youngquist in *Wood Handbook—Wood as an Engineering Material*, General Technical Report FPL–GTR–113, US Department of Agriculture Forest Service, Forest Products Laboratory, Madison, WI, USA, 1999, p.10.

3. *Paper and Composites from Agro-Based Resources*, Eds., R.M. Rowell, R.A. Young and J.K. Rowell, CRC Lewis Publishers, Boca Raton, FL, USA, 1996.

4. T.Q. Li and M.P. Wolcott, *Polymer Engineering and Science*, 2005, **45**, 4, 549.

5. *Polymer Alloys and Blends: Thermodynamics and Rheology*, Ed., L.A. Utracki, Hanser Publishers, Munich, Germany, 1990.

6. C. Lacroix, M. Grmela and P.J. Carreau, *Journal of Non-Newtonian Fluid Mechanics*, 1999, **86**, 37.

7. U. Yilmazer and D.M. Kalyon, *Journal of Rheology*, 1989, **33**, 1197.

8. M. Bousmina and R. Muller, *Rheologica Acta*, 1996, **35**, 369.

9. J. Greener and J.R.G. Evans, *Journal of Rheology*, 1998, **42**, 697.

10. *Rheology of Polymeric Systems: Principles and Applications*, P.J. Carreau, D.C.R. de Kee and R.P. Chhabra, Hanser/Gardner Publications, Cincinnati, OH, USA, 1996.

11. M. Muasher and M. Sain, *Polymer Degradation and Stability*, 2006, **91**, 5, 1156.

12. N.M. Stark and L.M. Matuana, *Journal of Applied Polymer Science*, 2003, **90**, 10, 2609.

13. S.M. Zabihzadeh, *BioResources*, 2010, **5**, 1, 316.

14. N.M. Stark and L.M. Matuana, *Journal of Applied Polymer Science*, 2004, **94**, 6, 2263.

15. N.M. Stark and L.M. Matuana, *Polymer Degradation and Stability*, 2004, **86**, 1, 1.

16. D. Ndiaye, E. Fanton, S. Morlat-Therias, L. Vidal, A. Tidjani and J-L. Gardette, *Composites Science and Technology*, 2008, **68**, 13, 2779.

17. W.C. Feist and D.N-S. Hon in *The Chemistry of Solid Wood*, American Chemical Society, Washington, DC, USA, 1984, p.401.

18. U. Müller, M. Ratzsch, M. Schwanninger, M. Steiner and H. Zöbl, *Journal of Photochemistry and Photobiology B*, 2003, **69**, 2, 97.

19. K.K. Pandey, *Polymer Degradation and Stability*, 2005, **90**, 1, 9.

20. K.K. Pandey, *Polymer Degradation and Stability*, 2005, **87**, 2, 375.

21. S. Girois, P. Delprat, L. Audouin and J. Verdu, *Polymer Degradation and Stability*, 1997, **56**, 2, 169.

22. I. Yakimets, D. Lai and M. Guigon, *Polymer Degradation and Stability*, 2004, **86**, 1, 59.

23. L.M. Matuana and D.P. Kamdem, *Polymer Engineering and Science*, 2002, **42**, 8, 1657.

24. N.M. Stark and S.A. Müller, *Wood and Fiber Science*, 2008, **40**, 2, 271.

25. L.M. Matuana, D.P. Kamdem and J. Zhang, *Journal of Applied Polymer Science*, 2001, **80**, 11, 1943.

26. N.M. Stark and L.M. Matuana, *Polymer Degradation and Stability*, 2007, **92**, 10, 1883.

27. N.M. Stark *Journal of Applied Polymer Science*, 2006, **100**, 4, 3131.

28. N.M. Stark, L.M. Matuana and C.M. Clemons, *Journal of Applied Polymer Science*, 2004, **93**, 3, 1021.

29. N.M. Stark and L.M. Matuana, *Polymer Degradation and Stability*, 2006, **91**, 5, 3048.

30. J.S. Fabiyi, A.G. McDonald, M.P. Wolcott and P.R. Griffiths, *Polymer Degradation and Stability*, 2008, **93**, 8, 1405.

31. D.T. Cooper and D.F. Van Haverbeke, *Eastern Cottonwood*, 2011. *http://www.na.fs. fed.us/pubs/silvics_manual/volume_2/populus/deltoides.htm*

32. M. Likon, M. Remskar, V. Ducman and F. Svegl, *Journal of Environmental Management*, 2013, **114**, 158.

33. A. Dufresne and M.R. Vignon, *Macromolecules*, 1998, **31**, 2693.

34. *Polymer Handbook*, Eds., J. Brandrup and E.H. Immergut, John Wiley and Sons, New York, NY, USA, 1989.

35. S.K. Wisniewska, J. Nalaskowski, E. Witka-Jezewska, J. Hupka and J.D. Miller, *Colloids and Surfaces B Biointerfaces*, 2003, **29**, 131.

36. T.T. Lim and X. Huang, *Industrial Crops and Products*, 2007, **26**, 125.

37. T.T. Lim and X. Huang, *Chemosphere*, 2007, **66**, 955.

38. M.A. Abdullah, A. Rahmah and Z. Man, *Journal of Hazardous Materials*, 2010, **177**, 683.

39. G. Bianchi in *Waxes: Chemistry, Molecular Biology and Function*, The Oily Press, Dundee, UK, 1995, p.175.

40. E.A. Baker in *The Plant Cuticle*, Academic Press, London, UK, 1982, p.139.

41. A.P. Tulloch in *Chemistry and Biochemistry of Natural Waxes*, Elsevier, New York, NY, USA, 1976, p.235.

42. R.S. Rengasamy, D. Das and C. Praba Karan, *Journal of Hazardous Materials*, 2011, **186**, 1, 526.

43. H.L. Hergert in *Lignins: Occurrence, Formation, Structure and Reactions*, Wiley-Interscience, New York, NY, USA, 1971, p.267.

44. B. Mohebby, *Journal of Agricultural Science and Technology*, 2008, **10**, 253.

45. M.O. Adebajo and R.L. Frost, *Spectrochimica Acta*, 2004, **A60**, 449.

46. C.J. Venverloo, *Holzforschung*, 1971, **1**, 25, 18.

47. K. K. Pandey, *Journal of Applied Polymer Science*, 1999, **71**, 1969.

48. X. Colom, J. Canavate, P. Pages, J. Saurina and F. Carrasco, *Journal of Reinforced Plastics and Composites*, 2000, **19**, 818.

49. F.W. Langkilde and A. Svantesson, *Journal of Pharmaceutical and Biomedical Analysis*, 1995, **13**, 409.

50. R.G. Zhbankov, S.P. Firsov, E.V. Korolik, P.T. Petrov, M.P. Lapkovski, V.M. Tsarenkov, M.K. Marchewka and H. Ratajczak, *Journal of Molecular Structure*, 2000, **555**, 85.

51. S.A. Jabarin and E.A. Lofgren, *Journal of Applied Polymer Science*, 1994, **53**, 4, 411.

52. *Handbook of Material Weathering*, Ed., G. Wypych, ChemTec Publishing, Toronto, ON, Canada, 1995.

53. A. Tidjani, *Polymer Degradation and Stability*, 2000, **68**, 3, 465.

54. C. David, M. Trojan, A. Daro and W. Demarteau, *Polymer Degradation and Stability*, 1992, **37**, 3, 233.

55. M. Kaci, T. Sadoun and S. Cimmino, *International Journal of Polymer Analysis and Characterization*, 2001, **6**, 5, 455.

56. X. Colom, F. Carrillo, F. Nogues and P. Garriga, *Polymer Degradation and Stability*, 2003, **80**, 543.

57. S.S. Fernando, P.A. Christensen, T.A. Egerton and J.R. White, *Polymer Degradation and Stability*, 2007, **92**, 12, 2163.

58. L. Guadagno, C. Naddeo, V. Vittoria, C. Giovanni and C. Cagnani, *Polymer Degradation and Stability*, 2001, **72**, 1, 175.

59. D.W. van Krevelen in *Properties of Polymers*, 3rd Edition, Elsevier Science Publishers, Amsterdam, Holland, 1990.

60. P.R. Hornsby, *Advances in Polymer Science*, 1999, **139**, 155.

61. W. Macosko in *Rheology: Principles, Measurements and Applications*, Wiley-VCH, New York, NY, USA, 1994.

62. T.Q. Li and M.P. Wolcott, *Composites Part A: Applied Science and Manufacturing*, 2004, **35**, 303.

63. R.N. Darie, M. Bercea, M. Kozlowski and I. Spiridon, *Cellulose Chemistry and Technology*, 2011, **5**, 1, 127.

64. S. Mohanty and S. K. Nayak, *Polymer Engineering and Science*, 2007, **47**, 10, 1634

65. S. Mir, T. Yasin, P.J. Halley, H.M. Siddiqi and T. Nicholson, *Carbohydrate Polymers*, 2011, **83**, 414.

66. M. Mosneagu, R.N. Darie, S.T. Aradoaei and I. Spiridon in *Proceedings of the 6th International Conference on Electrical and Power Engineering (EPE)*, Iasi, Romania, 28-30th October, 2010, p.383.

67. J.M. Dealy and R.G. Larson, *Structure and Rheology of Molten Polymers: from Structure to Flow Behaviour and Back Again*, Hanser Gardner Publications, Munich, Germany, 2006.

11 Effect of Physicochemical Conditions on Biomass-based Biocomposites

Manju Kumari Thakur, Vijay Kumar Thakur and Amar Singh Singha

11.1 Introduction

The use of waste biomass-based materials such as lignocellulosic natural fibres as sustainable materials is rapidly increasing because of their potential to substitute for certain petroleum-derived synthetic polymer materials [1-3]. Lignocellulosic natural fibres play a fundamental role in our life and society due to their wide range of properties. A number of fibres such as sisal, *Grewia optiva*, cotton, jute, flax, *Hibiscus sabdariffa*, pine needles or hemp are available all around the globe and are extensively produced and used in everyday life, as well as for industrial applications [4-7]. Biomass-based products and by-products have gained attention as a source of lignocellulosic fibers for use in a number of applications, especially as reinforcement for composites [8-11]. Biomass-reinforced polymer composites have demonstrated a marked improvement in the physical properties of polymeric materials for various applications [12-16]. Furthermore, the increasing concerns regarding shortages of landfill sites, persistence of nonbiodegradable petroleum-based polymers in the environment, and the dependence on depleting non-renewable petroleum resources have resulted in research into the utilisation of biomass-based materials as novel low cost materials for the future [17-20]. Biomass-based biocomposites, which are a combination of polymeric matrices (synthetic or natural) and biomass as reinforcement, have emerged as promising alternatives to synthetic fibre-reinforced composites because they offer a wide variety of advantages, such as low density, renewability, biodegradability, and better cost-performance ratio [21-24]. Indeed the growing global energy crisis and current environmental concerns have led to extensive research, which is primarily aimed at the production and use of biomass-based products and by-products as a source of fibres to reinforce polymeric matrices for composites [25-30].

Biomass-based lignocellulosic fibres offer a number of advantages over their synthetic counterparts, such as efficient biodegradability, non-abrasiveness, low cost, low specific density, eco-friendliness and non-toxicity. Their lack of toxicity brings healthier working conditions in processing and production facilities and their low specific density is an important advantage for composite applications in fields such as the automotive industry, where the use of light-weight materials in vehicles will reduce their fuel consumption. A number of lignocellulosic natural fibres such as flax, jute, kenaf, and sisal, have been extensively studied as reinforcement in polymer composites [21-30]. The easy availability and the continuous supply of biomass-based cellulosic fibres in large quantities have raised the demand for new fibres which can potential be used along with the fibres mentioned above. The use of inexpensive biomass-based lignocellulosic fibres can lower the cost of traditional composites to a considerable extent as the fibres obtained from biomass are between five and ten times cheaper than other materials which are currently used as reinforcements.

Among the various fibers found abundantly in Himachal Pradesh, in India, *Hibiscus sabdariffa* is one which is most frequently used by the local people for various household applications, owing to its high mechanical strength and physical characteristics. The bast fibres obtained from the *Hibiscus sabdariffa* plant have the potential to be directly or indirectly used for a number of applications, such as in green biocomposites.

In natural lignocellulosic *Hibiscus sabdariffa* fibres, cellulose is the main component and the elementary unit of a cellulose macromolecule is anhydro-D-glucose, which contains three hydroxyl groups (-OH). Due to the presence of -OH groups, all natural fibres have been found to be hydrophilic in nature and this is one of the obstacles to the successful application of these fibres in industry. Because, as a result of the presence of these groups; natural fibres become susceptible to degradation by moisture and water. In this chapter, we report some of our preliminary studies on the effects of physicochemical conditions on *Hibiscus sabdariffa* fibre-reinforced biocomposites [29]. A brief description of the synthesis process is given in the experimental section below.

11.2 Preparation of *Hibiscus sabdariffa*-based Biocomposites for Physicochemical Studies

Hibiscus sabdariffa fibres were obtained from local sources in the Himalayan region of India. The unwanted materials were separated from the *Hibiscus sabdariffa* fibres using standard methods. After purification, *Hibiscus sabdariffa* fibres were completely dried in a hot air oven and were subsequently converted into short fibres according to the standard method reported in our earlier studies [29]. The polymer used as the matrix in the present research work was a urea-formaldehyde (UF) resin which was synthesised by the method reported in the literature [29]. For the preparation of the biocomposites, dried short cellulosic fibres (3mm in length) were mixed thoroughly with the polymer resin at different loadings (10, 20, 30 and 40 wt%) using a mechanical stirrer. Subsequently, the composite mixtures were poured into specially made moulds and biocomposite sheets of size 150 × 150 × 5.0 mm were prepared by a compression moulding technique [29].

11.3 Physicochemical Studies

The effect of different physicochemical conditions on the synthesised biocomposites was determined in the following manner:

The effects of different solvents such as carbon tetrachloride (CCl_4), isobutanol (C_4H_9OH), methanol (CH_3OH), and water (H_2O) on the swelling behaviour of the *Hibiscus sabdariffa* fibre-reinforced biocomposites were evaluated by studying the swelling for particular time intervals. Biocomposite specimens were dried in an oven for a specified time at a particular temperature and then cooled in a desiccator for further analysis. Known weights (W_i) of the dried biocomposite samples with different loadings were immersed in 100 ml of different solvents at room temperature for 15 days. After the stipulated time period, the *Hibiscus sabdariffa* biocomposite samples were filtered and the excess solvents were removed with the help of filter paper. The final weights (W_f) of the composite samples were noted and the increase in initial weight was used to calculate the percentage swelling as shown in **Equation 11.1**:

$$\text{Percentage swelling } (P_S) = \frac{W_f - W_i}{W_i} \times 100 \qquad (11.1)$$

Hibiscus sabdariffa fibre-reinforced biocomposite samples were also subjected to moisture absorbption studies at different relative humidity (RH) levels in the range of 20 to 90%. The study was carried out by placing a known weight (W_i) of the dried *Hibiscus sabdariffa* biocomposite samples in a humidity chamber set at a particular humidity level for about 12 h, followed by measuring the final weight (W_f) of the biocomposite samples exposed to a particular relative humidity (RH). The percentage moisture absorption was calculated as shown in **Equation 11.2**:

$$\text{Percentage moisture absorbance } (\%M_{abs}) = \frac{W_f - W_i}{W_i} \times 100 \qquad (11.2)$$

Chemical resistance studies of the different biocomposite samples were also carried out in acidic and basic solutions by immersing the specific samples in 100 ml of 1 N sodium hydroxide (NaOH) and 1 N hydrochloric acid (HCl) for different intervals of time (24-144 h). These biocomposite samples were filtered out after the specified time periods, dried and weighed. The percentage chemical resistance (P_{cr}) was then calculated in terms of weight loss as shown in **Equation 11.3**:

$$\text{Percentage chemical resistance } (P_{cr}) = \frac{T_i - W_{aci}}{T_i} \times 100 \quad (11.3)$$

where:

T_i = Initial weight of biocomposites

W_{aci} = Weight after certain interval

After the swelling studies, thermogravimetric analysis (TGA) of the biocomposite samples with optimum (30 wt%) loading was carried out in a nitrogen atmosphere on a thermal analyser (Perkin Elmer) at a heating rate of 10 °C/min to characterise the decomposition of the samples.

11.4 Results and Discussion

The up-to-date literature from different sources clearly specifies that different biomass-based materials are, to some extent, affected by weathering conditions which severely hinders their practical utility. Indeed when biomass-based products are used outdoors, most of them are easily degraded by weathering [21-30], so in order to develop biomass-based materials with long-term durability, it is particularly important to study their weathering behaviour. As specified earlier, the main problems associated with biomass-based polymeric materials are their dimensional instability due to moisture/water absorption and adsorption, as well as poor chemical resistance and the fact that they are easily degraded by microorganisms. A number of studies have been carried out to examine the effects of different weathering conditions on biocomposites from different sources and this is a subject of much deliberation these days. In order to successfully commercialise the biocomposite materials procured from different kinds of biomass, analysis of their physicochemical properties is essential in order to determine the overall usefulness of biocomposites.

When immersed in different solvents, *Hibiscus sabdariffa* fibre-reinforced UF biocomposites with different fibre loadings exhibit different swelling behaviours as shown in **Figure 11.1**.

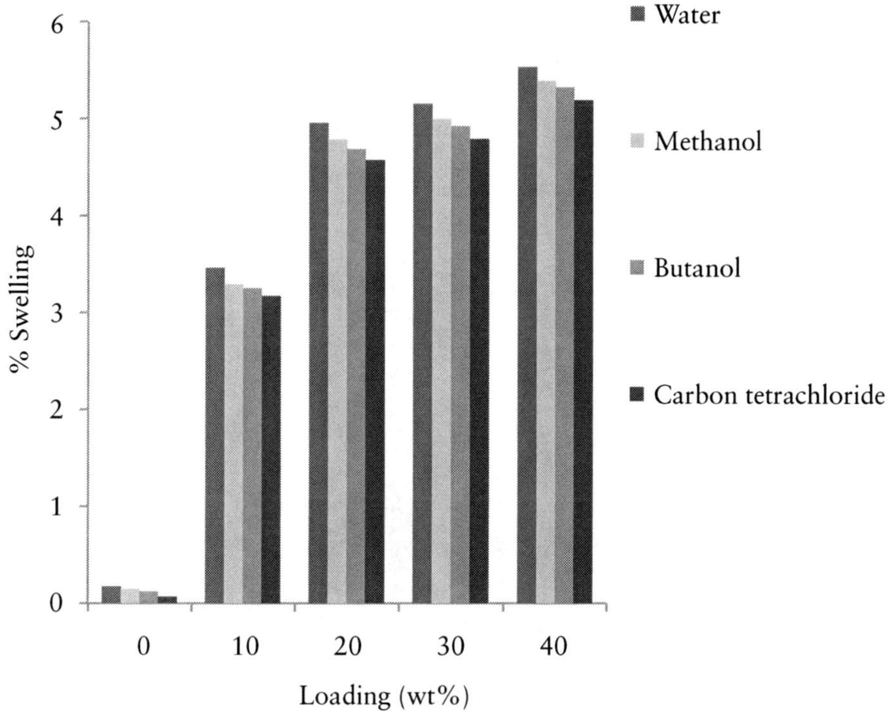

Figure 11.1 Swelling behaviour of biocomposites in different solvents

The swelling behaviour of *Hibiscus sabdariffa* fibre-reinforced biocomposites follows the trend: H_2O >CH_3OH > C_4H_9OH >CCl_4 in the different solvents. Swelling of the biocomposites has been found to increase with the increase in *Hibiscus sabdariffa* fibre loading due to the superior affinity of water for the higher number of -OH groups present in the fibre-reinforced biocomposites. The moisture absorption behaviour of the biocomposites at different RH levels is shown in **Figure 11.2.**

From the moisture absorption study, it has been observed that moisture absorption increases with an increase in RH level from 20-90% and with an increase in fibre loading, due to the presence of -OH groups in the biocomposites as explained earlier.

The chemical resistance behaviour of the biocomposites suggests that the resistance towards chemicals decreases with an increase in fibre loading (**Figure 11.3**).

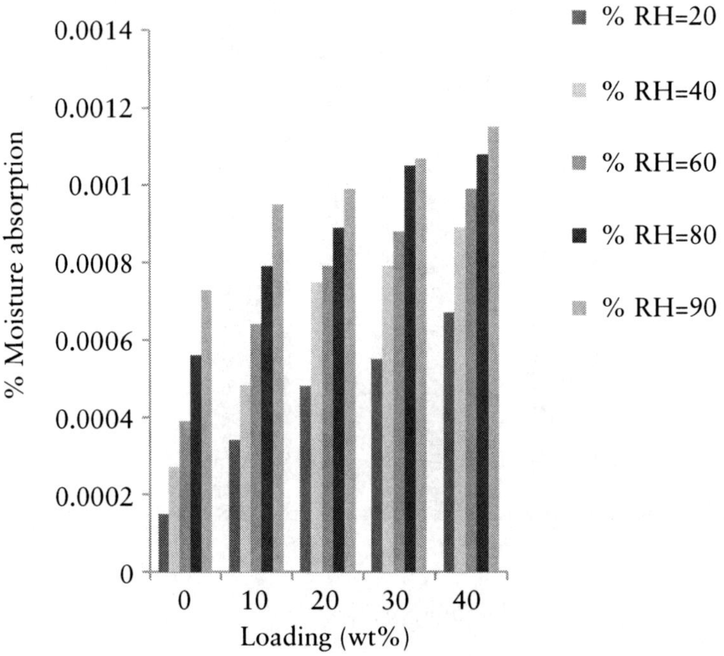

Figure 11.2 Moisture absorption behaviour of biocomposites at different RH levels

The decrease in chemical resistance with increasing fibre loading can be attributed to the higher number of hydrophilic -OH groups in *Hibiscus sabdariffa* fibres at higher loading. Composites with a higher number of -OH groups are vulnerable to chemical attack, resulting in decreased resistance towards the chemicals.

The thermal behaviour of the *Hibiscus sabdariffa* fibre-reinforced biocomposites was studied using TGA. **Table 11.1** shows the results of thermal degradation of the polymer resin and the fibre-reinforced composite with optimum loading (30 wt%) after swelling in water. It has been observed that the thermal stability decreases after swelling.

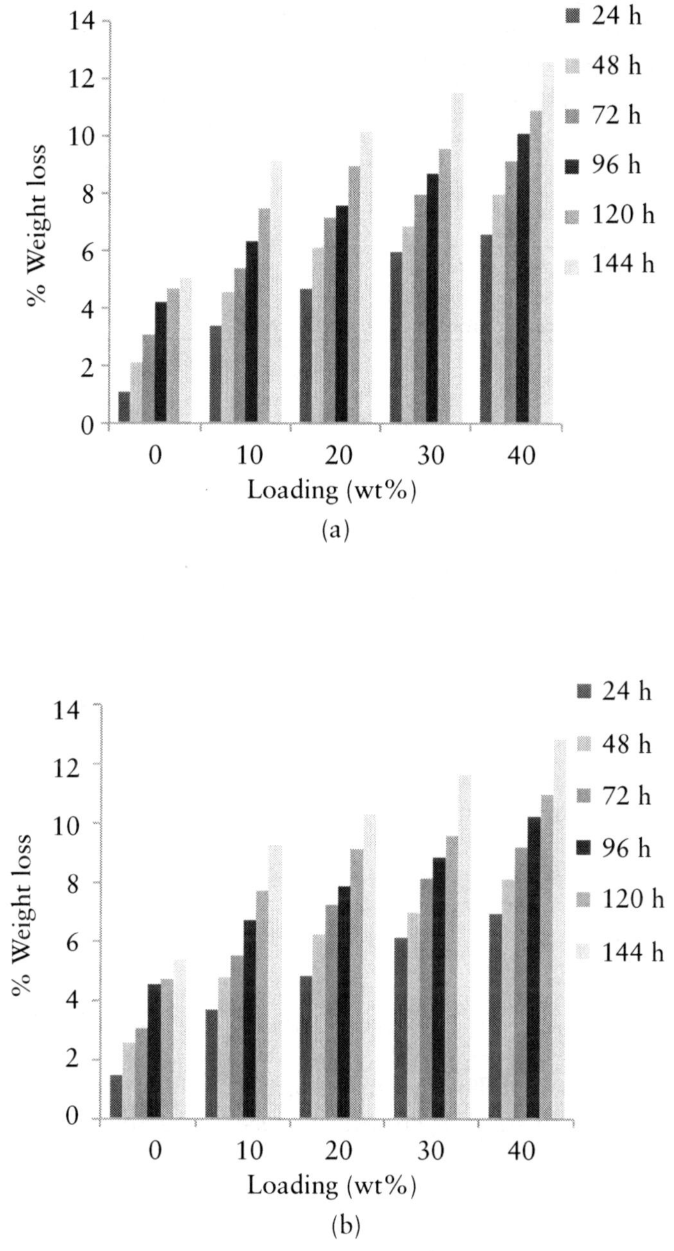

Figure 11.3 (a) Chemical resistance behaviour of biocomposites at 1 N HCl concentration, and (b) chemical resistance behaviour of biocomposites at 1 N NaOH concentration

Table 11.1 Thermogravimetric analysis of UF polymer resin and short fibre-reinforced composites with optimum loading (30 wt%)

Sample specification	Sample name	IDT (°C)	% wt. loss	FDT (°C)	%wt. loss	Final residue (wt%)
1	UF resin	221	28	963	87.5	12.5
2	SF-Rnf-UF	205	35	751	93.8	6.2

IDT: Initial decomposition temperature
FDT: Final decomposition temperature
SF-Rnf: Short fibre-reinforced composites

11.5 Conclusions

Despite some shortcomings, biomass-based biocomposite materials represent a suitable renewable alternative to synthetic fibre-reinforced composites. The structure and chemical composition of different kinds of biomass strongly affect the overall properties of the resulting biocomposites. In the present study, *Hibiscus sabdariffa* fibre-reinforced biocomposites have been subjected to weathering behaviour analysis. The results suggest that the amount of hydrophilic groups in terms of fibre loading greatly affects the physicochemical properties of the biocomposites and could be used as standard parameters for assessing end applications of these biocomposites in everyday life.

References

1. L. Yu, K. Dean and L. Li, *Progress in Polymer Science*, 2006, **31**, 576.

2. A.K. Rana, A. Mandal, B.C. Mitra, R. Jacobson, R. Rowell and A.N. Banerjee, *Journal of Applied Polymer Science*, 1998, **69**, 329.

3. J.A. Youngquist in *Wood Handbook—Wood as an Engineering Material*, General Technical Report FPL–GTR–113, US Department of Agriculture Forest Service, Forest Products Laboratory, Madison, WI, 1999, p.10.

4. A.S. Singha and V.K. Thakur in *Green Polymer Materials*, Studium Press LLC, Houston, TX, USA, 2011.

5. V.K. Thakur, A.S. Singha and M.K. Thakur, *International Journal of Polymeric Materials and Polymeric Biomaterials*, 2013, **62**, 226.

6. R.M. Rowell, R.A. Young and J.K. Rowell in *Paper and Composites from Agro-Based Resources*, CRC Lewis Publishers, Boca Raton, FL, USA, 1997.

7. C.K. Williams and M.A. Hillmyer, *Polymer Reviews*, 2008, **48**, 1, 1.

8. W. Helbert, J.Y. Cavaillé and A. Dufresne, *Polymer Composites*, 1996, **17**, 604.

9. V.K. Thakur, M.K. Thakur and R.K. Gupta, *Carbohydrate Polymers*, 2013, **97**, 18.

10. M. Matsuo, C. Sawatari, Y. Iwai and F. Ozaki, *Macromolecules*, 1990, **23**, 3266.

11. G. Siqueira, J. Bras and A. Dufresne, *Biomacromolecules*, 2009, **10**, 425.

12. J. Bras, D. Viet, C. Bruzzese and A. Dufresne, *Carbohydrate Polymers*, 2011, **84**, 211.

13. M.N. Anglès and A. Dufresne, *Macromolecules*, 2000, **33**, 8344.

14. G. Siqueira, J. Bras and A. Dufresne, *BioResources*, 2010, **5**, 727.

15. G. Siqueira, H. Abdillahi, J. Bras and A. Dufresne, *Cellulose*, 2010, **17**, 289.

16. V.K. Thakur, A.S. Singha and M.K. Thakur, *Advances in Polymer Technology*, 2013, **32**, S1, E427.

17. V.K. Thakur, A.S. Singha and M.K. Thakur, *Journal of Polymers and the Environment*, 2012, **20**, 1,164.

18. V.K. Thakur, A.S. Singha and M.K. Thakur, *Journal of Polymers and the Environment*, 2012, **20**, 2, 412.

19. V.K. Thakur, A.S. Singha and M.K. Thakur, *International Journal of Polymer Analysis and Characterization*, 2012, **17**, 6, 401.

20. X.M. Dong, J.F. Revol and D.G. Gray, *Cellulose*, 1998, 5, 19.

21. S. Beck-Candanedo, M. Roman and D.G. Gray, *Biomacromolecules*, 2005, **6**, 1048.

22. M.A.S. Azizi Samir, F. Alloin, M. Paillet and A. Dufresne, *Macromolecules*, 2004, **37**, 4313.

23. P.K. Ray, A.C. Chakraborty and S.B. Bandyopadhyay, *Journal of Applied Polymer Science*, 1976, **20**, 1765.

24. *International Encyclopedia of Composites*, Volume 4, Ed., S.M. Lee, VCH Publisher, New York, NY, USA, 1990.

25. V.K. Thakur and A.S. Singha, *International Journal of Polymer Analysis and Characterization*, 2011, **16**, 6, 390.

26. V.K. Thakur and A.S. Singha, *Bulletin of Material Science*, 2010, **33**, 3, 257.

27. V.K. Thakur and A.S. Singha, *Polymer-Plastics Technology and Engineering*, 2010, **49**, 11, 1101.

28. V.K. Thakur and A.S. Singha, *Polymer-Plastics Technology and Engineering*, 2010, **49**, 7, 694.

29. A.S. Singha and V.K. Thakur, *Polymer-Plastics Technology and Engineering*, 2009, **48**, 4, 482.

30. A. K. Bledzki and J. Gassan, *Progress in Polymer Science*, 1999, **24**, 2, 221.

12 Obtaining and Utilisation of Rice Husk Ash as a Filler of Polymers or Adsorbent for Oil Spill Clean-up

Sevdalina Turmanova, Svetlana Genieva and Lyubomir Vlaev

12.1 Introduction

A number of reviews and monographs [1-8] have been dedicated to rice husk (RHK) and the products obtained from its thermal degradation under different conditions. The production of rice, one of the major food crops in the world, generates one of the major wastes of the world, namely, RHK. It is reported that about 0.23 tons of RHK are generated per ton of rice produced. Efforts to utilise it have been handicapped by its tough, woody, abrasive nature, low nutritive properties, and resistance to degradation, great bulk and high ash content. Such efforts have resulted in minor usage, mostly in low-value applications in agricultural areas or as fuel. For these reasons, RHK is a serious technical waste problem and an ecological pollutant. However, because of the high silica content in RHK, its utilisation has been significantly widened in the past few decades. At present, RHK and especially rice husk ash (RHA), which is obtained after controlled burning of RHK, are the raw materials for the production of a series of silica-based materials [2, 9-14]. Recently, RHA has been successfully used as a sorbent of heavy metal ions, dyes and pigments from aqueous solutions [15] and as a support for heterogeneous catalysts used in different reactions with organic compounds [16].

In the last two decades, RHK and the products of its thermal degradation have often been used as fillers in paper, paint [3], polymers [17], polymeric composites [18] rubber [19], cement [20], adhesives and fertilisers [3]. Controlled burning of raw rice husk (RRH) in air yields two types of ash, namely white rice husk ash (WRHA) and black rice husk ash (BRHA). Both of these rice husk ashes have been used as fillers in polyethylene (PE) [21], polypropylene (PP) [22, 23], polystyrene [24] and other polymers [25]. Tremendous opportunities exist for more exhaustive research on RHA/polymer composites, which could, in the future, lead to the development of organic-inorganic hybrid materials with specific properties.

One of the most severe ecological problems nowadays is the contamination of the environment, particularly water basins, with heavy metal ions [26, 27], dyes, pigments [15, 28], petroleum and petroleum products [29, 30]. The adsorption technique has been found to be an economical and effective treatment method for removal of these components because this method is clean and produces no sludge [15, 26, 30].

The aim of the present work is to discuss the possibility of using RRH and the products of its thermal degradation under different conditions as a filler for polymer composites or as an adsorbent for cleaning up spilled crude oil or other oil products.

12.2 Chemical Composition, Constituent, Structure and Properties of Raw Rice Husk

According to the statistical data of the Food and Agriculture Organization, the global annual paddy production in 2003 was reported to be 582 million tons, cultivated in 171 million hectares of land. Rice husks (or rice hulls) are the fibrous, hard outermost coverings of rice grains. They are generated at 145 million tons per year, as an abundantly available by-product resulting from the vigorous development of the rice milling industry, and they account for one-fifth of the annual gross rice production throughout the world [31]. Global paddy production reached 628 million tons in 2005 with an additional 1% increase in 2006 [32]. These rice husks are not of commercial interest and cause serious pollution problems. It is necessary, then, to consider the use of this residue in polymer formulations with a clearly positive effect on the environment. In nature, RHK is tough, insoluble in water, woody and characterised by its inherent abrasion resistance behaviour and silica (SiO_2)-cellulose structural arrangement. Govindarao [33] has analysed all the data reported on the chemical composition of rice husks from various countries, including the Western world and Asia, and gives an average composition on a dry basis as organic matter 80 wt% and ash 20 wt%. The organic part is composed of approximately 42.8 wt% α-cellulose, 22.5 wt% lignin, 32.7 wt% hemicellulose and about 2 wt% other organic matter. Hemicellulose (xylan) is a mixture of D-xylose (17.52 wt%), L-arabinose (6.53 wt%), methylglucoronic acid (3.27 wt%) and D-galactose (2.37 wt%) [3, 34]. The chemical analyses of the inorganic part of rice husks showed that the main component is amorphous silica and small amounts of some oxides of alkali metals, alkaline earth metals, aluminium and iron.

12.3 Technologies Available for the Thermal Degradation of Rice Husk

There are many publications [7, 35, 36] in which different types and designs of combustor system for the thermal degradation of rice husks under dynamic conditions are described. Detailed analysis of the data in the literature showed that products of high quality and economic advantage may be obtained using a bubbling fluidised bed reactor for burning or pyrolysis of RRH [7, 37-39]. On the basis of the results obtained from a study carried out using laboratory equipment we built a pilot plant for the pyrolysis of RRH with a throughput capacity of 100 kg RRH/h [40], the scheme of which is presented in **Figure 12.1**.

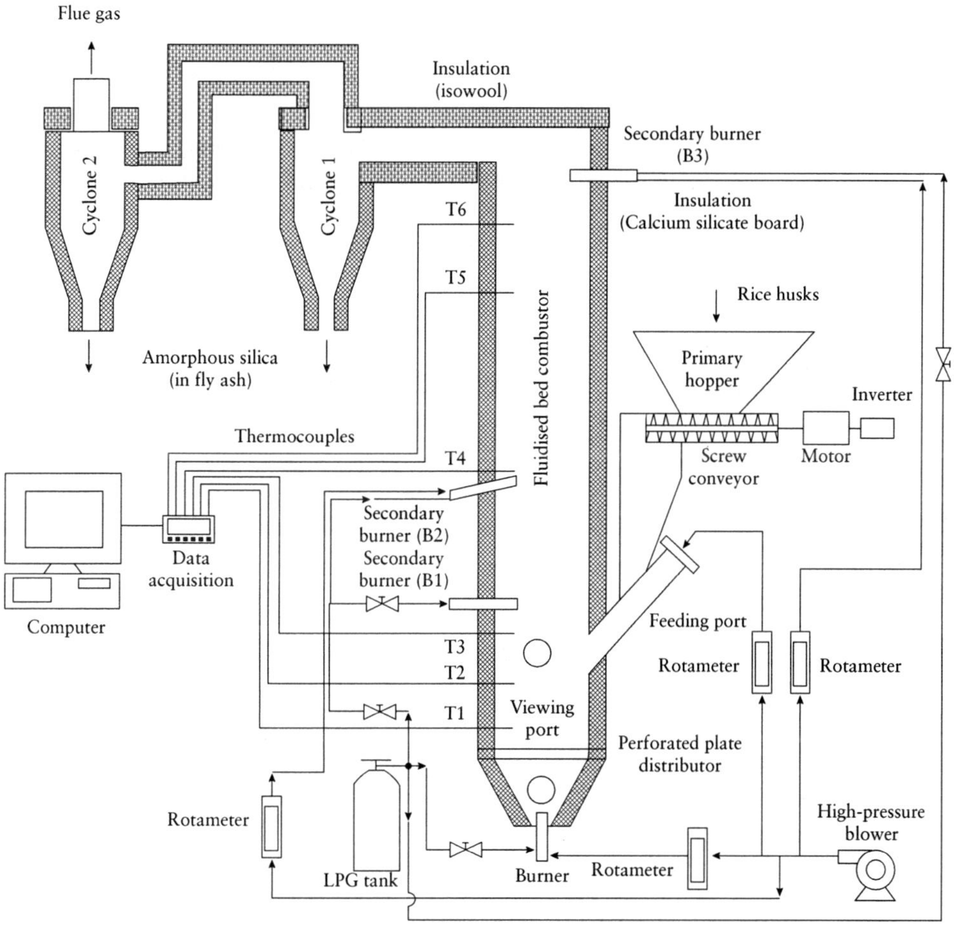

Figure 12.1 Scheme of the 0.3 m inner diameter fluidised bed combustor system with liquid petroleum gas (LPG) burner. Reproduced with permission from S. Genieva, S. Turmanova, A. Dimitrov, P. Petkov and L. Vlaev, *Journal of Thermal Analysis and Calorimetry*, 2012, **110**, 111. ©2012, Springer [40]

Metal coils are installed inside the two hot cyclones and inside the uppermost part of the fluidised bed combustor. Inside them, deionised water flows in reverse direction to the exhaust gases. The water is heated and vaporises, which results in dry steam. Thus a triple effect is achieved: the ecological problem of the utilisation of waste RHK is solved, and black or white rice husk ash, which are commercial products with different applications, and dry steam are obtained.

12.4 Physicochemical Characteristics of the Products of Thermal Degradation of Rice Husk

The main physicochemical characteristics of RHK, and the products of its thermal degradation in air under different conditions (which are used as fillers for polymers), are the morphology, crystalline or amorphous state, reactive functional groups on the surfaces, thermal stability and pore structure. Some of them are presented in **Table 12.1**, and those of Aerosil A200 (AR) (Degussa AG, Germany) are included for comparison [41].

Table 12.1 Physicochemical characteristics of RHK, WRHA, BRHA and AR

Parameter	RHK	WRHA	BRHA	AR
SiO_2, wt%	20.2	94.2	54.0	100
Moisture, wt%	7.1	0.8	0.6	<0.1
Mean particle size, μm	80	50	60	20
Surface area, m^2/g	<1	228	241	273
True density, g/cm^3	1.47	2.2	1.8	2.2

The main components of RHK are in lemma and palea form, which interlock tightly with each other [3, 34, 42, 43]. The outer epidermis of RRH is well organised and has a corrugated structure. The outer surface of the lemma is highly ridged, and the ridged structures have a linear profile. The epidermal cells of the lemma are arranged in linear ridges and furrows, and the ridges are punctuated with prominent globular protrusions [44]. Scanning electron micrographs show that the structure of RRH is globular [45].

After combustion of RHK in air, the morphology of the WRHA obtained tends to maintain its original shape although the product is brittle and loose when carefully pinched with the fingers. The hard residue formed consists of almost pure SiO_2 (94.2 wt%). As it has been pointed out [46], the rate of the combustion process depends strongly on the diffusion rate of air in RHK and the combustion products out of the bulk of the spherical globules of amorphous SiO_2.

The structure of BRHA is also globular but, due to the lower percentage of volatile products released, the solid residue contained significantly more carbon (42.6 wt%). In this case, therefore, the initial globular structure of RHK was also preserved due to the high thermal stability of SiO_2. X-ray diffraction analysis on the RHK before and after thermal decomposition showed that all samples are amorphous and no reflexes on the X-ray patterns, characterising the cristobalite or tridymite phase [45]. The WRHA and BRHA are highly porous materials with large internal specific surface area. Figure 12.2 shows the typical nitrogen adsorption-desorption isotherms of BRHA and WRHA and pore size distribution curves of both samples [40].

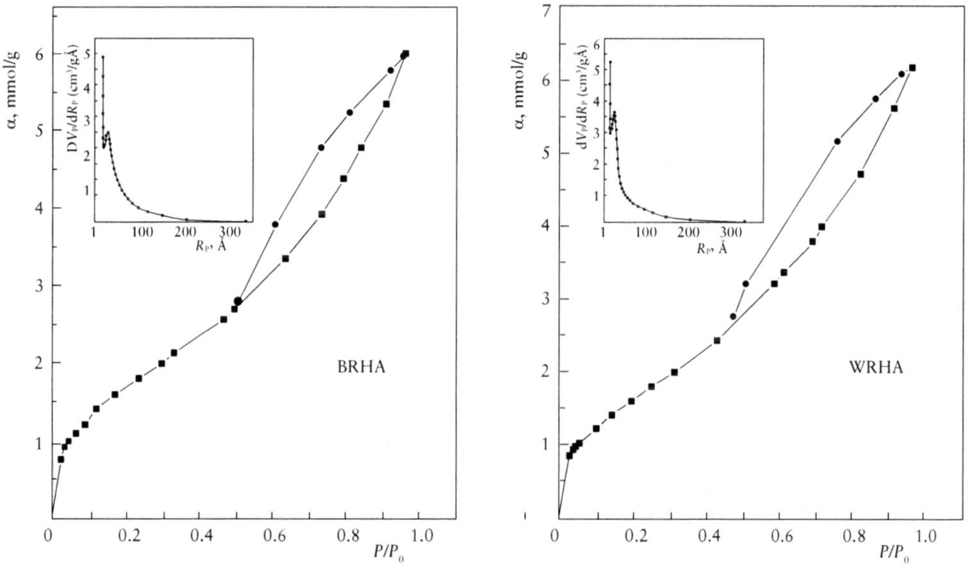

Figure 12.2 Adsorption isotherms of BRHA and WRHA and pore size distribution curves, where: α is the adsorption capacity, V_p is the pore volume, R_p is the pore radius, d*V*p/d*R*p is the distribution of pore volume with respect to pore radii, and P/P_0 is the relative presure. Reproduced with permission from S. Genieva, S. Turmanova, A. Dimitrov, P. Petkov and L. Vlaev, *Journal of Thermal Analysis and Calorimetry*, 2012, **110**, 111. ©2012, Springer [40]

As can be seen from **Figure 12.2**, the adsorption isotherms of BRHA and WRHA are of Type IV and each had a hysteresis loop (associated with capillary condensation) from Type H1 according to The International Union of Pure and Applied Chemistry classification [47, 48]. According to these observations, both the samples were mainly microporous (mean radius 1.8-2.0 nm) with narrow pore size distribution. It can be seen that the adsorption isotherms and the curves of pore volume distribution of BRHA and WRHA are identical. This is supported by the fact that the specific surface area of both samples was in the order of 230-240 m^2/g, and the pore volume was 0.22-0.25 cm^3/g. This was considered to be a sound basis on which to expect that both products could be used as adsorbents, although it should be kept in mind that the polar surface of WRHA will adsorb predominantly polar molecules while the non-polar surface of BRHA will adsorb mostly non-polar molecules.

12.5 Utilisation of Rice Husk and the Products of its Thermal Degradation as Fillers for Polymers

Using RHA as a filler in certain polymers gives composites with different properties, as well as better economics [2, 49, 50]. The addition of fillers to polymers is a fast and cheap method to modify the properties of the base polymer. Each filler brings its own characteristics to the matrix and, consequently, to the properties of the composite. Therefore, composites based on the same filler may have quite different properties depending on the nature of the matrix, the physical and chemical composition and compounding conditions [51-53]. There are many publications in the literature which report the use of natural fibres as replacements for man-made fibres in fibre-reinforced composites [2, 3, 43, 49-51, 54]. Yang and co-workers [54-56] studied the morphological and mechanical properties of RHK flour/PP composites filled with 10-40 wt% RHK at different testing conditions and temperatures. The authors observed decreased tensile strength and increased tensile modulus with increasing filler content. The composites showed brittleness at break and plastic deformation at higher temperatures. Advantages of natural fibres over man-made fibres include low density, low cost, recyclability and biodegradability [50]. These advantages make RHK a potential replacement for glass fibre in composite materials, which can be applied in the plastics, automotive and packaging industries to cut down on material cost. A serious problem with RHK fibres is their polar character, which creates incompatibility with most polymer matrices, and their relatively high moisture sorption. Therefore, chemical treatments should be employed to modify the fibre surface properties. As mentioned above, raw rice husks contain cellulose, hemicellulose, lignin, waxes and water soluble substances. The large amount of hydroxyl groups in cellulose imparts hydrophilic properties to natural fibres when used to reinforce hydrophobic matrices; the result is a very poor interface and poor resistance to moisture absorption. Hemicellulose is strongly bound to cellulose fibrils, presumably by hydrogen bonds. Lignin polymers are amorphous, highly complex and mainly aromatic compounds consisting of phenylpropane units and have low water sorption properties. The hydrophilic nature of natural fibres leads to a poor fibre/matrix interface with many polymers, so this often reduces their potential as reinforcing agents. Chemical modifications may be carried out in order to optimise the surface properties of the fibres. Chemical modifications may activate hydroxyl groups or introduce new moieties that can interlock effectively with the matrix. Generally, chemical coupling agents are molecules which possess two functions; the first function is to react with the hydroxyl groups of cellulose and the second one is to react with the functional groups of the matrix. In this respect, silanes, maleated ethylene, maleated propylene and a few acrylic-grafted linear polymers [21, 50, 54, 55] are very often used as coupling agents. They may reduce the number of cellulose hydroxyl groups in the fibre/matrix interface. In the presence of moisture, the organosilica part of the RHK leads to the formation of silanol groups. Then the silanol groups react with the hydroxyl groups of the fibre and form stable covalent bonds to the cell wall of the RHK that are chemisorbed onto the fibre surface. Therefore, the hydrocarbon chains provided by the application of silane restrain the swelling of the fibre by creating a crosslinked network due to covalent bonding between the matrix and the fibre. Yet many chemical treatments have achieved various levels of success in improving fibre strength and fibre-matrix adhesion in natural fibre-reinforced compounds, but this additional operation adds to the cost of production. This problem may be partially overcome by using BRHA produced from thermal degradation of rice husks.

12.6 Polypropylene Composites Filled with Rice Husk Ash

PP is a major thermoplastic stereoregular polymer, widely used in many industries, e.g., construction, in the automotive industry, computer components, wires and cables, home appliances, food packaging, and medical applications [57]. The mechanical, thermal, chemical and other properties of the plastic are improved by the addition of fillers. The mechanical properties of composites are the most important because they can be influenced by parameters such as the type of filler, type of matrix, filler concentration, filler dispersion, filler particle size, and the interaction between the filler and matrix. Extensive studies have been carried out during the last decade on the use of WRHA or BRHA as filler materials in PP and PE. Fuad and co-workers [58-61] investigated their use as fillers and the properties of the composites obtained. The mechanical properties of the composites were of comparable values to those of the PP composites prepared by filling with commercial Neuburg silica [61]. The flexural modulus and density of PP and PE composites increased with an increase in

RHA loading, whereas their tensile strength, elongation at break and impact strength decreased.

Most polymers absorb water in a humid atmosphere and when immersed in water. The amount of water absorbed by non-polar polymers containing a filler depends mainly on the nature of the filler. The moisture absorption by composites containing natural fillers has several adverse effects on their properties, and thus affects their long-term performance. In many of the applications of PP and its composites, the material is exposed to water and atmospheric moisture for long periods. This can lead to loss of adhesive strength due to weakening of the product at the fibre/matrix interface. This damage results from the diffusion of water molecules throughout the polymer chains causing plasticisation, chain rupture, and chemical degradation.

Knowledge of water and water vapour sorption in composites and in polymer matrices is considered to be of the utmost importance [47, 62-65]. Therefore, it is important to study in detail the water absorption behaviour in order to evaluate not only the consequences that the water absorption may have, but also how this water uptake can be minimised in some way. When the filler is a hydrophilic material, the addition of this filler into a PP matrix (which is hydrophobic) changes the water and water vapour sorption properties of PP and makes the composite a water and water vapour-sorbing material. In this case, the filled PP composite can be used as a desiccating packaging material. The kinetics and thermodynamics of water absorption as well as the comparison of water absorption and mechanical behaviours of PP composites filled with RHA have been studied [64-66]. The results reported in these papers concern studies on the kinetics and thermodynamics of the absorption of water by PP composites filled with different loadings (1, 3, 5, 10 or 20 wt%) of different fillers (RRH, WRHA, BRHA or AR) when immersed in water.

The thermal stability, and thermal and oxidative degradation of virgin PP and filled PP composites must be considerred when assessing their suitability for different applications. For a number of applications, it is necessary to increase the thermal stability of virgin PP by introducing certain reinforcing additives. The introduction of reinforcing additives increases the tensile strength of the composite with respect to the polymer matrix but, depending on the nature of the filler, the thermal stability of the composite may either increase or decrease. Incorporation of rigid inorganic particles is a promising approach to simultaneous improvement of both stiffness and toughness of plastics. For example, PP composites used for insulation of electrical power distribution and transmission cables operating at temperatures higher than 70 °C and under high electric fields must be heat resistant because the excessive heat generated in the cable is one of the main causes of dielectric failure [67]. On the other hand, PP constitutes one of the main components of plastic waste from industrial and domestic refuse. In order to recycle plastic waste to obtain energy or low molecular hydrocarbons, other additives, for example solids containing Lewis and Brønsted acid centres have a catalytic effect, and are used to decrease the thermal stability of the composites [67]. Therefore, comprehension of the kinetics of its thermal degradation of PP is of great interest for both theory and practice. Incorporation of RHK and RHA fillers into PP affected some of the thermal properties of the composites [57]. Addition of the BRHA filler raised the thermal degradation temperature whilst maintaining the oxidative stability. The degradation temperature of the WRHA composites was found to be independent of the filler loading but the oxidative stability deteriorated at higher filler content. Differential scanning calorimetry studies indicated that both WRHA and BRHA fillers act as weak nucleating agents and increase the degree of crystallinity of PP by a small margin [3]. Incorporation of RHA fillers was found to influence the thermal expansion coefficient and some of the mechanical properties of the composites. A new thermogravimetric analysis technique has been described for the computation of filler content in PP/RHA composites [68, 69].

Kinetic studies of the oxidative thermal degradation of composite materials obtained by filling PP with dispersed fillers of rice husks and AR are presented in **Table 12.2.**

Table 12.2 Kinetic parameters of non-isothermal degradation of PP composites filled with different fillers at a loading of 20 wt%

Parameter	Sample				
	PP	PP/RRH	PP/BRHA	PP/WRHA	PP/AR
E, kJ/mol	66.7	71.5	63.5	108.7	112.7
A, 1/min	1.45×10^4	7.52×10^4	9.39×10^3	4.89×10^7	6.33×10^8
$\Delta S^{\neq}$, J/mol K	–214.9	–200.9	–218.4	–147.3	–125.1
$\Delta H^{\neq}$, kJ/mol	60.7	65.8	57.6	102.8	107.4
$\Delta G^{\neq}$, kJ/mol	216.1	205.0	212.2	207.8	187.2

The highest values of the activation energy E and frequency factor A in the Arrhenius equation were observed for the composites filled with WRHA and AR, which have the same chemical content and porous structure. The change of Gibbs free energy $\Delta G^{\neq}$ reflects the total energy content of the system and showed the possibility of the formation of the activated complex from the reagents. This energy is influenced by two thermodynamic parameters, the changes of enthalpy ($\Delta H^{\neq}$) and entropy ($\Delta S^{\neq}$) of activated complex formation. The change of activation enthalpy shows the energy differences between the activated complex and the reagents. If this difference is small, the formation of the activated complex is favoured because the potential energy barrier is low. The change of reaction entropy reflects how near the system is to its own thermodynamic equilibrium. Low activation entropy means that the material has just passed through some kind of physical or chemical ageing process, bringing it to a state near its own thermodynamic equilibrium. In this situation, the material shows little reactivity, increasing the time necessary to form the activated complex. On the other hand, when high activation entropy values are observed, the material is far from its own thermodynamic equilibrium. In this case, the reactivity is high and the system can react faster to produce the activated complex, which resulted in the short reaction times observed [67].

RHK and RHA are used as fillers in different polymer composites. Siriwardena and co-workers [22, 25] studied the mechanical and morphological properties and water absorption of WRHA-filled PE/ethylene-propylene-diene terpolymer thermoplastic elastomer composites. Incorporation of WRHA improved the tensile modulus and flexural modulus and decreased the tensile strength, elongation at break, tear strength, and toughness of these composites [70]. WRHA-filled composites and moulded composites exhibited lower water uptake compared to silica-filled composites. All vulcanised composites showed lower water uptake than the non-vulcanised composites. After an immersion period in water, the tensile properties of the non-vulcanised composites were almost unaffected, while the vulcanised composites exhibited an increase in their tensile properties.

Ismail and co-workers [23, 24, 71, 73] investigated the effect of compatibilising agents on the mechanical properties, water absorption and morphological properties of PP/natural rubber blends filled with WRHA. The incorporation of WRHA improved the tensile modulus but decreased the tensile strength, elongation at break and stress at yield. Water adsorption increased with filler loading. However, the presence of compatibilising agents affected the water resistance of the composites. Studies of this kind may make it possible to design the controlled syntheses of composites filled with natural ingredients, combining the properties of the polymeric phase (matrix) and fillers to obtain products with properties which are suitable for practical applications. In conclusion, it may be believed that RHA/polymer composites could, in the future, lead to the development of organic-inorganic hybrid materials with specialised properties.

12.7 Utilisation of the Products of Thermal Degradation of Rice Husk as an Adsorbent of Crude Oil or Diesel Fuel

According to data presented by the United States National Academy of Sciences, the annual amount of oil and oil products spilled into the seas and oceans from various sources is about 20 million tonnes [74]. The sources of contamination are most often spills of oil and oil products by accidents with tankers or underwater pipes, as well as offshore oil platforms and oil refineries. Another of the marine industry's primary environmental concerns is the overboard discharge of liquid wastes. The discharge of wastewater must meet the more stringent waste disposal regulations imposed by local, national and international authorities. Recently shipping companies have been severely fined for the release of oily wastewater into the environment. A major source of overboard discharge is bilge water, a wastewater that accumulates in the bilges, which are in the very bottom of a ship. Cleaning up the consequences of oil and bilge water spills is quite a difficult task and the effectiveness of the activities depends to great extent on the choice of the most suitable method in each particular situation [75-77].

The kinetic curves of adsorption of diesel fuel at 25 °C onto BRHA and WRHA respectively are shown in **Figure 12.3** [40].

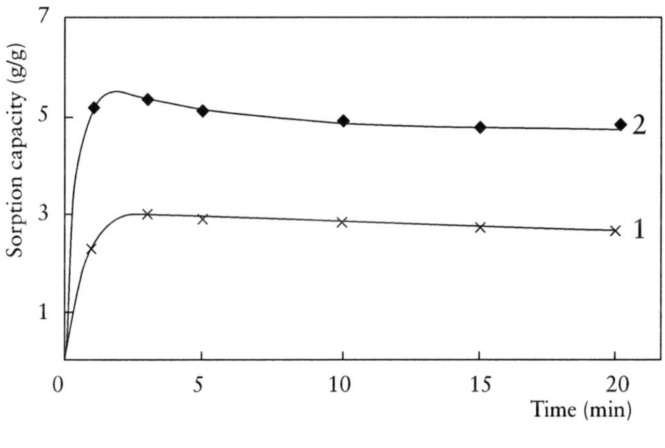

Figure 12.3 Kinetic curves of adsorption of diesel fuel onto WRHA (1) and BRHA (2). Reproduced with permission from S. Genieva, S. Turmanova, A. Dimitrov, P. Petkov and L. Vlaev, *Journal of Thermal Analysis and Calorimetry*, 2012, **110**, 111. ©2012, Springer [40]

As can be seen from **Figure 12.3**, the adsorption process occurred very quickly on both adsorbents to reach adsorption equilibrium after about 2-3 min. The observed rapid and high level of adsorption at the beginning of the process may be explained by the existence of the strong non-compensated forces on the pure surface of the adsorbent [78], which promotes the selective adsorption of some of the diesel fuel components.

The equilibrium sorption capacity of diesel fuel onto BRHA was estimated to be about 5 g/g, whilst that onto WRHA was only 2.8 g/g. It therefore follows that since the BRHA surface is non-polar and that of WRHA is polar, the curve representing the adsorption of diesel fuel onto BRHA was above that for WRHA, as the main components of diesel fuel have non-polar molecules [40].

At the same time the kinetic curve of water adsorption onto WRHA lies above that for BRHA which means that the BRHA surface is indeed hydrophobic [75] and, therefore, it would adsorb predominantly the non-polar hydrocarbon molecules contained in crude oil and diesel fuel. Whilst the thermodynamic equilibrium of water adsorption onto WRHA was reached in quite a short period (after about 2 min), for BRHA it was reached after about 10 min. Also, the maximal adsorption capacity of BRHA for water turned out to be nearly three times lower than that of WRHA. The lower water adsorption capacity of BRHA was the reason for preferring it as an adsorbent for collecting spills of oil and oil products. Our adsorption experiments [75] carried out with water-oil emulsion showed that the adsorption of water at these conditions is very slow and it took 10-12 days for the adsorbent to become saturated, grow heavier and precipitate spontaneously to the bottom, leaving the water surface clear of diesel fuel. This observation indicates that crude oil or oil products spilled in open basins can be cleaned up using BRHA as a collector.

Since the bilge waters contain different hydrocarbons different hydrocarbons we have studied the sorption capacity of some of them on BRHA and WRHA, respectively. In **Table 12.3** the maximal sorption capacity of BRHA and WRHA for these compounds at 25 °C are presented for comparison [40].

As can be seen from **Table 12.3** the sorption capacity of BRHA for water is less than that of WRHA, because BRHA is more hydrophobic. In all cases, the sorption capacity of BRHA for the hydrocarbons, diesel fuel and crude oil studied is higher than those of WRHA. This tendency may be explained by the non-polar character of the surface of BRHA. Moreover, there is a clear dependence between sorption capacity for hydrocarbons and ther boiling points. Based on the results presented, it may be concluded that BRHA is a more appropriate adsorbent for cleaning up bilge water containing crude oil and different hydrocarbons. Because BRHA saturated with these hydrocarbons has a high calorific value, it can be burnt in incinerators, industrial ovens or steam generators. In this way we abtain a double benefit - not only ecological but also economic.

Table 12.3 Sorption capacity of BRHA and WRHA at 298 K

Adsorbate	Boiling point, T (K)	Surface tension, σ_{298} (mN/m)	Density, d_{298} (g/cm^3)	BRHA, α (g/g)	WRHA α (g/g)
Water	373.1	72.81	0.997	0.82	1.23
Benzene	353.2	29.08	0.874	2.15	1.32
Toluene	383.6	29.01	0.865	2.78	1.82
n-Octane	399.1	21.62	0.703	3.38	2.86
m-Xylene	411.2	28.91	0.868	3.58	3.12
Diesel fuel	Large temperature interval	30.03	0.828	5.02	2.78
Crude oil	Large temperature interval	29.62	0.892	6.22	2.98

α – adsorption capacity

Adapted from S. Genieva, S. Turmanova, A. Dimitrov, P. Petkov and L. Vlaev, *Journal of Thermal Analysis and Calorimetry*, 2012, **110**, 111 [40].

12.8 Conclusions

Rice husk is an important by-product of the rice milling process and is a major waste product of the agricultural industry. It has now become very useful as a source of raw biomass material for manufacturing value-added silicone composite products. The method of controlled thermal treatment of RRH helps to solve the disposal and pollution problems of the rice milling industry and provides an excellent raw material for the preparation of composites for different applications. The present study was an attempt to establish the possibility of obtaining BRHA and WRHA *via* pyrolysis of RRH in a pilot plant fluidised bed reactor under different conditions. The obtaining of RHA/polymer composites can lead in the future to the production of organic-inorganic hybrid materials with specific properties, and BRHA may be used successfully as a cheap adsorbent for effective removal of crude oil and diesel fuel from water surfaces and for cleaning bilge water.

The abundance of waste RRH from the paddy milling industry, as well as its interesting complex behaviours, are a prerequisite for success in obtaining cheap and useful products and gives new alternatives for its applications as fillers of polymers and adsorbents for heavy metal ions, dyes, pigments and hydrocarbons. Hence, the production of value-added materials from RHK not only facilitates the utilisation of an abundantly available agro-waste, but also reduces environmental pollution and has an economic effect.

Acknowledgements

The research was realised by the financial support of the National Science Fund, Ministry of Education and Science, Bulgaria, Project 'Young scientists-2011', MU03/031, entitled 'Utilization of the products of pyrolysis of rice husk waste'.

References

1. S.E. Bailey, T.J. Olin, R.M. Bricka and D.D. Adrian, *Water Research*, 1999, **33**, 2469.

2. L. Sun and K. Gong, *Industrial and Engineering Chemistry Research*, 2001, **40**, 5861.

3. S. Chandrasekhar, K.G. Satyanarayana, P.N. Pramada, P. Raghavan and T.N. Gupta, *Journal of Materials Science*, 2003, **38**, 3159.

4. S. Babel and T.A. Kurniawan, *Journal of Hazardous Materials*, 2003, **B97**, 219.

5. T.H. Liou, *Carbon*, 2004, **42**, 785.

6. G. Crini, *Bioresource Technology*, 2006, **97**, 1061.

7. M. Rozainee in *Production of Amorphous Silica from Rice Husk in Fluidized Bed System*, Research Volume Number. 74526, Faculty of Chemical Engineering and Natural Resource Engineering, Universiti Teknologi Malaysia, Malaysia, 2007, p.118.

8. L.T. Vlaev, S.Ch. Turmanova and S.D. Genieva in *Pyrolysis: Types, Processes and Industrial Sources and Products*, Eds., W.S. Donahue and J.C. Brandt, Nova Science Publishers, New York, NY, USA, 2009, p.267.

9. P. Mishra, A. Chakraverty and H.D. Banerjee, *Journal of Materials Science*, 1986, **21**, 2129.

10. U. Kalapathy, A. Proctor and J. Shultz, *Bioresource Technology*, 2002, **85**, 285.

11. D. Kalderis, S. Bethanis, P. Paraskeva and E. Diamadopoulos, *Bioresource Technology*, 2008, **99**, 6809.

12. N. Sekar and T. Viruthagiri, *Chemical Engineering World*, 2005, **40**, 81.

13. E.S.M. Seo, M. Andreoli and R. Chiba, *Journal of Materials Processing Technology*, 2003, **141**, 351.

14. K.C. Nandi, D. Mukherjee, A.K. Biswas and H.N.A. Acharia, *Solar Energy Materials*, 1991, **22**, 161.

15. U.R. Lakshmi, V.C. Srivastava, I.D. Mall and D.H. Lataye, *Journal of Environmental Management*, 2009, **90**, 710.

16. F.W. Chang, W.Y. Kuo and H.C. Yang, *Applied Catalysis A: General*, 2005, **288**, 53.

17. D.S. Chaudhary, M.C. Jollands and F. Cser, *Advances in Polymer Technology*, 2004, **23**, 147.

18. S. Siriwardena, H. Ismail, U.S. Ishiaku and M.C.S. Perera, *Journal of Applied Polymer Science*, 2002, **85**, 438.

19. W. Arayapranee, N. Naranong and G.L.N. Rempel, *Journal of Applied Polymer Science*, 2005, **98**, 34.

20. S. Asavapisit and N. Ruengrit, *Cement and Concrete Composites*, 2005, **27**, 782.

21. S. Panthapulakkal, M. Sain and S. Law, *Polymer International*, 2005, **54**, 137.

22. S. Siriwardena, H. Ismail and U.S. Ishiaku, *Journal of Reinforced Plastics and Composites*, 2003, **22**, 1645.

23. H. Ismail, H.B. Hong, C.Y. Ping and H.P.S. Khalil, *Journal of Thermoplastic Composite Materials*, 2003, **16**, 121.

24. H. Ismail, Z. Mohamad and A.A. Bakar, *Polymer-Plastics Technology and Engineering*, 2003, **42**, 81.

25. S. Siriwardena, H. Ismail and U.S. Ishiaku, *Polymer-Plastics Technology and Engineering*, 2002, **41**, 419.

26. J. Wongjunda and P. Saueprasearsit, *Environmental Research Journal*, 2010, **4**, 244.

27. N. Gupta, A.K. Kushwaha and M.C. Chattopadhyaya, *Journal of the Taiwan Institute of Chemical Engineers*, 2012, **43**, 125.

28. C-Y. Chen, J-C. Chang and A-H. Chen, *Journal of Hazardous Materials*, 2011, **185**, 430.

29. D. Deschamps, H. Caruel, M-E. Borredon, C. Bonnin and C. Vignoles, *Environmental Science and Technology*, 2003, **37**, 1013.

30. D.H. Lataye, I.M. Mishra and I.D. Mall, *Chemical Engineering Journal*, 2009, **147**, 139.

31. P.M. Stefany, D. Garcia, J. Lopez and A. Jimenez, *Journal of Thermal Analysis and Calorimetry*, 2005, **81**, 315.

32. F. Yao, Q. Wu, Y. Lei and Y. Xu, *Industrial Crops and Products*, 2008, **28**, 63.

33. V.M.H. Govindarao, *Journal of Scientific and Industrial Research*, 1980, **39**, 495.

34. A.A.M. Daifullah, B.S. Girgis and H.M.H. Gad, *Materials Letters*, 2003, **57**, 1723.

35. S. Eiamsaard, Y. Kaewkohkiat, W. Lelaphatikul, C. Thianpong and P. Promvonge, *Communications in Heat and Mass Transfer*, 2008, **35**, 139.

36. I.M. Yusof, N.A. Farid, Z.A. Zaina and M. Azman, *Journal of Applied Sciences*, 2008, **8**, 622.

37. S. Huang, S. Jing, J. Wang, Z. Wang and Y. Jin, *Power Technology*, 2001, **117**, 232.

38. M. Fang, L. Yang, G. Chen, Z. Shi, Z. Luo and K. Cen, *Fuel Processing Technology*, 2004, **85**, 1273.

39. M. Rozainee, S.P. Ngo, A.A. Salema, K.G. Tan, M. Ariffin and Z.N. Zainura, *Bioresource Technology*, 2008, **99**, 703.

40. S. Genieva, S. Turmanova, A. Dimitrov, P. Petkov and L. Vlaev, *Journal of Thermal Analysis and Calorimetry*, 2012, **110**, 111.

41. S.Ch. Turmanova, A.S. Dimitrova and L.T. Vlaev, *Oxidation Communications*, 2008, **31**, 465.

42. C. Real, M.D. Alcala and J.M. Criado, *Journal of the American Ceramic Society*, 1996, **79**, 2012.

43. B.D. Park, S.G. Wi, K.H.A. Lee, P. Singh, T.H. Yoon and Y.S. Kim, *Biomass and Bioenergy*, 2003, **25**, 319.

44. L. Vlaev, I. Markovska and L. Lyubchev, *Thermochimica Acta*, 2003, **406**, 1.

45. S.D. Genieva, S.Ch. Turmanova, A.S. Dimitrova and L.T. Vlaev, *Journal of Thermal Analisys and Calorimetry*, 2008, **93**, 387.

46. M. Patel, A. Karera and P. Prasannak, *Journal of Materials Science*, 1987, **22**, 2457.

47. S.J. Gregg and K.S.W. Sing in *Adsorption, Surface Area and Porosity*, 2nd Edition, Academic Press, London, UK, 1982, p.252.

48. P. Tanev and L. Vlaev, *Journal of Colloid and Interface Science*, 1993, **160**, 110.

49. J.E. Moreland in *Proceedings of the 34th Annual Conference of the Society of the Plastics Industry, Reinforced Plastics/Composites Institute*, Society of the Plastics Industry, New York, NY, USA, 1972, Section 14-A.

50. X. Li, L.G. Tabil and S. Panigrahi, *Journal of Polymers and the Environment*, 2007, **15**, 25.

51. D.N. Saheb and J.P. Jog, *Advances in Polymer Technology*, 1999, **18**, 4, 351.

52. D.S. Chaudhary, M.C. Jollands and F. Cser, *Silicon Chemistry*, 2002, **1**, 281.

53. D.S. Chaudhary, M.C. Jollands and F. Cser, *Advances in Polymer Technology*, 2004, **23**, 147.

54. H.S. Yang, H.J. Kim, J. Son, H.J. Park, B.J. Lee and T.S. Hwang, *Composite Structures*, 2004, **63**, 305.

55. H.S. Yang, H.J. Kim, H.J. Park, B.J. Lee and T.S. Hwang, *Composite Structures*, 2007, **77**, 45.

56. H.S. Yang, M.P. Wolcott, H.S. Kim, S. Kim and H.J. Kim, *Polymer Testing*, 2006, **25**, 668.

57. J. Prachayawarakorn and N. Yeambunying, *Songklanakarin Journal of Science and Technology*, 2005, **27**, 343.

58. M.Y.A. Fuad, M, Jamaludin, Z.A.M. Ishak and A.K.M. Omar, *International Journal of Polymeric Materials*, 1993, **19**, 75.

59. M.Y.A. Fuad, M.J. Zaini, Z.A.M. Ishak and A.K.M. Omar, *Journal of Elastomers and Plastics*, 1994, **26**, 252.

60. M.Y.A. Fuad, Z. Ismail, Z.A.M. Ishak and A.K.M. Omar, *European Polymer Journal*, 1995, **31**, 885.

61. M.Y.A. Fuad, Z. Ismail, M.S. Mansor, Z.A.M. Ishak and A.K.M. Omar, *Polymer Journal*, 1995, **27**, 1002.

62. M. Razavi-Nouri, F.J. Dogouri, A. Oromiehie and A.E. Langroudi, *Iranian Polymer Journal*, 2006, **15**, 757.

63. S. Mishra and J. Verma, *Journal of Applied Polymer Science*, 2006, **101**, 2530.

64. L. Vlaev, S. Turmanova and A. Dimitrova, *Journal of Polymer Research*, 2009, **16**, 151.

65. S. Turmanova, A. Dimitrova and L. Vlaev, *Polymer-Plastics Technology and Engineering*, 2008, **47**, 809.

66. M. Jacob, K.T. Varughesee and S. Thomas, *Biomacromolecules*, 2005, **6**, 2969.

67. S.Ch. Turmanova, S.D. Genieva, A.S. Dimitrova and L.T. Vlaev, *Express Polymer Letters*, 2008, **2**, 133.

68. M.Y.A. Fuad, M.J. Zaini, M.J. Jamaludin, Z.A.M. Ishak and A.K.M. M. Omar, *Journal of Applied Polymer Science*, 1994, **51**, 1875.

69. P. Carniti and A. Gervasini, *Thermochimica Acta*, 2001, **379**, 51.

70. S. Siriwardena, H. Ismail and U.S. Ishiaku, *Polymer-Plastics Technology and Engineering*, 2001, **40**, 519.

71. H. Ismail and F.L. Chung, *International Journal of Polymeric Materials*, 1999, **43**, 301.

72. H. Ismail, L. Mega and H.P.S. Khalil, *Polymer International*, 2001, **50**, 606.

73. H. Ismail, J.M. Nizam and H.P.S. Khalil, *Polymer Testing*, 2001, **20**, 125.

74. I.A. Murashev and G.N. Semanov in *Methods of Testing and Estimation of Ability of Some Materials to Sorb Petroleum and Petroleum Products*, Trudiy CNIIMFA, USSR, 1978, **237**, 43. [In Russian].

75. L. Vlaev, P. Petkov, A. Dimitrov and S. Genieva, *Journal of the Taiwan Institute of Chemical Engeneers*, 2011, **42**, 957.

76. M-M. Li, H-C. Pan, S-L. Huang and M. Scholz, *Chemical Engineering and Technology*, 2013, **36** (4), 673.

77. A. Dimitrov, S. Genieva, P. Petkov and L. Vlaev, *Water, Air, and Soil Pollution*, 2012, **223**, 5087.

78. A.A. Al-Majed, A.R. Adebayo and M.E. Hossain, *Journal of Environmental Management*, 2012, **113**, 213.

13 The Potential of Lignin in Biocomposites

Fei-ling Pua and S.M. Sapuan

13.1 Introduction

The term 'biomass' includes all of the living matter on Earth. Biomass resources which have the potential to be renewable resources for biocomposite production will not compete directly with food crops for either livestock or man, because they include, for example, wood, sawdust, wood waste, herbaceous plants, agricultural waste, waste paper and sludge from pulp and paper mills. Besides, biomass which is abundant in large quantities and renewable is suitable to be used as low-cost sustainable and ethical renewable raw materials, and can be used as a renewable replacement for fuels and petrochemicals. According to Goyal and co-workers [1], the utilisation of biomass as raw material for conversion into any useful feedstock depends on its physical properties and chemical composition. The main constituents of biomass are cellulose, hemicellulose and lignin. **Table 13.1** shows different types of biomass resources and examples of these.

Table 13.1 Types of biomass resources

Types of biomass resources	Examples
Natural wastes	Agricultural waste, crop residues, waste wood from mills (sawmill, pulp and paper mill waste) and urban organic waste.
Forest products	Wood, logging residues, trees, shrubs, wood residues like dead branches and hollow logs, sawdust and bark.
Energy crops	Short rotation woody crops, herbaceous plants, grasses, starch crops, sugar crops and oilseed crops.

Adapted from H.B. Goyal, D. Seal and R.C. Saxena, *Renewable and Sustainable Energy Reviews*, 2008, **12**, 504 [1] and J. Chumpoo and P. Prasassarakich, *Energy and Fuels*, 2010, **24**, 2071 [6].

Lignin is traditionally regarded as a waste material in pulp and paper mills [4, 5]. It is a by-product of the pulping industry and has been used as a fuel in boilers in the pulping process in pulp and paper mills. Studies have shown that only about 1-2% of lignin, which is isolated from pulping liquors, has been used for the production of specialty products [7, 8]. Around 5 million metric tons of lignin are produced by industry every year. According to the published literature, some of the industrial lignin material can be used for the production of steam and energy as a heating fuel [9-12]. Commercial pulping methods such as the sulfite and kraft processes can produce high quality pulps. However, the non-cellulose fractions including hemicelluloses and lignin have, until recently only been burned to generate energy. The conversion of lignin into high-value chemical feedstocks such as carbon fibre, phenols and adhesives has been widely studied in the past few decades [13].

Lignin is a renewable resource which consists of aromatic and cyclohexyl compounds. It can be converted into different valuable substances which can be used for applications including as a polymer filler, to prepare artificial vanillin and vanillic acid, compatibilisers for polymer blends, activated carbon adsorbents for wastewater treatment, ion-exchange agents and dispersing agents. It can be used to produce polyols due to its hydroxyl content [14, 15].

Utilisation of waste lignin from the pulp and paper industry has gained attention recently as it is an abundant and traditional discarded waste from these industries in every country. The utilisation of lignin could have a great impact on the green and renewable energy industry and also the forest product industry. Due to the lack of effective conversion and modification methods, until now most of the lignin has been used largely as a low-grade fuel by the pulp and paper industry [16]. It has been reported that kraft pulping contributes approximately 80% of the world's chemical pulp production. Black liquor is a by-product of all types of pulping processes and contains around 30-34% of lignin. Black liquor is burned to provide or generate energy for mill operations, and white liquor, the strongly alkaline solution used in the first stage of the kraft process, can be recovered from it [17].

The properties and composition of lignin differ from plant to plant, depending on the soil conditions age of the plant and the growing environment. The composition of lignin also differs depending on the industrial pulping processes from which the lignin was obtained. The common pulping methods include kraft pulping, soda pulping and sulfite pulping. Until now, lignin has not been utilised significantly. However we can find some limited application of lignin in certain areas, for example as polymer additives, surfactants, ultraviolet (UV) stabilisers and base materials for the production of chemicals, dyes, fertilisers or adhesives [18, 19]. There are a few studies on the conversion of this waste into useful feedstocks. Some preliminary studies have reported on the utilisation of lignin as a filler or reinforcing phase in polymeric systems [7, 20, 21].

Lignin is an amorphous, polydisperse polymer which is biosynthesised by free radical polymerisation of sinapyl, coniferyl, coumaryl and coniferyl alcohols, which contain phenol, guaiacol (2-methoxy phenol) and syringol (2,6-dimethoxy phenol) and aromatic units, respectively. These are incorporated into lignin in the form of phenylpronoids: H-lignin (*p*-hydroxyphenyl), G-lignin (guaiacyl) or S-lignin (syringyl) [8]. The methoxylated phenylpropane unit consists of a number of functional groups such as phenolic, hydroxyl, carboxyl, benzyl alcohol, methoxyl, and aldehyde groups. In natural lignin, the positions on the aryl ring para- and ortho- to the hydroxyl groups are usually occupied by alkoxy or alkyl substituents [1, 3]. Lignins have different functional groups depending on the source of origin and processing [22]. Different functional groups in lignin will affect its effectiveness as a filler in polymer composites. The different functional groups in lignin will determine the solubility parameter, thermal stability, and compatibility with the polymer matrix.

13.1.1 Biocomposites

A composite material is made by combining two or more different materials artificially or mechanically to give a unique combination of properties [23, 24, 28]. Sapuan and co-workers [29] have shown that a composite material has a chemically distinct phase distributed within a continuous phase. The matrix is the continuous phase, normally a thermoplastic or thermosetting polymer, but could be a ceramic or metal, while the distributed phase is called the reinforcement. This reinforcement can be in various forms i.e., particles, short fibres, continuous fibres or laminate and the reinforcement materials are normally glass, carbon and aramid fibres. Nowadays, natural fibres play a major role as potential replacements for these. Today, the term 'composites' refers to the highly engineered combinations of polymer matrix and reinforcing materials such as natural fibres. The purpose is to achieve properties which the individual components themselves do not possess. The filler (also known as the binder) is the component that is bound by the matrix phase [25].

Fibre and particle composites are the two major classes of composites. A fibre composite is one in which the reinforcement or the filler is in form of fibres. The fibres used may be either continuous or discontinuous in form, short fibres, and chopped fibres and so on [26]. Fibres are defined as a class of materials that are in continuous filaments or are in discrete elongated pieces. On the other hand, composites in which particles of various shapes and sizes are distributed randomly within the matrix are known as particle composites [27].

Composites have been utilised for a long time to solve technological problems. Composite materials (otherwise known as engineering materials) are manufactured and designed for different applications such as sporting goods, aerospace parts, automotive components, consumer goods, or applied in the oil and marine industries. Of all materials, composite materials have the potential to replace widely used steel and aluminum because many show better performance..

According to Mazumdar [30], the reinforcing fibre provides the fibre composite with strength and stiffness, while the matrix gives rigidity and environmental protection. Callister [31] has claimed that the matrix provides two important functions which are to hold the reinforcement phase in place and to deform and distribute the stress to the reinforcement under applied loads. Ray and Rout [32] have reported that the plant fibre itself is a composite. Plant fibres are natural composites in which rigid, crystalline cellulose microfibrils reinforce an amorphous lignin and hemicellulose matrix [33]. Nowadays, plant fibres are used in order to replace synthetic fibres to produce various type of composites material that can be used in many applications.

Replacing synthetic fibre-reinforced synthetic polymer composites with biocomposites is helping to solve the growing environmental problems caused by conventional plastics, which are non-biodegredable, and disposing of these by burning generates toxic by-products. A biocomposite can be defined as a multiphase material formed by a matrix and reinforcement of any natural plants or fibres. Recently, the market for biocomposites has been increasing significantly due to the increase in oil price and the need to protect the environment. Biocomposites are environmentally friendly and are sometimes termed 'green composites'. They are biodegradable and can be a partial solution to many global environmental problems [34]. Natural fibres can serves as fillers or reinforcements and examples include fibres from various plants such as cotton, jute, flax, kenaf, coir (coconut husk), hemp, bamboo, rice husk, sugar palm, wood, oil palm, banana pseudostem, pineapple leaf, abaca, henequen, sugarcane bagasse or sisal, from animal products such as shells, skins and feathers, or fibres from recycled material such as wood or food crops. The fibres could also be arranged in a variety of forms (unidirectional, bidirectional, multidirectional, randomly distributed, short or continuous fibres) or used in the form of powders. Meanwhile, the choice of matrix can be thermoplastics and thermosetting polymers. Biopolymer or bioplastic matrices are derived from renewable resources such as vegetable oils, starches or cellulose. A wide number of combinations of fibres, fibre arrangements, and manufacturing processes are now available to produce the best materials which can be applied in our daily life. Therefore, the selection of composite materials can be very complicated.

Fibre-reinforced polymers are the main biocomposites. Polymer resin and fibres are the two main components usually used to form these. Fibre-reinforced polymer resins combine the high physical and mechanical performance of the fibres and the appearance, bonding and physical properties of the matrix polymers. At present, most of the matrix polymers are synthetic polymers, such as polystyrene (PS), polyethylene (PE), polyvinyl chloride and polypropylene (PP), which are mostly derived from fossil fuels. Examples of synthetic fibres are E-glass, S-glass, aramid, nylon, rayon and boron. The polymer matrix, in a composite, acts as an adhesive agent to hold the fibres in place and the fibres provide strength and stiffness to the structure [33].

13.1.2 Types of Lignin

According to Doherty and co-authors [35], lignocellulose materials are natural composites which are composed of three main components: cellulose, hemicellulose and lignin.

According to Ivan [36] and El Mansouri and Salvado [2], all types of lignin contain similar functional groups. Softwood lignin is built of guaicyl propane units which include a methoxyl group bonded to the third carbon atom of the aromatic ring. Meanwhile, hardwood lignin is built of guaicyl propane units and syringyl propane units. Syringyl propane units have methoxyl groups which are bonded to the third and the fifth carbon atom of the aromatic ring. The main functional groups in natural lignin include methoxyl, phenolic hydroxyl, primary and secondary aliphatic hydroxyl, ketone and aldehyde groups. The proportion of those functional groups depends on the method used to isolate the lignin and the plant species from which it came. Functional group analysis can be used to determine the structure of lignin.

In term of industry specifics, lignin can be categorised into two major categories. The first category is conventional or sulfur-containing lignin, including kraft and sulfonated lignin, which have been available and known for many years. These types of lignins are mainly obtained from softwoods. Another category is non-sulfur lignin which is obtained from other pulping processes, such as soda lignin, organosolv lignin, steam explosion lignin, hydrolysis lignin (mainly from biofuel production), and lignin obtained by oxygen delignification. Most of this lignin is not yet commercially exploited. Most of the lignin extracted from lignocellulosic materials in the pulp and paper industry is burned to generate energy and recover chemicals. Nevertheless, a small portion is used to produce automotive brakes, biodispersants, wood panel products, phenolic resins, polyurethane foams, epoxy resins for printed circuit boards, and surfactants [2, 7].

Methoxyl groups ($-OCH_3$) are found in lignin from all plants. The number of methoxyl groups depends on the isolation method and the species of plant. The methoxyl group contents of softwood and hardwood are different. According to a previous study on five different technical lignins [2], the methyl and methylene groups were present as shown by absorbance at bands 3000-2800 cm^{-1}. The type of spectrometer used was Fourier transform infrared (FTIR). The intensity of the absorbance bands was substantially higher for soda/anthraquinone lignin and organosolv lignin. However, the absorbance bands were relatively lower in intensity for kraft lignin and sulfonated lignins. There are primary aliphatic hydroxyl groups, secondary aliphatic and phenolic hydroxyl groups present in lignin. According to El Mansouri and Salvado [2], kraft lignins and soda/anthraquinone lignins contain the highest phenolic hydroxyl contents followed by organosolv lignin. Meanwhile, kraft lignin has the highest aliphatic hydroxyl content but the content was relatively low for other samples. Kraft and soda/anthraquinone lignins have higher contents of carboxylic groups compared with the other lignins. The report indicates that these two types of lignins (kraft and soda/anthraquinone lignins) were highly degraded during the pulping process. In addition, strongly modified or oxidised lignins can also contain other types of carbonyl-containing groups. Wang and co-workers [37], in their FTIR analysis of lignin microstructure, showed that lignin contains various types of carbonyl groups and aromatic compounds, such as syringyl ring with C-O stretching (1329 cm^{-1}), guaicyl ring with C-O stretching (1270 cm^{-1}), C-C, C-O, C=O stretching (1220 cm^{-1}), and aromatic C-H deformation (1031 cm^{-1}).

13.1.2.1 Kraft Lignin

According to Norgren and Edlund [38], the natural lignin structure will change more when wood is treated chemically during kraft pulping. During the pulping process, the aryl-ether bonds in the lignin macromolecule are cleaved, the molecular weight decreases and the number of phenolic groups per mass unit increases due to the conditions in the digester. This makes the kraft lignin more soluble than native wood lignin. Besides, the solubility of kraft lignin in water is influenced by pH, ionic strength and reaction temperature.

According to Vishtal and Kraslawski [12], kraft lignin is the lignin produced from the sulfate process (also known as kraft pulping). This process is used for about 85% of total world lignin production. Approximately 90-95% of the lignin contained in the wood is dissolved in an aqueous solution of sodium sulfide and sodium hydroxide during the cooking process. Most of the kraft lignin is used as low grade burning fuel. In the current pulp and paper mills, incineration or direct burning of black liquor (which contains lignin) produces energy and part of the extracted dissolved lignin is use for further conversion into valuable products such as adsorbents. About 100,000 tonnes of solid kraft lignin are produced yearly [2, 7].

Black liquor is obtained during end of the kraft pulping process. It consists 30-40%

of lignin and other inorganic compounds like extractives. Lignin is extracted from the black liquor which collects during the pulping process. Kraft lignin has high contents of inorganic or mineral matter which usually range from 6 to 15 wt% on a dry-and-ash-free basis. Lignin is well-known as a good feedstock for the production of low cost porous carbon materials [39, 40]. There are some studies in which the lignin is used as the precursor to prepare porous activated carbon adsorbents. Kraft lignin contains different features from other types of lignin. It contains an increased amount of phenolic hydroxyl groups due to extensive cleavage of β-aryl bonds during the cooking process. Some biphenyl and other condensed structures are also formed in the lignin structure. Kraft lignin has a high content of ash, however the ash can be removed *via* washing or treatment with dilute acid solution [2, 12].

13.1.2.2 Soda Lignin

Soda lignin is the major component of the black liquor from the soda pulping process. The soda pulping process is a chemical process to produce wood pulp with sodium hydroxide as the cooking chemical. This process gives pulp with a lower tear strength than other chemical pulping process such as kraft pulping. From the environmental and biomass resource points of view, the soda lignin generated in the pulping process has to be recovered to avoid pollution. Currently, most of the recovered soda lignin is burned to generate energy. Only 1% of the lignin is used for producing higher-value products. Therefore, it is important to explore applications for this abundant biomass. However, the main obstacle for efficient utilisation of lignin is its poor reactivity due to the condensed structure of lignin.

According to Vishtal and Kraslawaski [12], soda lignin is produced from the soda or soda-anthraquinone pulping processes. The difference between kraft and soda pulping is the sulfur-free medium of the cooking liquor in the latter process. Extraction or separation of lignin is one alternative to reduce or solve the environmental problems and fully utilise the waste from pulp and paper mills. Soda lignin may originate from many different plant sources and process variations. Soda lignins all have a low molecular weight, high phenolic hydroxyl content and relatively low (but variable) glass transition temperature [7]. Soda lignin from non-woody plants contains more *p*-hydroxyl units and carboxyl groups. Soda lignins contain high silicate and nitrogen contents and they are suitable for polymer synthesis, e.g., phenolic resin synthesis [12, 41].

13.1.2.3 Other Lignins

There are other pulping processes besides kraft and soda pulping. The organosolv pulping process is a technique which uses an organic solvent to solubilise the lignin and hemicellulose of the fibres. It is an environmental friendly alternative to kraft pulping because the organic solvent can be easily recovered by distillation. According to Vishtal and Kraslawaski [12], the organic solvents which are used in the organosolv pulping process are acetic acid, formic acid, ethanol and organic peroxy acids. Lignin is separated from organosolv pulping process *via* solubilisation (i.e., lignin is recovered from the solvent). Solubilisation makes it possible to obtain less modified lignin materials. Thus, organosolv lignin can achieve higher homogeneity than other lignins. The major properties of organosolv lignin are low molecular weight and high chemical purity. According to Lora and Glasser [7], organosolv lignins are hydrophobic and have low solubility in water. However, they contain several reactive side chains which are useful for further chemical reactions.

There are more technical lignins or other types of lignin in the market. An example is sulfonated lignin, which is a by-product of the sulfite pulping of wood. During the delignification process, lignin is sulfonated, degraded and solubilised.

'Ionic liquid lignins' is a new development whereby the lignin is extracted from wood using ionic liquids. According to Binder and co-workers [42], the lignin obtained *via* fractionation with ionic liquids is not yet available on any industrial scale. However, the lignin obtained from the ionic liquid has some characteristics which are similar to those of organosolv lignin and so can be used for the same applications.

13.2 Applications of Lignin in Polymer Composites

13.2.1 Lignin/Epoxy

Lignin has found many potential applications in polymer composites. Doherty and co-workers [35] have reviewed the potential value-added applications of lignin as components of polymer composites. The review discusses epoxy/lignin composites, starch/lignin blends, protein/lignin blends, polyolefin/lignin blends, synthetic rubber/lignin blends and other lignin/polymer blends or polymer composites. Blending of two or more types of materials will provides better material properties to achieve specific goals.

There are two common applications of lignin which are as a filler for composites or as a constituent of polymer blends, and as a precursor to obtain chemical feedstocks for the synthesis of polymers such as polyurethanes and epoxy [3]. One of the main examples of the lignin/polymer blend is lignin/epoxy resin blends. According to previous studies [3, 43, 44], several co-monomers and curing reagents have been used to obtain lignin-derived epoxy resins. Simionescu and co-workers [3] reported that the lignin/epoxy and epoxy resins in binary mixtures, which contain different compositions of lignin and epoxy resin, show good compatibility after a crosslinking reaction at elevated temperature. It was reported that a lignin/epoxy composite with additives (dibutyl phthalate, titanium dioxide, talc and chalk) with good mechanical and electrical properties was obtained. Lignin/epoxy composite adhesive films can be used in the electronics industry.

Wood and co-authors [45] reported on the role of lignin in the fabrication of hemp/epoxy composites. The resulted composites were made from an epoxy resin reinforced with hemp fibre, with addition of kraft lignin. The incorporation of lignin improved the impact, tensile and flexural strength of composites. Lignin has shown the potential to be an alternative to other chemical treatments to improve the fibre-matrix adhesion in composite fabrication. Since lignin can work as a compatibiliser and impart reinforcement properties to the the composite, it will turn waste lignin from the paper mill into value-added products in the composite industry. Sun and co-workers [46] carried out a study on the curing kinetics and mechanism of a lignin-based epoxy/anhydride resin system. The cure rate and cure kinetics of the resin are two of the criteria which will affect the mechanical properties of the cured epoxy. Lignin-based polymers can been considered as another way in which to use this paper mill waste on a large industrial scale.

Sasaki and co-workers [47] investigated the production of different types of cured epoxy resins using the lignin from steam-exploded bamboo. The results showed that bamboo lignin could be successfully used as a resource material for epoxy resins. The lignin-based epoxy resins were compared with petroleum-based epoxy resins and it was found that the thermal decomposition properties if the lignin-based epoxy resins exceeded the dip solder resistance. Moreover, the lignin-based epoxy resin showed 70% of the flexural strength of the petroleum-based epoxy resin. This indicates that lignin has potential applications in the electrical and electronics industries.

Besides bamboo lignin, oil palm empty fruit bunch (EFB) lignin can also be used a as a curing agent for epoxy composites. Khalil and co-workers [48] reported on the preparation of an epoxy/lignin composite reinforced with EFB fibre. The epoxy which was cured with a fixed amount of lignin (25% lignin) was shown to be a better matrix material and succeeded in achieving the optimum value compared with other formulations (15, 20 and 30 wt% lignin contents). Based on the results of mechanical, morphological and thermal analyses, the EFB/lignin-epoxy composite material is potentially useful for applications in the automotive industry, and predominantly for vehicle interiors.

13.2.2 Lignin/Thermoplastic Composites

Biocomposites or biobased composite materials have gained attention over the past two decades. Thermoplastics have become one of the synthetic polymer choices for matrix selection. Alexy and co-workers [49] reported on the fabrication of lignin/thermoplastic composites. The effect of blending lignin with PE and PP on the physical properties was investigated. Lignin was used as filler in both PP and PE. The obtained results show that lignin acts as a good stabiliser, mainly in the PP blend. Lignin acts as an initiator of the degradation process under long-term heat stress. Meanwhile in blends with PE, lignin has no practical effect on the processing stability, but it can be the initiator of the degradation process after short-term exposure of the samples to UV light. The investigation clarifies the interaction between lignin and different types of thermoplastic matrix, mainly with regard to the degradation process.

A few previous studies had investigated the interaction between lignin and thermoplastics. For example, Morandim-Giannetti and co-workers [50] revealed the potential of Acetosolv lignin as an additive in PP/natural fibre composites. The thermal, mechanical and morphological properties of PP/coir composites were studied in the report. The effect of lignin in the presence and absence of maleic anhydride-grafted PP on the properties of composites was studied. The study indicated that incorporation of lignin in composite fabrication will improve the initial thermal degradation temperature and oxidation induction times. Lignin can be the potential antioxidant and also the barrier to thermal degradation of composite materials.

Pucciariello and co-workers [51] used lignin from straw to prepare lignin/thermoplastic blends with low density polyethylene (LDPE), linear LDPE, high density polyethylene and atactic PS. The results obtained showed that lignin can acts as a stabiliser against UV radiation for thermoplastic polymer blends. The study was aimed at finding new potential applications for straw lignin, due to the large abundance of this resource in the Baslicata region of Italy. The incorporation of lignin into thermoplastic material increases the modulus but decreases the tensile strength and the elongation break of the material. However, poor compatibility between lignin and the synthetic polymer can be overcome *via* improved mixing techniques or the addition of a compatibiliser or compatibilising agent.

In another study, a single screw extruder was used to compound and blend a thermoplastic material, polyethylene terephthalate (PET) with lignin. The thermal properties and the crystalline structure were studied and the morphological results showed a good dispersion of lignin particles in the thermoplastic matrix material. The study concluded that the presence of lignin particles in the composite or blend produced an enhancement of the crystallinity and crystal dimensions of the PET matrix material [52].

Hardwood kraft lignin was blended with synthetic polymers including polyethylene oxide (PEO), polyvinyl alcohol (PVA), PET and PP [53]. Intermolecular interaction in the blends of lignin with synthetic polymers was analysed by FTIR and thermal analysis. The study revealed that the lignin/PEO blend and lignin/PET blend were miscible, whereas the lignin/PP and lignin/PVA blends were immiscible. Infrared analysis revealed that the basis for the miscibility of lignin and thermoplastic blend was most probably due to the hydrogen bonding interactions between the hydroxyl hydrogens of lignin and the accepting sites of the thermoplastic. The investigation showed that strong hydrogen bonds are formed when lignin is blended with PEO, but no hydrogen bonds are formed in the PET/lignin blends.

According to previous study, lignin can be used to chemically modify a thermoplastic matrix before it is incorporated in to the blending process [54]. An esterified lignin was blended with LDPE with maleic anhydride-grafted LDPE as a compatibiliser. The purpose of esterification was to enhance the hydrophobicity of lignin. Thermogravimetic analysis revealed a better thermal stability for the compatibilised lignin/LDPE blends. The crystallinity of the blends increased slightly after the addition of the esterified lignin filler.

13.2.3 Lignin-based Biodegradable Composites

Biodegradable polymers or composites which can be easily degraded have recently started to attract the attention of both the public and researchers. The application of biodegradable polymers and renewable agro-wastes as packaging materials is one of the options available to solve the problem of solid waste from inert polymers. There are several polymers of synthetic aliphatic polyesters and natural polymers which are being used as biodegradable materials [55, 56]. Poly(lactic acid) (PLA), poly caprolactone, PVA, polybutylene succinate (PBS) and starch/polymer blends are the most promising examples of biodegradable matrix polymers used in mulch films [57]. These polymers have good potential as biodegradable matrices in environmental friendly composites [58]. PVA/starch blend plastics are some of the most popular of the biodegradable plastics. PVA is widely used in agricultural mulch film or biodegradable packaging [59, 60]. A lignin-based biodegradable composite from PBS was prepared and studied by Sahoo and co-workers [18]. The report revealed that the addition of 65 wt% of lignin material in to PBS improved the tensile and flexural properties of the composite. However, the study found that incorporation of 1 wt% of a compatibiliser such as methylene diphenyl diisocyanate into the lignin-based composite can enhanced the flexural strength, impact strength and tensile strength simultaneously. This indicated that lignin can play the role of a reinforcing filler in PBS to improve the mechanical and thermomechanical properties of the composite materials.

Fernandes and co-authors [61] studied the thermal and photochemical stability of PVA/modified lignin blends. A biodegradable composite film from a kraft lignin derivative blended with PVA was obtained by a solution casting method. The resulting composite film showed a greater thermal stability than the pure PVA film and the thermal stability was increased after UV radiation. The study reveals that lignin or lignin derivatives are potential resources for composite materials which need good thermal stability. Spiridon and co-workers [62] have reported on the preparation and characterisation of composite films containing adipic acid-modified starch microparticles as the filler and glycerol-plasticised starch as the matrix, which were reinforced by lignin. These were prepared *via* the casting method. The effect of lignin incorporation into the matrix was investigated. Two types of lignin were used in this study: lignin extracted from beech wood, and industrial kraft lignin isolated using the LignoBoost process. The thermal stability and surface water resistance of the composite were improved after the incorporation of lignin. The tensile strength was increased compared with the composite material without any lignin reinforcement. The composite containing lignin also showed improved hydrophobicity and a reduction of the water absorption in a high humidity atmosphere. The study proposed a simple, low-cost route to prepare a new biodegradable modified starch reinforced lignin composite material. The potentially biodegradable film can be used in the packaging industry or for agricultural mulching.

Li and co-workers [63] have studied the thermal and mechanical properties of biodegradable blends of poly(L-lactic acid) (PLLA) with lignin. FTIR analysis revealed the existence of an intermolecular interaction between the matrix, (PLLA) and lignin. The result of mechanical property tests showed that the maximum strength and the elongation break start to decrease when the lignin content is about 20 wt%, meanwhile the Young's modulus remained almost constant when the lignin content is about 20 wt%. The study showed that approximately 20 wt% of lignin content with PLLA is the optimum level. The blends of lignin with PLLA were thought to be promising low cost, environmentally friendly materials.

In addition, Graupner [64] revealed another potential use of lignin in biodegradable composites. This report showed that the adhesion between cotton fibre and PLA matrix can be improved by the presence of lignin. Moreover, incorporation of lignin increased the Young's modulus and tensile strength of the composite materials. Thus, lignin could be used as a natural adhesion promoter for lignin-free fibres such as cotton in a PLA matrix if a good tensile strength and stiffness is required for the product specification.

13.3 Conclusions

Lignin as an abundant natural resource and, traditionally, a waste material has properties which make it suitable for many applications. It can play a role as an additive or stabiliser in the composite fabrication process or can partly replace petroleum-based components. The presence of lignin in polymer plends or composites improves their thermal stability. The importance of making use of biomass and biowaste materials is likely to increase in this decade.

Acknowledgements

The authors wish to acknowledge the Ministry of Education, Malaysia and Universiti Putra Malaysia for providing financial support through the Research University Grant Scheme (project number: RUGS: UPM/700/2/1/RUGS-05-03-11-1443RU) for this research work.

References

1. H.B. Goyal, D. Seal and R.C. Saxena, *Renewable and Sustainable Energy Reviews*, 2008, **12**, 504.

2. N.E. El Mansouri and J. Salvado, *Industrial Crops and Products*, 2007, **26**, 116.

3. C.I. Simionescu, V. Rusan, M.M. Macoveanu, G. Cazacu, R. Lipsa, C. Vasile, A. Stoleiru and A. Ioanid, *Composite Science and Technology*, 1993, **48**, 317.

4. W. Thielemans, E. Can, S.S. Morye and R.P. Wool, *Journal of Applied Polymer Science*, 2002, **83**, 323.

5. D. Stewart, *Industrial Crops and Products*, 2008, **27**, 202.

6. J. Chumpoo and P. Prasassarakich, *Energy and Fuels*, 2010, **24**, 2071.

7. J.H. Lora and W.G. Glasser, *Journal of Polymers and Environment*, 2002, **19**, 39.

8. C.A. Mullen and A.A. Boateng, *Fuel Processing Technology*, 2010, **91**, 1446.

9. S. Rastogi and N. Dwivedi, *Plant Science*, 2008, **173**, 3, 264.

10. C. Mai, O. Milstein and A. Huttermann, *Journal of Biotechnology*, 2000, **79**, 2, 173.

11. D. Mohan, C.U. Pittman and P.H. Steele, *Journal of Colloid and Interface Science*, 2006, **297**, 2, 489.

12. A. Vishtal and A. Kraslawaski, *Bioresources*, 2011, **6**, 3, 3547.

13. H. Zhang, H. Deng, L. Lin, Y. Sun, C. Pan and S. Liu, *Bioresource Technology*, 2010, **101**, 2311.

14. J. Huang and K. Zhang, *Polymer*, 2002, **43**, 2287.

15. S. Sarkar and B. Adhikari, *European Polymer Journal*, 2001, **37**, 1391.

16. M. Tian, J. Wen, D. MacDonald, R.M. Asmussen and A. Chen, *Electrochemistry Communications*, 2010, **12**, 527.

17. E.A. Borges da Silva, M. Zabkova, J.D. Araujo, C.A. Cateto, M.F. Barieiro, M.N. Belgacem and A.E. Rodrigues, *Chemical Engineering Research and Design*, 2009, **87**, 1276.

18. S. Sahoo, M.O. Seydibeyoglu, A.K. Mohanty and M. Misra, *Biomass and Bioenergy*, 2011, **35**, 4230.

19. M.N.S. Kumar, A.K. Mohanty, I. Erickson and M. Misra, *Journal of Biobased Materials and Bioenergy*, 2009, **3**, 1.

20. G. Cazacy, M.C. Pascu, L. Profire, A.I. Kowarsk, M. Mihaes and C. Vasile, *Industrial Crops and Products*, 2004, **20**, 261.

21. H. Hatakeyama, A. Nakayachi and T. Hatakeyama, *Composites Part A: Applied Science and Manufacturing*, 2005, **36**, 698.

22. A.U. Buranov and G. Mazza, *Industrial Crops and Products*, 2008, **28**, 237.

23. S.M. Sapuan in *Concurrent Engineering for Composites*, UPM Press, Serdang, Malaysia, 2010, p.9.

24. S.M. Sapuan in *Bahan Komposit Berasaskan Polimer*, UPM Press, Serdang, Malaysia, 2000, p.1. [In Malay]

25. D.L.C. Deborah in *Composite Materials: Science and Applications*, Springer Verlag, London, UK and New York, NY, USA, 2010, p.1.

26. J.M. Berthelot in *Composite Materials: Mechanical Behaviour and Structural Analysis*, Springer Verlag, London, UK and New York, NY, USA, 1998, p.1.

27. M. Madhujit in *Mechanics of Composite Materials and Structures*, Universities Press, India, 2004, p.1.

28. M.R. Rowell in *Polymers and Other Advanced Materials: Emerging Technologies and Business Opportunities*, Plenum Press, New York, NY, USA, 1995, p.659.

29. S.M. Sapuan, A.R. Rahmat, Y.S. Ismail and M.Y. Hassan in *Glosari Bahan Komposit*, Pusat Penerbitan Universiti (UPENA) UiTM, Shah Alam, Malaysia, 2006, p.183. [In Malay]

30. S.K. Mazumdar in *Composites Manufacturing: Materials, Product, and Process Engineering*, CRC Press, Boca Raton, FL, USA, 2002, p.1.

31. W.D.J. Callister in *Material Science & Engineering*, John Wiley & Sons Inc, Oxford, UK, 2002, p.627.

32. D. Ray and J. Rout in *Natural Fibers, Biopolymers and Biocomposites*, CRC Press, Boca Raton, FL, USA, 2005, p.291.

33. A.K. Mohanty, M. Misra, L.T. Drzal, S.E. Selke, B.R. Harte and G. Hinrichsen in *Natural Fibers, Biopolymers and Biocomposites*, CRC Press, Boca Raton, FL, USA, 2005, p.1.

34. M.S. Huda, A.K. Mohanty, L.T. Drzal, E. Scut and M. Misra, *Bioresource Technology*, 2005, **101**, 8489.

35. W.O.S. Doherty, P. Mousavioun and C.M. Fellows, *Industrial Crops and Products*, 2011, **33**, 259.

36. B. Ivan *Characterization of Natural and Technical Lignins using FTIR Spectroscopy*, Luleå University of Technology, Sweden, 2007. [M.Sc. Thesis]

37. S. Wang, K. Wang, Q. Liu, Y. Gu, Z. Luo, K. Cen and T. Frasson, *Biotechnology Advances*, 2009, **27**, 562.

38. M. Norgren and H. Edlund, *Colloids and Surfaces A: Physiochemical and Engineering Aspects*, 2001, **194**, 239.

39. N.E. El Mansouri and J. Salvado, *Industrial Crops and Products*, 2006, **24**, 8.

40. V. Fierro, E.A.V. Torn, A. Celzard and D. Montan, *Journal of Hazardous Materials*, 2007, **149**, 126.

41. K. Wormeyer, T. Ingram, B. Saake, G. Brunner and I. Smirnova, *Bioresource Techno*logy, 2011, **102**, 8, 4157.

42. J.B. Binder, M.J. Gray, J.F. White, C.Z. Zhang and J.E. Holladay, *Biomass and Bioenergy*, 2009, **33**, 9, 1122.

43. Y. Ebata, inventor and assignee; Japanese Patent, JP 210 816, 2004.

44. S. Hirose, H. Hatakeyema and M. Funabashi, inventors and assignees; Japanese Patent, JP 2002 284 791, 2002.

45. B.M. Wood, S.R. Coles, S. Maggs, J. Meredith and K. Kirwan, *Computer Science and Technology*, 2011, **71**, 1804.

46. G. Sun, H. Sun, Y. Liu, B. Zhan, N. Zhun and K. Hu, *Polymer*, 2007, **48**, 330.

47. C. Sasaki, M. Wanaka, H. Takagi, S. Tamura, C. Asada and Y. Nakamura, *Industrial Crops and Products*, 2013, **43**, 757.

48. H.P.S.A. Khalil, M.M. Marliana, A.M. Issam and I.O. Bakare, *Materials and Design*, 2011, **32**, 2604.

49. P. Alexy, B. Kosikova and G. Podstranska, *Polymer*, 2000, **41**, 4901.

50. A.A. Morandim-Giannetti, J.A.M. Agnelli, B.Z. Lancas, R. Magnabosca, S.A. Casarin and S.H.P. Bettini, *Carbohydrate Polymers*, 2012, **87**, 2563.

51. R. Pucciariello, V. Villani, C. Bonini, M. D'Auria and T. Vetere, *Polymer*, 2004, **45**, 4159.

52. M. Canetti and F. Bertini, *Composites Science and Technology*, 2007, **67**, 3151.

53. J.F. Kadla and S. Kubo, *Composites Part A: Applied Science and Manufacturing*, 2004, **35**, 395.

54. R.R.N. Sailaja and M.V. Deepthi, *Materials and Design*, 2010, **31**, 4369.

55. R. Jayasekara, I. Harding, I. Bowater, G.B.Y. Christie and G.T. Lonergan, *Polymer Testing*, 2004, **23**, 17.

56. T. Ishigaki, Y. Kawagoshi, M. Ike and M. Fujite, *World Journal of Microbiology and Biotechnology*, 1999, **15**, 321.

57. P. Halley, R. Rutgers, S. Coombs, J. Ketterls, J. Gralton, G. Christie, M. Jenkins, H. Beh, K. Griffin, J. Ranjiny and G. Lonerga, *Starch*, 2001, **53**, 362.

58. A.P. Mathew, O. Kristiina and S. Mohini, *Journal of Applied Polymer Science*, 2004, **97**, 2014.

59. T. Iwanami and T. Uemura, Japan *Journal of Polymer Science and Technology*, 1993, **50**, 767.

60. C. Bastioli, V. Bellotti L.D. Giudice and G. Gilli, *Journal of Environmental Polymer Degradation*, 1993, **1**, 181.

61. D.M. Fernandes, A.A.W. Hechenleitner, A.E. Job, E. Radovanocic and E.A.G. Pineda, *Polymer Degradation and Study*, 2006, **91**, 192.

62. I. Spiridon, C-A. Teaca and R. Bodirlau, *Journal of Materials Science*, 2011, **46**, 3241.

63. J. Li, Y. He and Y. Inoue, *Polymer International*, 2003, **52**, 949.

64. N. Graupner, *Journal of Material Science*, 2008, **43**, 5222.

14 Biopolymer Nanocomposites Reinforced with Nanocrystalline Cellulose

Mehdi Jorfi, Nasim Amiralian, Mahesh V. Biyani and Pratheep K. Annamalai

14.1 Introduction

The economic consequences of depleting petroleum resources, the demands from customers for high performance lightweight materials and a growing environmental awareness have ignited the incentive to use renewable resources for making new materials. By exploiting the synergistic effect of different properties from nanoparticles and polymers, the properties of biobased materials can be enormously improved. The utilisation of cellulose nanoparticles has gained increasing attention from researchers and industrialists due to the possibilities of creating advanced functional nanocomposite materials for a variety of applications. The inherent biodegradability of both cellulose nanoparticles and biopolymers has an added advantage from an environmental perspective. This chapter is intended to review the recent research activities and discuss the processing methods, structure-property relationships and potential applications of the biopolymer nanocomposites reinforced by cellulose nanocrystals or nanofibres.

Fuel economy, demands for lightweight high performance materials and environmental concerns have compelled researchers to develop new materials from alternative or renewable resources, in order to secure sustainable development [1, 2]. Renewability of resources depends on the availability and life cycle of the raw materials [3]. In recent decades, polymers from renewable bioresources have increasingly gained the attention of researchers. This is attributed not only to their renewability and environmental friendliness, but also to their processability and the range of tailorable properties [4-8]. As a biopolymer reinforcement, cellulose is particularly attractive due to its abundance, renewability, cost effectiveness, biodegradability, potential for enhancing properties and its chemical functionalities. Research activities on reinforcing biobased polymers with cellulose nanoscale particles/fibres have been accelerated around the globe, for developing 'green materials' [9-14]. This chapter discusses the recent trends and important developments in this field.

14.2 Biopolymers

The polymers which are derived or extracted from most widely available biorenewable resources (plants, marine animals and microorganisms) are called 'biopolymers'. During the growth cycles of organisms, polymers are produced as biomass or by-products [15, 16]. They are regarded as environmentally friendly materials not only due to their source but also, in most cases, to their inherent biodegradability [1, 17, 18]. It should be noted that there is a significant proportion of polymers produced from biological resources that are not 'biodegradable' according to the internationally accepted standards [19]. **Figure 14.1** represents the four types of biopolymers based on their pathways and biodegradability [3, 15, 19, 20].

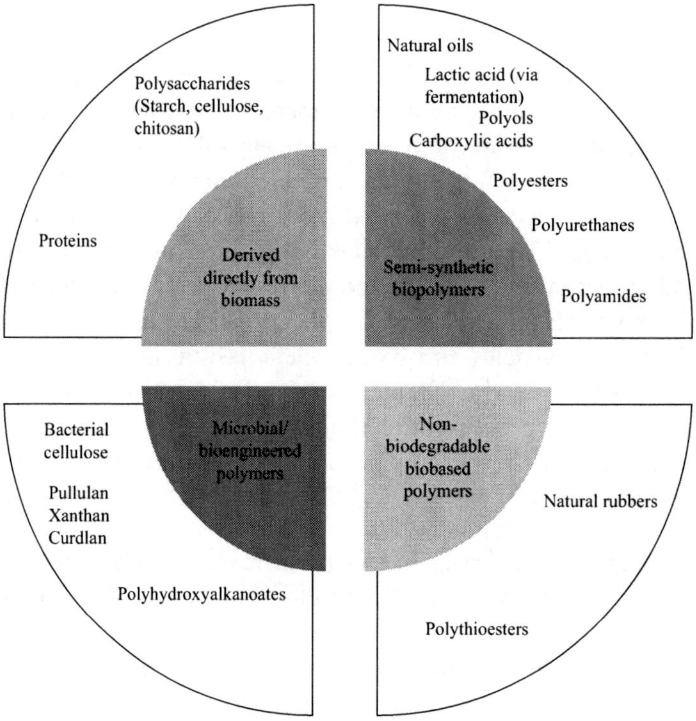

Figure 14.1 Polymers from biomass, a renewable resource. Adapted from P.K. Annamalai and R.P. Singh, in *Cellulose Fibers: Bio- and Nano-Polymer Composites*, Eds., S. Kalia, B.S. Kaith and I. Kaur, Springer-Verlag Berlin Heidelberg, Germany, 2011, p.519 [3]; R. Chandra and R. Rustgi, *Progress in Polymer Science*, 1998, **23**, 7, 1273 [15]; A. Steinbuchel, *Current Opinion in Biotechnology*, 2005, **16**, 6, 607 [19]; and D.L. Kaplan in *Biopolymers from Renewable Resources*, Springer-Verlag, Berlin, Germany 1998, p.1 [20]

The first type of polymers (such as polysaccharides, proteins and gums) are directly derived from the most abundant sources. The second type of polymers (polyesters, polyurethanes, polyamides and others) are synthesised by classical chemical synthesis methods using biobased monomers. These 'semi-synthetic biopolymers' have been of particular commercial interest due to their tailorability and ease of scaling up. The third type is typically generated by microorganisms (especially bacteria) as a by-product of their biochemical cycles. The composition and properties of these microbial polymers can be engineered by strictly varying the nutrient composition and applying biotechnological tools, especially using genetically modified organisms. These three types of polymers have been widely defined as 'biodegradable'. The fourth class consists of polymers that are still derived from biomass but widely demonstrated as 'non-degradable' ones. They include natural rubber (NR), microbial polythioesters and highly crosslinked gums. It should be mentioned that NR accounts for a large volume (68% in 2007) of the total commercial market for biopolymers [21].

All four classes of biopolymers have great potential in the growing commercial plastics market with the global production capacity of bioplastics at about 1,161 metric tonnes in 2011, and this is projected to increase up to 5,779 metric tonnes in 2016, as indicated by a recent market survey [22]. However, there appear to be significant obstacles in translating them into real applications due to their poor properties (mechanical properties, thermal stability, water absorption and barrier properties) in comparison to their synthetic counterparts. Thus, the current research activities are focused on improving the required properties and introducing new functional properties for improving their performance [8, 23]. To reinforce or modify the properties of these polymers, 'nanocomposite materials', in which at least one dimension of one component is in the nanometre range, are being developed. In the last decade, the interest in nanocomposites has been immensely increased, mainly due to the attainment of potentially unique property profiles as the result of clever mixing of polymer matrices (which are melt processable, have low density and other desirable properties) and nanoparticles (which contribute to the improvement in stiffness, thermal stability, higher surface area and interfacial volume). Incorporation of nanoparticles in the biobased polymers can enhance the physical, mechanical and thermal properties whilst retaining the inherent biodegradability. These biopolymer nanocomposites represent a class of versatile materials that can be employed in various applications ranging from tissue engineering, artificial bones, gene therapy, biosensors, sustained drug delivery and food packaging.

14.3 Cellulose Nanocrystals

During the biosynthesis of cellulose, which is chemically composed of poly(β-1,4-D-anhydroglucopyranose), highly crystalline nanoscale fibrils with outstanding mechanical properties are formed. They are bound by intra- and intermolecular hydrogen bonding and organised into elementary fibrils separated by non-crystalline regions (lignin, pectin, hemicellulose). By defibrillation and/or removing the other amorphous components, the highly crystalline nanofibrils can be recovered. They are typically derived from a variety of sources *via* acid hydrolysis, enzymatic hydrolysis, oxidation, mechanical treatment, ultrasonic treatment or combinations thereof. Depending on the method used and the sources, they can be obtained in a range of aspect ratios and morphologies. They are also categorised as cellulose nanocrystals (CNC), cellulose nanofibres (CNF) and micro-/nanofibrillated cellulose (MFC/NFC). **Figure 14.2** compares the different derivatisation methods, dimensions and morphologies of nanofibrils that can be obtained from them [24].

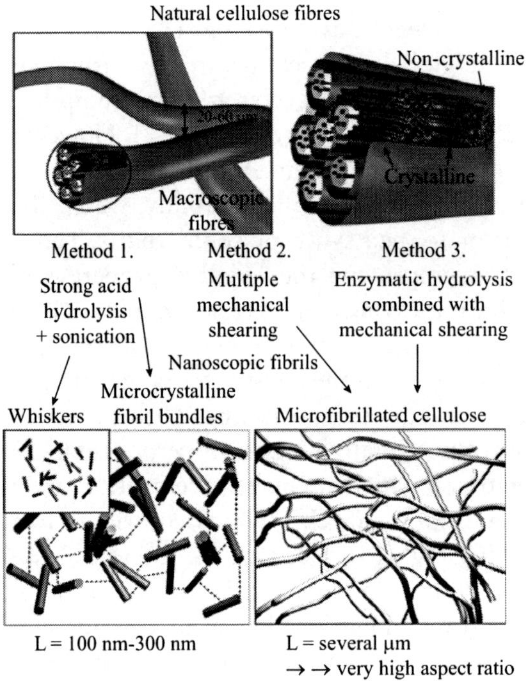

Figure 14.2 Schematic representation of derivatisation of macroscopic cellulosic fibres into nanoscale fibrils. L is the length and the two arrows represent higher aspect ratio. Reproduced with permission from M. Paakko, M. Ankerfors, H. Kosonen, A. Nykanen, S. Ahola, M. Osterberg, J. Ruokolainen, J. Laine, P.T. Larsson, O. Ikkala and T. Lindstrom, *Biomacromolecules*, 2007, 8, 6, 1934. ©2007, ACS [24]

Method 1 shows the conversion of cellulose macrofibres into highly crystalline and low aspect ratio fibrils (as aggregates) *via* aggressive acid hydrolysis and sonication [25-27]. Method 2 represents the disintegration or defibrillation of macrofibres into highly entangled and inherently connected fibrils and mechanically strong networks of MFC/NFC through the application of high shear forces [28-30]. Method 3 incorporates an additional enzymatic hydrolysis which produces a mixture of predominantly cellulose I fibrils (about 5 nm thickness) and fibril aggregates (about 10-20 nm thickness) [24]. Owing to their higher specific surface area and energy, the CNC exhibit a strong tendency to aggregate. Thus, their technological commercial potential may be limited because of the poor dispersion of nanocrystals (or the lack of control over it) in most of the organic media and polymer matrices [31, 32]. Therefore, achieving and understanding better dispersion of CNC in polymer matrices or other media is indeed important in this area of research. **Figure 14.3** represents the currently available pathways for the preparation of nanocomposites containing CNC.

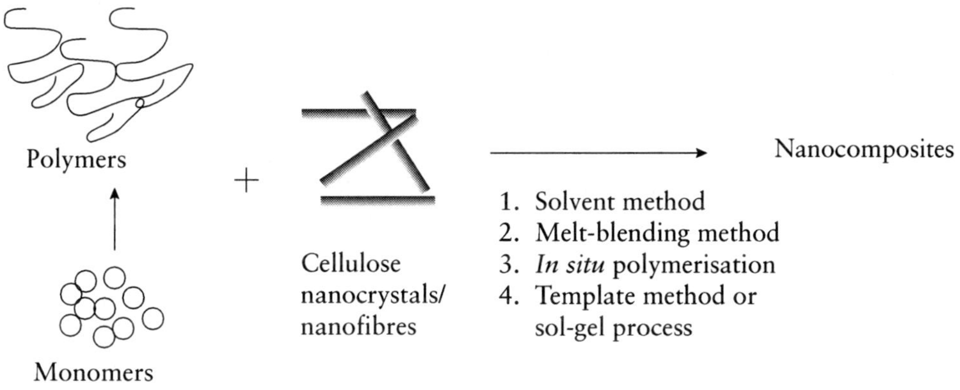

Figure 14.3 Schematic representation of processing pathways of CNC-based nanocomposites

The first one relies on a '*solvent method*' where CNC are first dispersed in a solvent in which the host polymer is also soluble or dispersible in latex form, followed by casting and drying (evaporation of solvent) [33-36]. The second method is a conventional '*melt blending*' method where CNC can be melt blended with any polymer whose softening or melting temperature is below 200 °C. To alleviate the poor dispersion, surfactants or compatibilisers are very often used, especially with most common non-polar or hydrophobic polymer matrices [37-39]. Sometimes, CNC are dispersed well in the monomers/precursors/prepolymers (with or without solvent) and subsequently polymerised/cured further to obtain the nanocomposites. This method is regarded as '*in situ polymerisation*' [40, 41]. The fourth method is based on a '*solvent exchange sol-gel process*' where a template percolating network of CNC is first formed and the polymer/monomer is then imbibed into it [31, 42, 43]. This method is known as the'*template approach*' or '*sol-gel method*' as the CNC template is formed as an organogel from a homogeneous dispersion of CNC by successive solvent exchange with the solvent in which cellulose nanocrystals are not dispersible. This versatile method becomes very important for achieving a high level of dispersion and transport properties where the percolating network of CNC plays a vital role.

14.4 Nanocomposites of Polymers Directly Derived from Biomass

14.4.1 Cellulose Nanocomposites

Owing to its insolubility in most common organic solvents, high crystallinity and high (theoretical) softening temperature, the melt processing of neat cellulose becomes difficult. Thus, cellulose derivatives (such as esters, ethers, xanthates and nitrate) are processed as semi-crystalline polymers, with the aid of different plasticisers and additives, and extruded to produce various grades of commercial cellulose plastics in pelletised form [18, 44]. These chemically modified cellulose derivatives were readily reinforced with CNC *via* solvent [45-47] and melt blending methods [48]. They have shown improved mechanical properties, indicating strong interactions owing to the chemical structural similarities between the filler and the matrix. To achieve better dispersion of CNC, the sol-gel template method has also been used in a recent study where bionanocomposites based on CNC and cellulose acetate butyrate were prepared using solvent exchange of CNC by the sol-gel method, followed by casting [49]. A better dispersion of CNC was confirmed, in both the polymer solutions (as indicated by strong flow birefringence), and in the solid nanocomposite films (as indicated by transparency and by scanning electron microscopy (SEM) analysis). Owing to their better dispersion and strong interactions, these highly transparent films exhibited a significant increase of about 83% and 70%, in the Young's modulus and strength, respectively, for nanocomposites with 12% w/w of CNC without affecting the strain values. Thus, preparing nanocomposites *via* the sol-gel route i.e., forming an organogel of CNC, dispersing and casting, can provide a promising route for obtaining cellulose nanocomposites with well-dispersed CNC, leading to improved mechanical properties whilst retaining transparency.

14.4.1.1 All-cellulose Nanocomposites

Another type of nanocomposite is prepared by partially dissolving the cellulose matrix and/or welding the nanocrystals (undissolved cellulose). The matrix and reinforcement phases of all-cellulose composites are completely compatible with each other, allowing efficient stress transfer and adhesion at their interface. These types of composites have been demonstrated to be highly tough, fully biobased, and biodegradable. The nanowelding of microfibres was recently demonstrated by fabricating all-cellulose nanocomposites *via* two methods (direct and indirect), using the ionic liquid 1-butyl-3-methylimidazolium chloride (BMIMCl) as a solvent. **Figure 14.4** shows a cross-sectional model of binding of nanofibres.

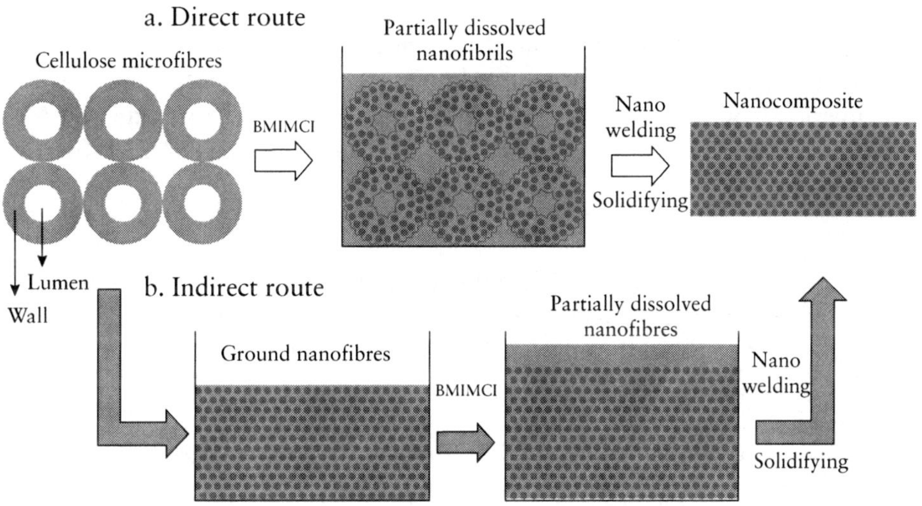

Figure 14.4 Cross-sectional schematic model of (a) direct and (b) indirect fabrication of nanocomposite from cellulose microfibre (CMF) by partial dissolving/nanowelding. Grey dots or filled circles represent undissolved nanofibrils of cellulose Iβ (reinforcement phase); continuous phase (grey coloured) in partially dissolved nanofibrils and nanocomposites are, BMIMCl and, the nanocrystalline cellulose (NCC) (matrix phase). Reproduced with permission from H. Yousefi, T. Nishino, M. Faezipour, G. Ebrahimi and A. Shakeri, *Biomacromolecules*, 2011, **12**, 11, 4080. ©2011, ACS [50]

The average diameter of the reinforcing component (undissolved nanofibrils) in the nanocomposite made directly from CMF (NCC microfibre) was 53 ± 16 nm.

Nanowelding of these fibres is apparent from the improved mechanical properties and imparted transparency (**Figure 14.5**) [50].

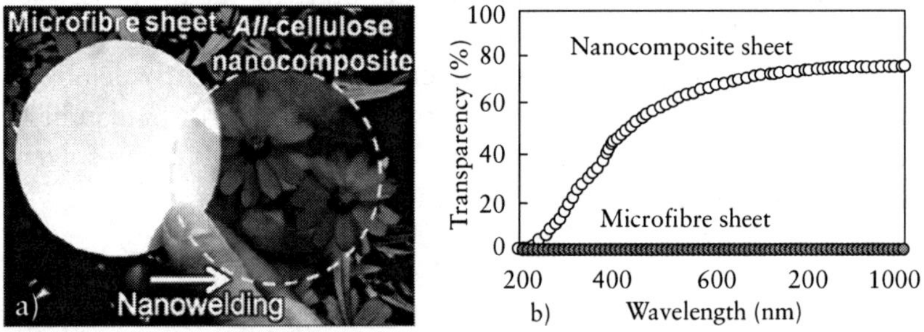

Figure 14.5 (a) Qualitative and (b) quantitative comparison of optical transparency before and after nanowelding by ionic liquid. Reproduced with permission from H. Yousefi, T. Nishino, M. Faezipour, G. Ebrahimi and A. Shakeri, *Biomacromolecules*, 2011, **12**, 11, 4080. ©2011, ACS [50]

Since films from both the methods exhibited similar morphologies and properties, it can be inferred that the direct preparation of all-cellulose nanocomposites from microfibres could be easier, faster and cheaper than using CNC/CNF as the starting material.

For the potential use in controlled release applications, all-cellulose nanocomposite gels were developed by a rapid thermally-induced phase separation, followed by a regeneration process (from sodium hydroxide/urea/water solution) [51]. It was demonstrated that CNC could act as the 'bridge' to facilitate the crosslinking of cellulose chains during the gel formation process. Moreover, they can significantly improve the dimensional stability and mechanical strength of the regenerated gels (**Figure 14.6**).

In vitro investigation under physiological conditions showed that the bovine serum albumin was released *via* a Fickian diffusion mechanism with a decrease in diffusion coefficient from 0.42×10^{-4} (for neat gel) to 0.21×10^{-4} mm^2/s (for gel with 50% w/w of CNC) where the CNC reinforced gel structure served as an effective diffusion barrier. Therefore these gels featuring membrane-matrix hybrid structures could be used for the controlled release of macromolecules.

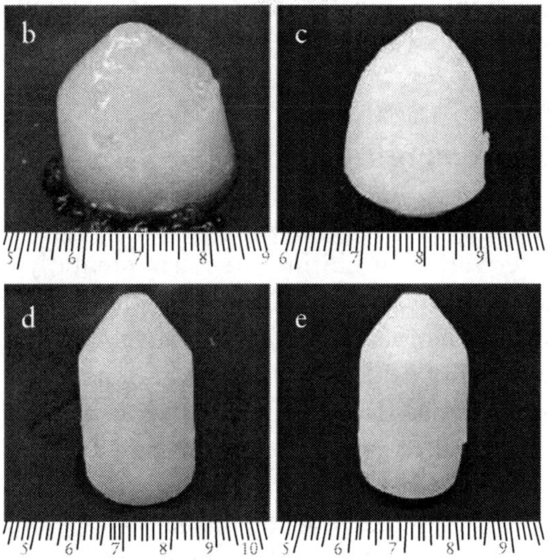

Figure 14.6 Photographs of cellulose gels before (left) and after (right) regeneration in water at 25 °C with different CNC contents: 0 (b, c) and 50% w/w (d, e). Reproduced with permission from Y.X. Wang and L.Y. Chen, *Carbohydrate Polymers*, 2011, **83**, 4, 1937. ©2011, Elsevier [51]

14.4.2 Starch Nanocomposites

Starch, which consists of linear amylose and branched amylopectin, is the most abundant edible polymer in nature. Though it is not truly a thermoplastic polymer, it can be melt-processed with the aid of water and/or other plasticisers using conventional plastics processing equipment [52-54]. Various approaches have been investigated to minimise or overcome the poor characteristics, especially retrogradation *via* moisture absorption [55]. It has been found that reinforcing starch with nanosized fillers is an effective method to obtain high performance starch-based biocomposites [52-54]. The nanocomposites reinforced with lignocellulosic nanocrystals or nanofibres like ramie [56], bamboo [57], wheat straw nanofibres [58] and also cotton seed [59] exhibited relatively good dispersion in the matrix and good adhesion between the matrix and fillers due to chemical similarities and the formation of three-dimensional networks as a result of filler-filler and filler-matrix interactions *via* intermolecular hydrogen bonding. The mechanical properties (tensile strength and Young's modulus) and water resistance of the nanocomposites showed substantial improvements compared to the pure thermoplastic starch [54, 55, 59, 60]. The nanocomposites also showed a higher storage modulus, glass transition temperature (T_g), and degradation temperature, a very good optical transparency and water resistance [61].

14.4.3 Chitosan Nanocomposites

Chitin is the most abundant nitrogen-rich polysaccharide in nature. Chitosan, the *N*-deacetylated derivative of chitin, is environmentally friendly, non-toxic, biodegradable, and antibacterial. By combining the benefits of chitosan and reinforcement with various nanoparticles, a wide range of materials can be produced which are applicable in many fields including medical, packaging, agricultural, and automotive. Since chitosan dissolves easily in water or acidic medium, the solvent method is typically adopted to prepare chitosan-based nanocomposites [62-65]. Fernandes and co-workers [65] have prepared nanocomposites with two different chitosans (one of which is soluble in acidic media and the other in water) *via* the solvent method by incorporating NFC. These highly transparent (transmittance of 20-90%) and flexible films (**Figure 14.7**) exhibited better mechanical properties, with a maximum enhancement of the Young's modulus of 78% and 150% for high molecular weight chitosan and water-soluble high molecular weight chitosan (WSHCH), respectively with 20 wt% of NFC; and of 200% and 320% for low molecular weight chitosan and water-soluble low molecular weight chitosan, respectively with 60 wt% of NFC.

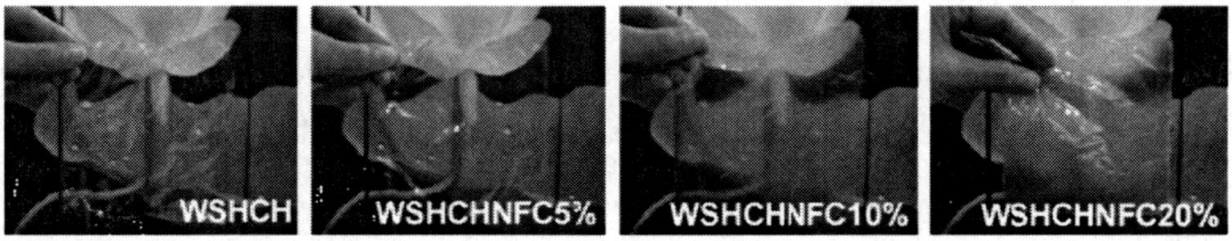

Figure 14.7 Photographs of WSHCH nanocomposite films with different loadings (0-20 wt%) of NFC. Reproduced with permission from S.C.M. Fernandes, C.S.R. Freire, A.J.D. Silvestre, C.P. Neto, A. Gandini, L.A. Berglund and L. Salmen, *Carbohydrate Polymers*, 2010, **81**, 2, 394. ©2010, Elsevier [65]

The nanocomposites showed increased thermal stability, with, for example, a significant increase in the initial degradation temperature from 227 °C to 271 °C [65]. These materials have great potential to be exploited in packaging applications where transparency, mechanical properties, thermal stability and retained biodegradability are required.

In a recent investigation [66], *in situ* grafting of acrylic acid onto chitosan in the presence of CNC (10% w/w) has led to superabsorbent hydrogel materials with faster equilibrium swelling conditions as well as an improved swelling capacity which was increased by approximately 100 units, from 381 to 486. The incorporation of CNC increased the average dimension of the pores to increase the porosity. These materials also displayed a functional pH/ionic responsive behaviour.

14.4.4 Other Polysaccharide Nanocomposites

From biomass, we can also obtain low molecular weight polysaccharides ranging from viscous liquids to solids with different functionalities (e.g., sulfates, carboxylates) such as carrageenans, aligin/alginates, hemicellulose, pectin and many others. Carrageenans, which are generally used as an alternative to gelatin, are a family of linear sulfated polysaccharides that are extracted typically from red seaweeds. They are available in three forms κ, λ and ι-carrageenan and are used in foods, cosmetics and pharmaceuticals. The nanobiocomposites prepared by incorporating CNC into a κ/ι-carrageenan hybrid matrix, exhibited the reduction in water vapour permeability of about 71% compared with the neat matrix [67]. They can be utilised for biodegradable food packaging. Edible films of fruit purees, with or without polysaccharides, have been developed. In a recent study, nanocomposites of acerola purée with alginate were prepared by plasticising with corn syrup and reinforcing with CNC from different origins (cotton, coconut-husk). The CNC improved the overall tensile and barrier (for water) properties of the films [68]. CNC-reinforced hydrogels of hemicellulose with a unique network structure were prepared from wood CNC coated with chemically modified hemicelluloses *via* the sol-gel method followed by *in situ* polymerisation [69].

14.4.5 Protein Nanocomposites

Proteins can be derived from animals (casein, whey and collagen) and plants (zein, soy and gluten). The proteins that have the potential for applications (as polymeric materials) are, for the most part, neither soluble nor fusible without degradation, so they are used in the form in which they are found in nature. Since proteins can be used in applications where biocompatibility with the cells is required, most of the protein-based nanocomposites for biomedical applications are prepared with biologically important nanoparticles [70]. Nanocomposites prepared from glycerol-plasticised sodium caseinate and CNC *via* the solvent-method have shown a significant increase in the mechanical properties as well as water uptake. However, they were less transparent and their surfaces were more hydrophilic than those of neat ones [71]. A similar increase in mechanical properties was observed by Phiriyawirut and co-workers [72] in a recent study on zein proteins reinforced with CNC.

14.5 Nanocomposites of Polymers Synthesised using Biobased Monomers

Polymers from biobased monomers can offer a range of materials where the classical chemical synthetic routes allow tailorability of functions and properties. Biobased monomers can be easily derived by classical chemical methods, i.e., hydrolysis, transesterification and fermentation of biomass. In addition to the monomers for polyesters, polyurethanes and polyamides, the surfactants, additives and plasticisers are generated from biomass [21, 73]. They are also known as 'semi-synthetic biopolymers'.

14.5.1 Polylactic Acid Nanocomposites

Poly(lactic acid) (PLA) is a linear aliphatic polyester whose monomer, lactic acid, is typically derived by fermentation from plant biomass such as corn, potato, cane sugar, rice, cotton and other starch-rich products. PLA is one of the most studied biodegradable polymers, mainly because of its high mechanical properties and easy processability compared to other biopolymers. It has great potential to replace petroleum-based plastics since it can be melt-processed as easily as conventional plastics [74]. Among its many potential uses, PLA has already found applications in textiles, the automotive and medical fields and in the packaging industry. However, its range of application is still limited by its high production cost, its brittleness and its low thermal stability [75].

With the aim of developing sustainable 'green composites', optimising the cost/performance balance and improving the mechanical and thermal properties, reinforcing the biobased PLA by renewable and biodegradable CNC will be an effective method. As PLA is soluble in many organic solvents and can be processed by conventional melt processing, reinforcing PLA with CNC or CNF has been widely demonstrated by solvent casting or solvent spinning) [76-78] and melt blending methods [37, 79, 80]. In these studies, the use of surfactants or chemical modification (e.g., silylation and acetylation) of surface hydroxyl groups has been observed to be effective for improving the dispersibility in common organic solvents like chloroform and *N,N*-dimethylformamide. The reinforcement of PLA by CNC has been demonstrated by a significant increase in the mechanical properties and heat distortion temperature

[81]. Biodegradable PLA/CNC nanocomposites were also made by electrospinning with suitable solvent mixtures [82, 83]. **Figure 14.8** shows the surface morphology of the electrospun mats after one month of *in vitro* degradation and schematically represents the possible recycling of PLA *via in vitro* degradation [83].

Figure 14.8 SEM images of electrospun mats of (a) neat PLA; (b) PLA/5% w/w CNC nanocomposites after one month of *in vitro* degradation; and (c) Schematic representation of PLA/CNC nanocomposites mats *via in vitro* degradation showing possible recycling method. Reproduced with permission from Q.F. Shi, C.J. Zhou, Y.Y. Yue, W.H. Guo, Y.Q. Wu and Q.L. Wu, *Carbohydrate Polymers*, 2012, **90**, 1, 301. ©2012, Elsevier [83]

14.5.2 Naturally Occurring Oil-based Polymer Nanocomposites

Naturally occurring oils which can be derived from both plant and animal sources are some of the most versatile renewable substrates and can be an ideal alternative chemical feedstock to petroleum ones. Hydrolysis and transesterification methods are used to produce mainly glycerol and fatty acids (or their alkyl esters) from natural oils [84]. Fatty acids can be used for the production of multiple functional group-containing molecules such as polyols, dicarboxylic acids, surfactants, resins, stabilisers, plasticisers, and others. Hydrolysis, epoxidation, hydroformylation, and metathesis are some of the other methods which may be employed to convert oils to useful chemicals and materials [73]. These bio-based polyols can be introduced into unsaturated polyester resins to make composite parts which have the potential to replace petrochemical-based polyols derived from propylene oxide in polyurethane formulations [85]. Sources of natural oils include soybean oil, lard, canola oil, algal oil, waste grease, and others.

Polyurethane is one of the most versatile polymeric materials with regard to both processing methods and mechanical properties. Due to increasing concerns about human health and environmental friendliness, organic and solvent-free (waterborne) polyurethane (WPU) as a nontoxic, non-flammable, and biodegradable material has been rapidly developed and used in many applications [86]. WPU formulations based on synthetic polyols have been demonstrated to be superior in mechanical and thermal properties after reinforcing by CNC, significantly even with very low CNC loadings [86, 87].

Recently, biobased polyurethane nanocomposites were prepared by dispersing CNC (0-3 wt%) in polyol obtained from castor oil (by alcoholysis) and curing further. Rheological properties (shear modulus and complex viscosity) of CNC incorporated polyol suspensions indicated a solid-like behaviour (independent of the frequency applied) even with the addition of only 0.5% w/w of CNC. This strong gel-like behaviour at very low concentration (0.5-1%) was correlated with the critical percolation threshold indicating the interfacial interactions by hydrogen bonding rendered by hydroxyl groups on the CNC surface [88].

A recent study on a biocompatible WPU based on castor oil/polyethylene glycol (PEG) filled with low level loadings of *Eucalyptus globulus* cellulose nanocrystals (ECN) showed a significant enhancement in mechanical properties. A high quality dispersion of ECN in the WPU matrix was achieved due to specific interactions with both the hard segments and soft segments in the microphase-separated WPU, causing shifts of the soft segment T_g and the hard segment melting temperature toward higher temperatures [89].

Growing concerns for environmental protection have accelerated the,, production of epoxidised substitutes derived from renewable resources to reduce the use of synthetically derived epoxy polymers. Soybean oil, which is one of the most abundant and lowest cost vegetable oils, can be particularly used for a wide range of biobased derivatives. Epoxidised soybean oil (ESO), a derivative of soybean oil which bears oxirane (epoxy) functional groups, shows excellent promise as an inexpensive, renewable material for industrial application at a reasonable cost [84, 90].

With a view to developing both new biodegradable and nanostructured materials, composites of ESO and bacterial cellulose (BC) have been prepared. The composites revealed high transparency and better mechanical properties than the neat matrix, due to the hierarchical network structure of BC in the ESO matrix [90]. The improvement in stiffness, high tensile strength and retained transparency of these BC-reinforced green materials can be potentially useful options for a wide range of industrial applications where toughness and biodegradability are required.

14.6 Nanocomposites of Microbial/Bioengineered Polymers

14.6.1 Polyhydroxyalkanoate Nanocomposites

Polyhydroxyalkanoates (PHA) are produced naturally by different types of microorganisms as an energy storage product. Different approaches, including the bioengineering of bacteria or plants and variation in the composition of nutrients, have been established to develop PHA [91, 92]. Since they can be produced with different comonomer ratios and a wide range of crystallinity, mechanical properties and retained biodegradability, they are versatile and have the potential to replace petroleum-based plastics in packaging, agricultural and biomedical applications [93]. One of the most widely investigated members of this family is poly-*R*-3-hydroxybutyrate (PHB), a hydrophobic polymer that is truly biocompatible and biodegradable. Because of its partially crystalline structure along with its plastic-like properties, PHB is a potential candidate for replacing the widely used non-degradable polymers such as polyethylene and polypropylene [94].

Bionanocomposites based on PHB and CNC were prepared by dispersing CNC in PEG plasticiser and subsequently incorporating the CNC/PEG suspensions in the matrix. The thermal properties of the nanocomposites indicated an enlargement in the processing window in comparison to the neat PHB. The nanocomposites showed a remarkable increase in the strain level (50 times greater than that of the neat PHB), without a significant loss of the tensile strength of the nanocomposites even at a low concentrations of CNC (up to 0.45% w/w). This promising behaviour was explained in terms of a considerable chain orientation promoted by the presence of CNC in the same direction of the applied load, which improved the shear flow of the polymer matrix. Reinforcement of PHB with CNC in the presence of a plasticiser has been found to improve the thermal processability and the mechanical properties. Using PEG as the plasticiser, the PHB matrix was reinforced with CNC by dispersing CNC in the

PEG plasticiser and then subsequently incorporating the CNC/PEG suspensions into the matrix. The thermal properties of the nanocomposites indicated an enlargement in the processing window in comparison to the neat PHB. The nanocomposites exhibited a significant increase in the elongation even with the small amount of CNC (0.45% w/w in the final nanocomposites), but without compromising the tensile strength significantly [95]. These results indicated that they could be potentially used in a variety of applications.

Interestingly, nanocrystalline BC which is also a bioengineered polymer (*vide infra*) has been used to reinforce PHB [96, 97]. The PHB/BC nanocomposites with a homogeneous dispersion of BC exhibited better mechanical properties and higher crystallinity than the isolated neat PHB [96]. According to Cai and co-workers [97], as PHB and BC have homogeneous nanosized spherulites and fibrils, their nanocomposites have shown high light transmission properties. The thermal stability of the transparent nanocomposites was found to improve from 195 °C (for neat PHB) to 250 °C (for nanocomposites). They believe that PHB/BC nanocomposites have good potential to be considered for various applications.

14.6.2 Bacterial Cellulose Nanocomposites

BC is a unique and emerging biopolymer which can be generated by microorganisms. BC is characterised by very long polymer chains, a degree of polymerisation in the range of 4000-10,000 repeating units, high crystallinity, high purity, an extremely large amount of water (in some cases more than 99%), and high tensile strength of individual CNF equivalent to that of steel or Kevlar [98-100]. It has gained increasing attention for use in multiple biomedical applications due to its high strength, purity, hydrophilicity, biocompatibility and an architecture that can be engineered structurally and chemically at the nano, micro and macroscale by altering the biosynthesis process. The main unique features of BC that set it apart from synthetic polymers and other cellulose types are the three-dimensional (3D) gel biosynthesised from simple low molecular weight substances, a control of porosity, the BC/water balance and surface morphology (coating) of BC, the direct control on the shape of 3D fibre network structure and *in situ* formation of composites (**Figure 14.9**).

Therefore, BC has proven to be a remarkably versatile bioengineered polymer and can be used in a wide variety of applications, such biomedical devices, electronics and acoustics. It has also been demonstrated for potential use in novel materials such as headphone and loudspeaker applications [101]. Detailed information about the BC-based materials literature has been previously reviewed elsewhere [98, 101-103].

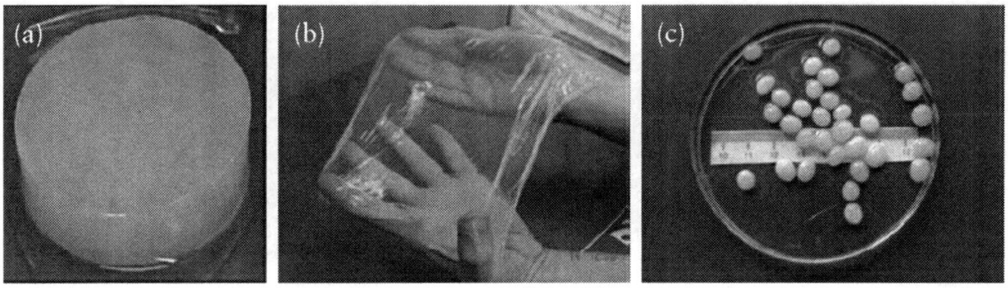

Figure 14.9 (a) BC fleece produced in reactor with a circular profile; (b) thin film of BC; and (c) BC spheres formed by agitated culture. Reproduced with permission from P. Gatenholm and D. Klemm, *MRS Bulletin*, 2010, **35**, 03, 208. ©2010 Cambridge University Press [102]

Owing to its crystalline nanostructure, BC has excellent mechanical properties (elastic modulus 15-35 GPa) which are quite different from those of common plant cellulose. BC-based composites have been processed with various organic matrices such as resins, polysaccharides, proteins, as well as inorganic substrates such as metals and metal oxides [98, 104]. BC films and BC-based composites are being used in acoustic diaphragms, paper products, and transparent films [104]. Detailed information about the preparation, characterisation and applications of BC–based composites (as well as MFC- and CNC-based composites) has been recently reviewed by Hubbe and co-workers [105]. Tajima and co-workers [106] produced BC composites with elastic moduli up to 90 GPa by adding carboxymethyl cellulose, methyl cellulose or PEG to the BC culture media.

14.6.2.1 Applications of Bacterial Cellulose based Nanocomposites

In situ modification and post-processing of BC-based composites with inorganic nanoparticles (titania, silica and silver) have also been explored for different applications. BC fleeces were used to prepare inorganic nanoparticles of different shapes with uniform distribution of size. The BC nanofibres acted as a nanotemplate for the formation and fixation of titania nanotubes [108].

Figure 14.10 shows some recent exciting developments in using BC for biomedical applications.

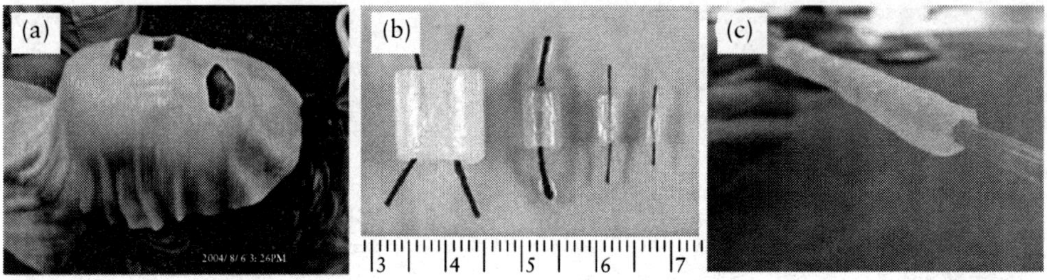

Figure 14.10 Medical applications of BC-based materials. (a) A never-dried BC membrane for wound healing. Reproduced with permission from W.K. Czaja, D.J. Young, M. Kawecki and R.M. Brown, *Biomacromolecules*, 2007, **8**, 1, 1. ©2007, ACS [103]; (b) BC tubes with different diameters created by a matrix technology for arterial grafting applications; and (c) a long BC tube used as a blood vessel implant. Reproduced with permission from P. Gatenholm and D. Klemm, *MRS Bulletin*, 2010, **35**, 03, 208. ©2010 Cambridge University Press [102]

One of best known clinical applications of BC is as a topical material in wound healing. Burns are very complex injuries, which cause extensive damage to skin tissue. The results of many studies indicate that topical applications of BC improve the healing process of burns and chronic wounds. **Figure 14.10a** shows a BC membrane used for treating patients with severe second degree burns [103]. Maneerung and co-workers [107] have prepared BC-based nanocomposites by impregnating silver nanoparticles into the three-dimensional networks of BC. They exhibited strong antimicrobial activity against both *Escherichia coli* (Gram-negative) and *Staphylococcus aureus* (Gram-positive), which are bacteria that are commonly found in infected wounds, indicating their potential for use in antimicrobial wound dressings [107]. Cardiovascular diseases are some of the top medical reasons causing the death of both men and women throughout the world . Every year, thousands of patients undergo heart by-pass surgery where the grafted vessels are drawn from the legs or thorax of the patients. Klemm and co-workers [101, 109-112] have developed prototypes of BC tubes (trade name BASYC, BActerial SYnthesised Cellulose) with different diameters which can be used for arterial grafting applications (**Figures 14.10b** and **c**). The walls of the BASYC tubes consist of BC loaded with 90% water or more. These authors have clearly demonstrated that these BC tubes have very good surgical handling, can be sterilised using standard methods and can be used to successfully replace the carotid arteries [110, 111]. Another area of development is the potential for BC use in tissue engineering. Thanks to its unique three-dimensional cellulose network, and mechanical properties, BC could potentially be an ideal scaffold material for tissue engineering [113-115]. In a recent investigation on the biocompatibility of BC in cell cultures, Watanabe and co-workers [113] showed that a serum-soaked BC membrane is an effective substrate for use in tissue engineering.

14.7 Nanocomposites of Non-degradable Biobased Polymers

The obvious example of this family of polymers is NR which is interestingly a carbon chain polymer, *cis*-1,4-polyisoprene, and accounts for about 40% of the global rubber consumption and 68% of the global production capacity of biobased polymers [21]. Nanocomposites of NR can be prepared by the solvent or casting method by dispersing the CNC in the latex of NR. NR was first reinforced with CNC extracted from the rachis of the edible date palm (*Phoenix dactylifera*) [116]. The stiffness of the nanocomposites was observed to improve especially above the T_g. The reinforcing effect was also found to be higher for nanocomposites with MFC compared to those with CNC, owing to the higher aspect ratio and possible entanglements of MFC [117]. The moisture sorption of NR nanocomposites was found to increase up to 5% CNC loading and to decrease further at higher loading. Interestingly, incorporation of CNC increased the rate of degradation of rubber in soil [118].

NR nanocomposites were also prepared by melt compounding after coagulating the CNC with latex (with and without curatives) [119]. However, research efforts are still required to optimise the processing conditions for exploiting the CNC-based nanocomposites with NR.

Another class of non-degradable biobased polymers is polythioesters, which are composed of mercaptoalkanoic acids and are typically synthesised by non-specific PHA synthases [19]. Reinforcing this matrix with CNC is yet to be documented, but an improvement in mechanical properties and degradability can be expected.

The enormous amount of literature that is currently available on cellulosic nanomaterials indicates that CNC or CNF with different aspect ratios and morphology can be obtained from various natural resources. In many cases, the solvent method is observed to be an effective method (at least as a first step) of dispersing CNC in polymer matrices. For achieving a high level of dispersion in nonpolar matrices, the use of a surfactant or surface modification of CNC or CNF is certainly required. The sol-gel template method has also demonstrated its uniqueness and versatility for incorporating CNC or CNF into most non-polar polymer matrices, at high quality of dispersion of nanocrystals even above their percolation threshold loading, as it allows to preform the organols containing a percolating network of nanofibrils.

To achieve global commercialisation of cellulose-reinforced biopolymer nanocomposites, processing them by conventional melt compounding methods still requires a great deal of attention in order to optimise the processing parameters without affecting the inherent properties of polymer and cellulose. Though these biobased polymer nanocomposites are considered to be biologically degradable, recyclability *via* both conventional (melt processing, pyrolysis and chemical recycling) and biological degradation still remains an issue of environmental concern and a thrust area for future research.

14.8 Concluding Remarks and Future Perspectives

The current trend of developing new materials from renewable resources is driven not only by environmental concern, regulations and demand for alternative sources to petroleum, but also by the advancements happening in the materials science and technology fields. The present contribution has appraised the current state of development of biomass-based nanocomposites based on nanocrystalline cellulose. Over the last decade, an increasing number of academic and industrial research activities have been developed worldwide in the area of NCC-reinforced biopolymeric materials. The processing strategies reported to date are essential for developing fully biobased polymer nanocomposites. Despite the constant developments in the field of cellulose-based nanocomposites, in order to make the transition from laboratory to industrial production, a favourable cost/performance ratio is still required. In order to achieve the true sustainability of cellulose-based nanocomposites, the inexpensive processing strategies, the degradability and recycling of nanocomposites are still to be addressed. Hence, exploring new materials based on biopolymers and cellulose nanocrystals or nanofibres with proven processability and sustainability may accelerate the adoption of these greener materials and these may then slowly replace the petroleum-based polymers and/or contribute to a more environmentally friendly world marketplace.

Acknowledgments

The authors gratefully acknowledge Professor Christoph Weder, the director of Adolphe Merkle Institute for his fruitful discussions and encouragement. They also acknowledge financial support received from the Swiss National Science Foundation (NRP 62: Smart Materials, Nr. 406240_126046), Adolphe Merkle Foundation, Australian Research Council (Discovery Grant No. DP0877161) and University of Queensland.

References

1. S. Mecking, *Angewandte Chemie-International Edition*, 2004, **43**, 9, 1078.

2. M. Patel, *Biodegradable Polymers and Plastics*, 2003, 83.

3. P.K. Annamalai and R.P. Singh, in *Cellulose Fibers: Bio- and Nano-Polymer Composites*, Eds., K. Susheel, B.S. Kaith and I. Kaur, Springer-Verlag Berlin Heidelberg, 2011, p.519.

4. A. Gandini, *Macromolecules*, 2008, **41**, 24, 9491.

5. C.K. Williams and M.A. Hillmyer, *Polymer Reviews*, 2008, **48**, 1, 1.

6. A. Gandini, *Green Chemistry*, 2011, **13**, 5, 1061.

7. L.Yu, K. Dean and L. Li, *Progress in Polymer Science*, 2006, **31**, 6, 576.

8. J.K. Pandey, A.P. Kumar, M. Misra, A.K. Mohanty, L.T. Drzal and R.P. Singh, *Journal of Nanoscience and Nanotechnology*, 2005, **5**, 4, 497.

9. M.A.S.A. Samir, F. Alloin and A. Dufresne, *Biomacromolecules*, 2005, **6**, 2, 612.

10. S. Kamel, *Express Polymer Letters*, 2007, **1**, 9, 546.

11. J.W. Rhim and P.K.W. Ng, *Critical Reviews in Food Science and Nutrition*, 2007, **47**, 4, 411.

12. M.J. John and S. Thomas, *Carbohydrate Polymers*, 2008, **71**, 3, 343.

13. S.J. Eichhorn, A. Dufresne, M. Aranguren, N.E. Marcovich, J.R. Capadona, S.J. Rowan, C. Weder, W. Thielemans, M. Roman, S. Renneckar, W. Gindl, S. Veigel, J. Keckes, H. Yano, K. Abe, M. Nogi, A.N. Nakagaito, A. Mangalam, J. Simonsen, A.S. Benight, A. Bismarck, L.A. Berglund and T. Peijs, *Journal of Materials Science*, 2010, **45**, 1, 1.

14. S. J. Eichhorn, *Soft Matter*, 2011, **7**, 2, 303.

15. R. Chandra and R. Rustgi, *Progress in Polymer Science*, 1998, **23**, 7, 1273.

16. G. Swift, *Abstracts of Papers of the American Chemical Society*, 1998, **216**, U1.

17. S. Karlsson and A.C. Albertsson, *Polymer Engineering and Science*, 1998, **38**, 8, 1251.

18. J. Simon, H.P. Muller, R. Koch and V. Muller, *Polymer Degradation and Stability*, 1998, **59**, 1-3, 107.

19. A. Steinbuchel, *Current Opinion in Biotechnology*, 2005, **16**, 6, 607.

20. D.L. Kaplan in *Biopolymers from Renewable Resources*, Ed., D.L. Kaplan, Springer-Verlag, Berlin, Germany, 1998, p.1.

21. A.H. Tullo, *Chemical and Engineering News*, 2008, **86**, 39, 21.

22. *http://en.european-bioplastics.org/market/*

23. S.S. Ray and M. Bousmina, *Progress in Materials Science*, 2005, **50**, 8, 962.

24. M. Paakko, M. Ankerfors, H. Kosonen, A. Nykanen, S. Ahola, M. Osterberg, J. Ruokolainen, J. Laine, P.T. Larsson, O. Ikkala and T. Lindstrom, *Biomacromolecules*, 2007, **8**, 6, 1934.

25. O. A. Battista, *Industrial and Engineering Chemistry*, 1950, **42**, 3, 502.

26. H. Ono, Y. Shimaya, K. Sato and T. Hongo, *Polymer Journal*, 2004, **36**, 9, 684.

27. D. Tatsumi, S. Ishioka and T. Matsumoto, *Journal of the Society of Rheology Japan*, 2002, **30**, 1, 27.

28. A.N. Nakagaito and H. Yano, *Applied Physics A: Materials Science and Processing*, 2004, **78**, 4, 547.

29. N. Lavoine, I. Desloges, A. Dufresne and J. Bras, *Carbohydrate Polymers*, 2012, **90**, 2, 735.

30. I. Siro and D. Plackett, *Cellulose*, 2010, **17**, 3, 459.

31. J.R. Capadona, O. van den Berg, L.A. Capadona, M. Schroeter, S.J. Rowan, D.J. Tyler and C. Weder, *Nature Nanotechnology*, 2007, **2**, 12, 765.

32. O. van den Berg, J.R. Capadona and C. Weder, *Biomacromolecules*, 2007, **8**, 4, 1353.

33. P. Hajji, J.Y. Cavaille, V. Favier, C. Gauthier and G. Vigier, *Polymer Composites*, 1996, **17**, 4, 612.

34. V. Favier, H. Chanzy and J.Y. Cavaille, *Macromolecules*, 1995, **28**, 18, 6365.

35. V. Favier, G.R. Canova, J.Y. Cavaille, H. Chanzy, A. Dufresne and C. Gauthier, *Polymers for Advanced Technologies*, 1995, **6**, 5, 351.

36. N. Ljungberg, C. Bonini, F. Bortolussi, C. Boisson, L. Heux and J.Y. Cavaille, *Biomacromolecules*, 2005, **6**, 5, 2732.

37. K. Oksman, A.P. Mathew, D. Bondeson and I. Kvien, *Composites Science and Technology*, 2006, **66**, 15, 2776.

38. A.L. Goffin, J.M. Raquez, E. Duquesne, G. Siqueira, Y. Habibi, A. Dufresne and P. Dubois, *Polymer*, 2011, **52**, 7, 1532.

39. A.J. de Menezes, G. Siqueira, A.A.S. Curvelo and A. Dufresne, *Polymer*, 2009, **50**, 19, 4552.

40. S. Z. Li, M.M. Xiao, A. Zheng and H.N. Xiao, *Biomacromolecules*, 2011, **12**, 9, 3305.

41. A.L. Goffin, J.M. Raquez, E. Duquesne, G. Siqueira, Y. Habibi, A. Dufresne and P. Dubois, *Biomacromolecules*, 2011, **12**, 7, 2456.

42. J.R. Capadona, K. Shanmuganathan, D.J. Tyler, S.J. Rowan and C. Weder, *Science*, 2008, **319**, 5868, 1370.

43. J.R. Capadona, K. Shanmuganathan, S. Trittschuh, S. Seidel, S.J. Rowan and C. Weder, *Biomacromolecules*, 2009, **10**, 4, 712.

44. K.J. Edgar, C.M. Buchanan, J.S. Debenham, P.A. Rundquist, B.D. Seiler, M.C. Shelton and D. Tindall, *Progress in Polymer Science*, 2001, **26**, 9, 1605.

45. J. Lu and L.T. Drzal, *Journal of Polymer Science, Part B: Polymer Physics Edition* 2010, **48**, 2, 153.

46. C. Bilbao-Sainz, J. Bras, T. Williams, T. Senechal and W. Orts, *Carbohydrate Polymers*, 2011, **86**, 4, 1549.

47. M. Grunert and W.T. Winter, *Journal of Polymers and the Environment*, 2002, **10**, 1-2, 27.

48. D. Bondeson, P. Syre and K.O. Niska, *Journal of Biobased Materials and Bioenergy*, 2007, **1**, 3, 367.

49. J.E. Ayuk, A.P. Mathew and K. Oksman, *Journal of Applied Polymer Science*, 2009, **114**, 5, 2723.

50. H. Yousefi, T. Nishino, M. Faezipour, G. Ebrahimi and A. Shakeri, *Biomacromolecules*, 2011, **12**, 11, 4080.

51. Y.X. Wang and L.Y. Chen, *Carbohydrate Polymers*, 2011, **83**, 4, 1937.

52. M.N. Angles and A. Dufresne, *Macromolecules*, 2001, **34**, 9, 2921.

53. A P. Mathew and A. Dufresne, *Biomacromolecules*, 2002, **3**, 3, 609.

54. A.P. Mathew, W. Thielemans and A. Dufresne, *Journal of Applied Polymer Science*, 2008, **109**, 6, 4065.

55. A.P. Kumar and R.P. Singh, *Bioresource Technology*, 2008, **99**, 18, 8803.

56. Y.S. Lu, L.H. Weng and X.D. Cao, *Carbohydrate Polymers*, 2006, **63**, 2, 198.

57. D.G. Liu, T.H. Zhong, P.R. Chang, K.F. Li and Q.L. Wu, *Bioresource Technology*, 2010, **101**, 7, 2529.

58. A. Alemdar and M. Sain, *Composites Science and Technology*, 2008, **68**, 2, 557.

59. Y.S. Lu, L.H. Weng and X.D. Cao, *Macromolecular Bioscience*, 2005, **5**, 11, 1101.

60. D.P.E.M. Teixeira, A.A.S. Curvelo, E. Corradini, M.N. Belgacem, A. Dufresne, *Carbohydrate Polymers*, 2009, **78**, 422.

61. Y. Chen, C.H. Liu, P.R. Chang, D.P. Anderson and M.A. Huneault, *Polymer Engineering and Science*, 2009, **49**, 2, 369.

62. K. Avik, R.A. Khan, S. Salmieri, S.L. Tien, B. Riedl, J. Bouchard, G. Chauve, V. Tan, M.R. Kamal and M. Lacroix, *Carbohydrate Polymers*, 2012, **90**, 4, 1601.

63. J.P. de Mesquita, C.L. Donnici, I.F. Teixeira and F.V. Pereira, *Carbohydrate Polymers*, 2012, **90**, 1, 210.

64. A.D. Liu and L.A. Berglund, *Carbohydrate Polymers*, 2012, **87**, 1, 53.

65. S.C.M. Fernandes, C.S.R. Freire, A.J.D. Silvestre, C.P. Neto, A. Gandini, L.A. Berglund and L. Salmen, *Carbohydrate Polymers*, 2010, **81**, 2, 394.

66. C. Spagnol, F.H.A. Rodrigues, A.G.B. Pereira, A.R. Fajardo, A.F. Rubira and E.C. Muniz, *Carbohydrate Polymers*, 2012, **87**, 3, 2038.

67. M.D. Sanchez-Garcia, L. Hilliou and J.M. Lagaron, *Journal of Agricultural and Food Chemistry*, 2010, **58**, 24, 12847.

68. H.M.C. Azeredo, K.W.E. Miranda, M.F. Rosa, D.M. Nascimento and M.R. de Moura, *LWT-Food Science and Technology*, 2012, **46**, 1, 294.

69. M.A. Karaaslan, M.A. Tshabalala, D.J. Yelle and G. Buschle-Diller, *Carbohydrate Polymers*, 2011, **86**, 1, 192.

70. K.K.W. Wong, T. Douglas, S. Gider, D.D. Awschalom and S. Mann, *Chemistry of Materials*, 1998, **10**, 1, 279.

71. M. Pereda, G. Arnica, I. Racz and N.E. Marcovich, *Journal of Food Engineering*, 2011, **103**, 1, 6.

72. M. Phiriyawirut and P. Maniaw, *Open Journal of Polymer Chemistry*, 2012, **2**, 2, 56.

73. D. Sengupta and R.W. Pike in *Handbook of Climate Change Mitigation*, Eds., W.-Y. Chen, J. Seiner, T. Suzuki and M. Lackner, Springer Science+Business Media, New York, NY, 2012, p.1043.

74. E.S. Takacs and J. Vlachopoulos, *Plastics Engineering*, 2008, **64**, 9, 28.

75. E. Fortunati, I. Armentano, Q. Zhou, A. Iannoni, E. Saino, L. Visai, L.A. Berglund and J.M. Kenny, *Carbohydrate Polymers*, 2012, **87**, 2, 1596.

76. M. Sanchez-Garcia and J. Lagaron, *Cellulose*, 2010, **17**, 5, 987.

77. A.H. Pei, Q. Zhou and L.A. Berglund, *Composites Science and Technology*, 2010, **70**, 5, 815.

78. P. Tingaut, T. Zimmermann and F. Lopez-Suevos, *Biomacromolecules*, 2010, **11**, 2, 454.

79. A.N. Frone, S. Berlioz, J.F. Chailan, D.M. Panaitescu and D. Donescu, *Polymer Composites*, 2011, **32**, 6, 976.

80. K. Oksman, A.P. Mathew and M. Sain, *Plastics Rubber and Composites*, 2009, **38**, 9-10, 396.

81. M. Jonoobi, J. Harun, A.P. Mathew and K. Oksman, *Composites Science and Technology*, 2010, **70**, 12, 1742.

82. C.H. Xiang, Y.L. Joo and M.W. Frey, *Journal of Biobased Materials and Bioenergy*, 2009, **3**, 2, 147.

83. Q.F. Shi, C.J. Zhou, Y.Y. Yue, W.H. Guo, Y.Q. Wu and Q.L. Wu, *Carbohydrate Polymers*, 2012, **90**, 1, 301.

84. R.A. Holser, *Industrial Crops and Products*, 2008, **27**, 1, 130.

85. A. Tullo, *Chemical and Engineering News*, 2007, **85**, 19, 14.

86. Y.X. Wang, H.F. Tian and L.N. Zhang, *Carbohydrate Polymers*, 2010, **80**, 3, 665.

87. X. Cao, H. Dong and C.C. Li, *Biomacromolecules*, 2007, **8**, 899.

88. V.M. Wik, M.I. Aranguren and M.A. Mosiewicki, *Polymer Engineering and Science*, 2011, **51**, 7, 1389.

89. Z.Z. Gao, J. Peng, T.H. Zhong, J. Sun, X.B. Wang and C. Yue, *Carbohydrate Polymers*, 2012, **87**, 3, 2068.

90. A. Retegi, I. Algar, L. Martin, F. Altuna, P. Stefani, R. Zuluaga, P. Ganan and I. Mondragon, *Cellulose*, 2012, **19**, 1, 103.

91. W.D. Luzier, *Proceedings of the National Academy of Sciences of the United States of America*, 1992, **89**, 3, 839.

92. P. Suriyamongkol, R. Weselake, S. Narine, M. Moloney and S. Shah, *Biotechnology Advances*, 2007, **25**, 2, 148.

93. H.P.S. Abdul Khalil, A.H. Bhat and A.F. Ireana Yusra *Carbohydrate Polymers*, 2012, **87**, 2, 963.

94. Z.J. Cai and G. Yang, *Materials Letters*, 2011, **65**, 2, 182.

95. P.S.D.O. Patricio, F.V. Pereira, M.C. dos Santos, P.P. de Souza, J.P.B. Roa and R.L. Orefice, *Journal of Applied Polymer Science*, 2013, **127**, 5, 3613.

96. H.S. Barud, J.L. Souza, D.B. Santos, M.S. Crespi, C.A. Ribeiro, Y. Messaddeq and S.J.L. Ribeiro, *Carbohydrate Polymers*, 2011, **83**, 3, 1279.

97. Z.J. Cai, G.A. Yang and J. Kim, *Current Applied Physics*, 2011, **11**, 2, 247.

98. D. Klemm, F. Kramer, S. Moritz, T. Lindström, M. Ankerfors, D. Gray and A. Dorris, *Angewandte Chemie International Edition*, 2011, **50**, 24, 5438.

99. H. Yano, J. Sugiyama, A.N. Nakagaito, M. Nogi, T. Matsuura, M. Hikita and K. Handa, *Advanced Materials*, 2005, **17**, 2, 153.

100. G. Guhados, W. Wan and J.L. Hutter, *Langmuir*, 2005, **21**, 14, 6642.

101. D. Klemm, B. Heublein, H-P. Fink and A. Bohn, *Angewandte Chemie International Edition*, 2005, **44**, 22, 3358.

102. P. Gatenholm and D. Klemm, *MRS Bulletin*, 2010, **35**, 03, 208.

103. W.K. Czaja, D.J. Young, M. Kawecki and R.M. Brown, *Biomacromolecules*, 2006, **8**, 1, 1.

104. R.J. Moon, A. Martini, J. Nairn, J. Simonsen and J. Youngblood, *Chemical Society Reviews*, 2011, **40**, 7, 3941.

105. M.A. Hubbe, O.J. Rojas, L.A. Lucia and M. Sain, *Bioresources*, 2008, **3**, 3, 929.

106. K. Tajima, M. Fujiwara, M. Takai and J. Hayashi, *Mokuzai Gakkaishi*, 1995, **41**, 8, 749. [In Japanese]

107. T. Maneerung, S. Tokura and R. Rujiravanit, *Carbohydrate Polymers*, 2008, **72**, 1, 43.

108. S. Yamanaka and J. Sugiyama, *Cellulose*, 2000, **7**, 3, 213.

109. D. Klemm, D. Schumann, F. Kramer, N. Heßler, M. Hornung, H-P. Schmauder and S. Marsch, *Advances in Polymer Science* 2006, **205**, 49.

110. D. Klemm, D. Schumann, U. Udhardt and S. Marsch, *Progress in Polymer Science*, 2001, **26**, 9, 1561.

111. J. Wippermann, D. Schumann, D. Klemm, H. Kosmehl, S. Salehi-Gelani and T. Wahlers, *European Journal of Vascular and Endovascular Surgery*, 2009, **37**, 5, 592.

112. D. Schumann, J. Wippermann, D. Klemm, F. Kramer, D. Koth, H. Kosmehl, T. Wahlers and S. Salehi-Gelani, *Cellulose*, 2009, **16**, 5, 877.

113. K. Watanabe, Y. Eto, S. Takano, S. Nakamori, H. Shibai and S. Yamanaka, *Cytotechnology*, 1993, **13**, 2, 107.

114. G. Helenius, H. Bäckdahl, A. Bodin, U. Nannmark, P. Gatenholm and B. Risberg, *Journal of Biomedical Materials Research Part A*, 2006, **76A**, 2, 431.

115. A. Svensson, E. Nicklasson, T. Harrah, B. Panilaitis, D.L. Kaplan, M. Brittberg and P. Gatenholm, *Biomaterials*, 2005, **26**, 4, 419.

116. A. Bendahou, Y. Habibi, H. Kaddami and A. Dufresne, *Journal of Biobased Materials and Bioenergy*, 2009, **3**, 1, 81.

117. A. Bendahou, H. Kaddami and A. Dufresne, *European Polymer Journal*, 2010, **46**, 4, 609.

118. J. Bras, M.L. Hassan, C. Bruzesse, E.A. Hassan, N.A. El-Wakil and A. Dufresne, *Industrial Crops and Products*, 2010, **32**, 3, 627.

119. P.M. Visakh, S. Thomas, K. Oksman and A.P. Mathew, *Composites Part A: Applied Science and Manufacturing*, 2012, **43**, 4, 735.

15 Cellulose Nanocrystals and Related Polymer Nanocomposites

Mariana Pereda and Alain Dufresne

15.1 Introduction

Over the last two decades natural fibres have emerged as low cost, lightweight and apparently environmentally superior realistic alternatives to fibres such as glass in composite applications. Lignocellulosic fibres are ubiquitous throughout the world in plants such as grasses, reeds, woody vegetation and in the stalks of all plants. The properties of low cost and low density are particularly attractive for automotive applications, because these will lead to improved fuel efficiency and reduced gas emissions. Composites reinforced by natural fibres are also likely to be environmentally superior because their production has a lower environmental impact compared to glass fibres, and end-of-life incineration results in recovered energy and carbon credit.

However, one of the main drawbacks of lignocellulosic fibres is the large variation in properties which are affected by climatic conditions, maturity of the source plant and type of soil. The elimination of macroscopic flaws by disintegrating the natural grown fibres and then separating the almost defect-free highly crystalline fibrils is one of the basic approaches for preventing this problem. Natural fibres display a hierarchical structure and show a multilevel organisation. Chemical (leading to the production of cellulose nanocrystals [CNC]), mechanical (leading to the production of nanofibrillated cellulose [NFC]), and enzymatic treatments or a combination of these can be applied to disintegrate natural fibres. The dimensions and morphology of the ensuing elongated nanometre-scale cellulosic particles depend on the nature, sequence and severity of these treatments as well as the source of cellulose. Nanometric reinforcements can be classified according to their morphology, i.e., spherical, rod-like and platelet-like nanoparticles that present three, two and one nanometre scale dimensions, respectively.

Nanocrystals occur as rod-like particles, but other descriptors have been also used in the literature to designate them. They are mainly referred to as whiskers [1, 2], nanowhiskers [3-8], CNC [9-13], nanocrystalline cellulose [14], cellulose crystallites [15] or microcrystals [16], despite their nanoscale dimensions.

In addition to the attractive properties of natural fibres, cellulose nanoparticles have generated significant attention from the scientific community because of their availability, biodegradability, biocompatibility and, more specifically, because of their impressive mechanical properties, nanoscale dimensions [17], and their low energy consumption and modest abrasivity during processing [18]. The remarkable mechanical properties of CNC (axial Young's modulus close to that of steel and Kevlar) resulting from their small number of defects make cellulose nanoparticles ideal candidates for the processing of reinforced polymer composites. On the other hand, their intrinsic nanoscale properties can be used for developing advanced nanomaterials, i.e., multiphase materials where at least one of the constituent phases has one dimension less than 100 nm. This field has attracted the attention, scrutiny and imagination of both scientific and industrial communities in recent years. Behind the push for nanocomposites, a large window of opportunity has opened to overcome the limitations of traditional micrometre-scale composites. Research in this field is literally exploding because of the intellectual appeal of building blocks on the nanometre scale and because the technical innovations permit the design and creation of new materials and structures with unprecedented flexibility, improved physical properties and significant industrial applications.

Sustainability and green issues continue to be top priorities for many businesses and individuals, stimulating the search for non-petroleum-based structural materials that suit their needs without being harmful to the environment. These very stiff nanoparticles are therefore suitable for producing green nanocomposite materials.

Such new high-value materials are the subject of continuing research and are commercially interesting in terms of new products for the pulp industry and agricultural sector.

15.2 Hierarchical Structure of Lignocellulosic Materials

Plant fibres are cellular hierarchical biocomposites made up of cellulose, hemicellulose, lignin, extractives and trace elements. Their structure spans many length scales, like many other biological tissues including bones and teeth, in order to provide maximum strength with a minimum of material. Wood, which is approximately 30-40 wt% cellulose with about half in nanocrystalline form and half in amorphous form, is a very well-known example (**Figure 15.1**). Metres describe the whole tree, centimetres describe the structures within the cross-section, millimetres describe the growth rings, tens of micrometres the cellular anatomy, micrometres describe the layer structure within the cell walls, tens of nanometres describe the configuration of the cellulose fibrils in a matrix (which is mainly composed of hemicellulose and lignin), and nanometres describe the molecular structures of cellulose, hemicellulose, and lignin and they are the typical distances between components that interact in wood. Therefore, wood, and other lignocellulosic materials, are cellulosic fibrillar nanocomposite materials.

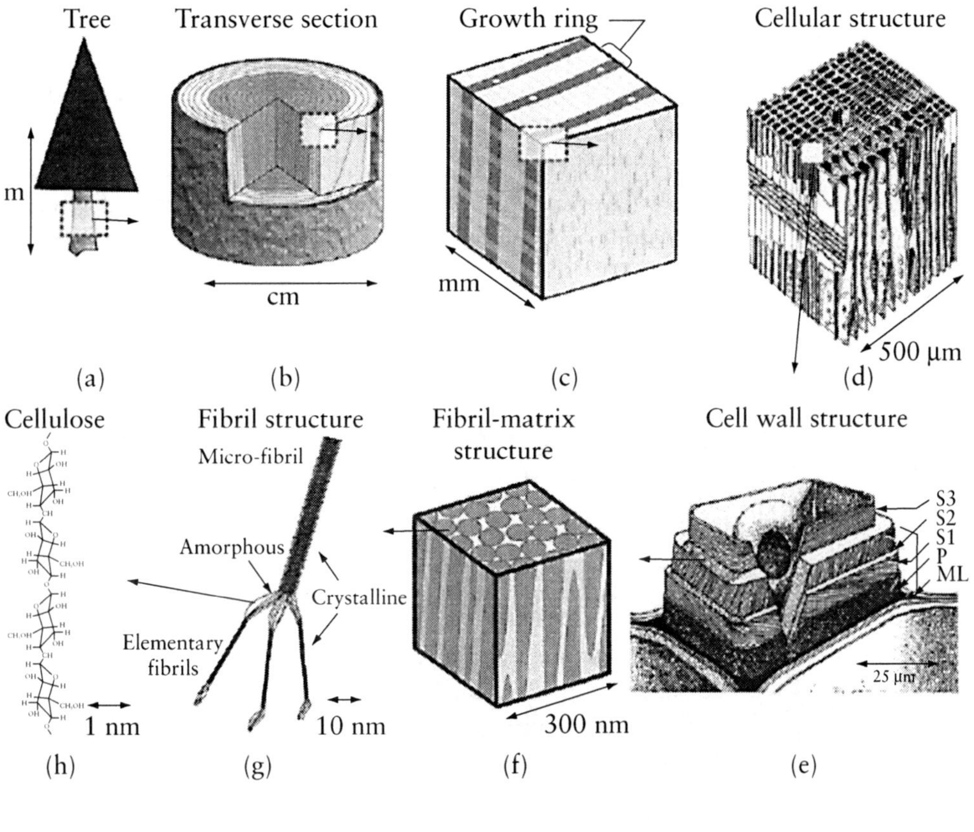

Figure 15.1 Hierarchical structure of wood: from tree to cellulose. S1, S2 and S3 are the three secondary layers (S1 = outer layer, S2 = middle layer, and S3 = inner layer), P is the primary wall, and ML is the middle lamella. Reproduced with permission from M.T. Postek, A. Vladár, J. Dagata, N. Farkas, B. Ming, R. Wagner, A. Raman, R.J. Moon, R. Sabo, T.H. Wegner and J. Beecher, *Measurement Science and Technology*, 2011, **22**, 024005. ©2011, IOP Science [19]

Plant cell wall architectures are nanodimensional structures composed of multiple elementary cellulose nanofibril arrangements, self-assembled in a manner similar to liquid crystals. Cell walls differ in their composition and orientation of cellulose fibrils, the microfibrillar (or spiral) angle being the angle at which the cellulose fibrils are oriented to the normal axis. In general, the fibre strength increases with increasing cellulose content and decreasing spiral angle. **Figure 15.1g** shows the elementary cellulosic fibrils, which contain both crystalline and amorphous segments and are hundreds to a thousand or more nanometres long. The cross-sectional dimension of these fibrils varies according to the origin of the cellulose, being about 3 nm for wood cellulose [20]. The cellulosic microfibrils consist of aggregates of elementary nanofibrils formed by aligned extended cellulose molecules, laterally stabilised by hydrogen bonds and containing both ordered and less-ordered regions [21]. Their structure consists of a predominantly crystalline cellulosic core, covered by a sheath of paracrystalline polyglucosan material surrounded by hemicelluloses [22]. These microfibrils are cemented by other polymers such as lignin and hemicelluloses and aggregate further to form lignocellulosic fibres. Depending on their origin, the microfibril diameters range from about 2 to 20 nm [23]. A model of the wood microfibril is presented in **Figure 15.2.**

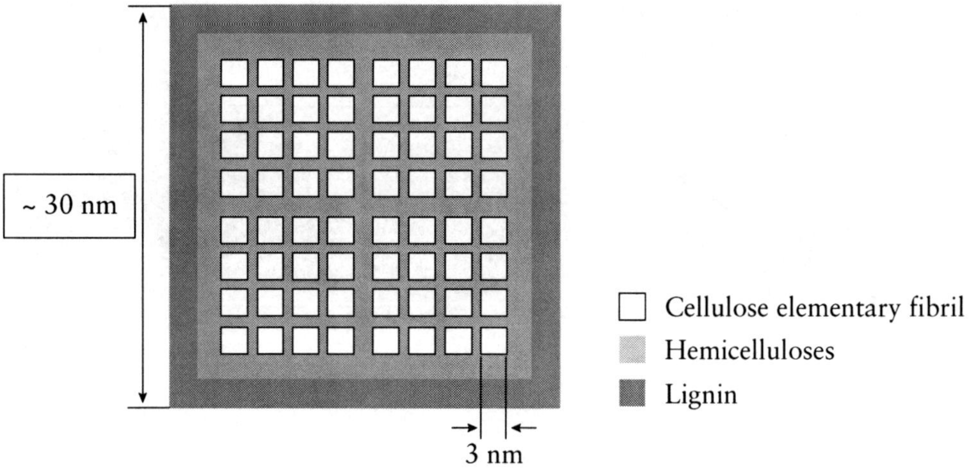

Figure 15.2 Model of the wood cellulose microfibril structure, consisting of elementary nanofibrils. Reproduced with permission from A. Dufresne in *Nanocellulose: From Nature to High Performance Tailored Materials*, Walter de Gruyter GmbH & Co. KG, Berlin, Germany and Boston, MA, USA, 2012, Chapter 1. ©2012, Walter de Gruyter GmbH & Co. [24]

CNC are liberated crystalline segments of elementary nanofibrils which are liberated after the removal of the amorphous segments. This hierarchical structure of natural fibres, based on their elementary nanofibrilar components, leads to the unique strength and high performance properties of different terrestrial and aquatic plant species. Indeed, the most important attribute of wood and other lignocellulosic materials is their mechanical properties, in particular, their unusual ability to provide high mechanical strength and high strength-to-weight ratio while allowing for flexibility to counter large dimensional changes due to swelling and shrinking. The mechanical properties of cellulose can be characterised by its properties in both the ordered (so-called crystalline) and disordered (so-called amorphous) regions of the molecule. The chain molecules in the disordered regions contribute to the flexibility and the plasticity of the bulk material, while those in the ordered regions contribute to the elasticity of the material. As they are almost defect-free (as mentioned previously), the modulus of CNC is close to the theoretical limit for crystalline cellulose, which gives it the potential to be even stronger than steel and similar in strength to Kevlar.

15.3 Crystalline Structure of Cellulose: Young's Modulus

Cellulose is a high molecular weight homopolysaccharide composed of β–1,4–linked-D-glucopyranose [25]. The basic chemical structure of cellulose is presented in **Figure 15.3.**

One of the characteristics of cellulose is that its monomers have three hydroxyl groups. These hydroxyl groups and their hydrogen bonding ability play a major role in crystalline packing and in governing the important physical properties of this highly cohesive material. These hydrogen bonds also hinder the free rotation of the rings along their linking glycosidic bonds, resulting in the stiffening of the chain.

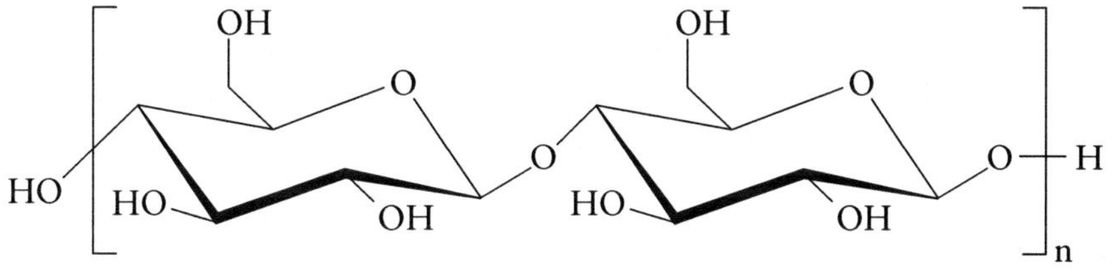

Figure 15.3 Basic chemical structure of cellulose

The ribbon-like character observed for cellulosic macromolecules allows adjacent cellulose chains to fit closely together in an ordered crystalline region. The degree of crystallinity and typical dimensions of cellulose nanofibrils are dependent on their origin, although the biosynthetic mechanism is the same for all organisms [26, 27]. Cellulose exists in several crystal forms, differing in unit cell dimensions and, possibly, in chain polarity. Natural cellulose is referred to as cellulose I, or native cellulose. In nature, cellulose occurs as slender and rod-like or thread-like nanoparticles, due to a linear association of crystallites. Cellulose I consists of a mixture of cellulose Iα (triclinic structure) and Iβ (monoclinic structure). This native cellulose can be converted irreversibly into the thermodynamically more favourable cellulose II polymorph by swelling in a concentrated sodium hydroxide aqueous medium (mercerisation process). Four other polymorphs of cellulose (III_I, III_{II}, IV_I and IV_{II}) can be prepared by different treatments [28]. These different crystalline forms differ in terms of geometry and orientation of the cellulose chains within their respective unit cells.

The elastic or Young's modulus of the crystalline region of cellulose is an important property of this material, especially with respect to the ultimate aim of exploiting its full potential in composite materials. It was first studied experimentally in 1962 by Sakurada and co-workers [29] from the crystal deformation of cellulose I using highly oriented bleached ramie fibres, and a value of 137 GPa was reported. The crystallite modulus of native cellulose along the chain has been calculated based on X-ray analysis of the molecular conformation and the force constants used in the vibrational analysis [30]. The calculated values were 172.9 and 70.8 GPa, when taking (or not taking) into account intramolecular hydrogen bonding, respectively. These values indicate the important role of intramolecular interactions on the determination of the crystallite modulus and chain deformation mechanism. The elastic modulus of the crystalline regions of cellulose polymorphs in the direction parallel to the chain axis was also measured by using X-ray diffraction analysis [31]. Values of 138, 88, 87, 58 and 75 GPa were reported for cellulose I, II, III_I, III_{II} and IV_I, respectively. This indicated that the skeleton conformation of the different polymorphs changed upon crystal transitions and that each polymorph had completely different mechanical properties from all the others. The crystal transition induced a skeletal contraction accompanied by a change in intramolecular hydrogen bonds, which is considered to result in a drastic change in the modulus value of the cellulose polymorphs. Indeed, cellulose I which showed the highest modulus value displayed the longest fibre identity period.

In contrast to the experimentally reported value of 137 GPa, the theoretical Young's modulus along the chain axis reported by Tashiro and Kobayashi [32] based on lattice-dynamic model treatment was 167.5 and 162.1 GPa for the forms I and II, respectively. The lower value observed for form II was ascribed to the lower force constant value of intramolecular hydrogen bonds, showing again the importance of this kind of interaction. On the other hand, the intermolecular hydrogen bonds were found to play a minor role. Anisotropy of the Young's modulus and linear compressibility in the planes perpendicular to the chain axis were also calculated. The two transverse moduli were 11 and 50 GPa. The elastic modulus of the cellulose I_β crystal was also calculated by the molecular mechanics simulation technique [33], and values ranging from 124 to 155 GPa were reported.

More recently, Raman spectroscopy has been used to measure the elastic modulus of nanocrystals prepared from tunicin, the cellulose extracted from tunicate –a sea animal– [34], yielding a value of 143 GPa. This result was clearly similar to those of 145 and 150 GPa reported for tunicate single microfibrils prepared by 2,2,6,6-tetramethylpiperidine-1-oxyl radical (TEMPO) oxidation and sulfuric acid (H_2SO_4) hydrolysis, respectively and measured by atomic force microscopy (AFM) using a three-point bending test [35]. These values are much higher than the Young's modulus of plant cellulosic fibres (e.g., 35-45 GPa for flax fibres). By comparing the experimentally measured force-distance curves with three-dimensional finite element calculations of tip indentation on CNC, Lahiji and co-workers [36] developed a procedure to calculate the transverse elastic modulus of CNC.

Table 15.1 Longitudinal (E_L) and transverse (E_T) moduli of crystalline cellulose

Material	Method	E_L (GPa)	E_T (GPa)	Reference
Bleached ramie fibres (cellulose I)	X-ray diffraction	137	-	[29]
Cellulose I	Calculation	172.9	-	[30]
Cellulose II		70.8	-	
Cellulose I	X-ray diffraction	122-135	-	[39]
Cellulose II		106-112	-	
Cellulose I	Calculation	167.5	11	[32]
Cellulose II		162.1	50	-
Purified ramie fibres (cellulose I)	X-ray diffraction	138	-	-
Cellulose II		88	-	-
Cellulose III_I		87	-	[31]
Cellulose III_{II}		58	-	-
Cellulose IV_I		75	-	-
Cellulose I_β	Calculation	124-155	-	[33, 34]
Cellulose I_β	Raman spectroscopy	143	-	-
TEMPO-oxidised cellulose I_β	AFM	145	-	[35]
Acid hydrolysed cellulose I_β		150	-	-
Wood	AFM		18-50	[36]

- Data is not available from the literature.

The influence of relative humidity (RH) on the stiffness of CNC was measured by comparing AFM measurements on the same nanoparticle under different humidity conditions (from 0.1 to 30% RH). A minimal effect of RH was reported, confirming the resistance of the cellulose crystals to water penetration. The different values reported for the stiffness of CNC are collected in **Table 15.1**. These values are comparable to those reported for aromatic ring polymers such as poly-*p*-phenylene terephthalamide (153-200 GPa) and poly-*m*-phenylene isophthalamide (88 GPa) [37] but they are much more lower than that of polyethylene (PE) (235 GPa), which possesses the maximum elastic modulus of the crystalline regions in the direction parallel to the chain axis [38]. However, taking into account the cross-sectional area of each individual molecule, the Young's modulus values of cellulose and PE are similar.

15.4 Cellulose Nanocrystals

CNC are particles with at least one dimension smaller than 100 nm [40, 41]. This scale shift increases the specific area of the particles, from approximately a few to a few hundred m^2/g [42, 43]. This, in turn, results in an increase in the interfacial area with the polymeric matrix and a decrease of the average interparticle distance in the matrix, allowing possible particle-particle interactions. On the other hand, their nanosize character and the resulting increase of surface energy make difficult to prepare a homogeneous dispersion of nanoparticles in a continuous medium.

When used as components of composites, nanoparticles have the ability to improve the properties of different matrix materials, even at low filler content, without any detrimental effect on impact resistance and plastic deformation. A reduction of gas diffusion (a barrier effect) is also likely to occur. Moreover, cellulosic nanoparticles are distinguished by their liquid crystal behaviour when suspended in water, showing birefringence when observed under polarised light.

15.4.1 Extraction Process: Acid Hydrolysis

The extraction or isolation of crystalline cellulosic regions, in the form of rod-like nanocrystals, is a simple process based on acid hydrolysis. Rånby [44] and Battista and co-workers [45] were the first researchers to produce stable suspensions of colloidal-sized cellulose crystals by H_2SO_4 hydrolysis. This process was later optimised by finding a more rapid and high-yield extraction method [46, 47].

Even if in most publications it is not mentioned, the extraction procedure starts by purification and bleaching processes to purify cellulose by removing non-cellulosic constituents, such as lignin and hemicelluloses. After this treatment, acid hydrolysis is performed on the bleached material under controlled conditions.

The amorphous regions of cellulose act as structural defects and facilitate the diffusion of acid into the cellulose fibres thereby helping the transverse cleavage of the cellulose microfibres into short nanocrystals. This transformation occurs by disrupting the amorphous regions surrounding and embedding the cellulose microfibrils whilst leaving the crystalline segments intact as schematised in **Figure 15.4**. It is attributed to the faster hydrolysis kinetics of the amorphous domains compared to the crystalline ones. Hydronium ions penetrate the cellulosic material in the amorphous domains promoting the hydrolytic cleavage of the glycosidic bonds and releasing individual crystallites.

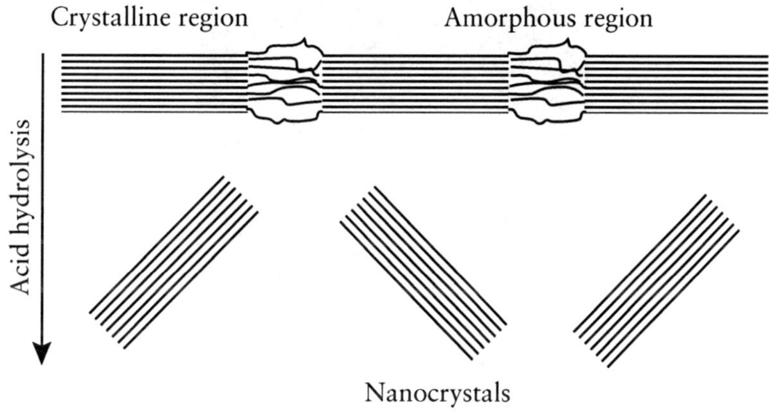

Figure 15.4 Acid hydrolysis breaks down disordered (amorphous) regions and isolates nanocrystals

After the acid hydrolysis, the resulting suspension is subsequently diluted with water and washed by successive centrifugations. Dialysis against distilled water is then performed to remove free acid from the dispersion. Disintegration of aggregates and complete dispersion of the nanoparticles is obtained by a sonication step. The dispersions are stored in a refrigerator after a filtration process aiming to remove residual aggregates. This general procedure has to be adapted, depending on the nature of the cellulosic substrate.

At the end, CNC are obtained as aqueous suspensions, whose stability depends on the dimensions of the dispersed species, particle size distribution and surface charge.

Figure 15.5 shows a picture taken between crossed nicols of a diluted CNC aqueous suspension obtained from capim dourado (golden grass). It shows the birefringent character of an acid-treated CNC suspension as reported by Marchessault and co-workers in 1959 [48]. This birefringence results from two origins: (i) a structural form anisotropy of cellulose and (ii) a flow anisotropy resulting from alignment of nanocrystals under flow because the suspension is shaken before observation.

Figure 15.5 Photograph of an aqueous suspension of capim dourado CNC (0.50 wt%) observed between crossed nicols showing the formation of birefringent domains. Reproduced with permission from G. Siqueira, H. Abdillahi, J. Bras and A. Dufresne, *Cellulose*, 2010, **17**, 289. ©2010, Springer [49]

15.4.2 Effect of Hydrolysis Conditions on the Properties of Cellulose Nanocrystals

The factors governing the products of the acid hydrolysis are the concentration of cellulose (the starting material), the acid type and concentration, the hydrolysis time and temperature, and the ultrasonic treatment time [50].

15.4.2.1 Hydrolysis Time

Dong and co-workers [51] were among the first researchers to study the effect of hydrolysis conditions on the properties of the resulting CNC. They proved that a longer hydrolysis time leads to shorter nanocrystals and also to an increase in their surface charge. Beck-Candanedo and co-workers [52] found by studying the hydrolysis of cellulose from softwood and hardwood pulp, that the reaction time is one of the most important parameters in this acid hydrolysis. They reported that very long reaction times result in the production of sugar molecules *via* the complete digestion of cellulose. In contrast, shorter reaction times yield only yield large undispersable fibres and aggregates.

15.4.2.2 Acid and Cellulose Concentration

The acid concentration was also found to affect the morphology of CNC prepared from sugar beet pulp as reported by Azizi Samir and co-workers [53]. The effect of acid-to-pulp ratio on the critical concentration and rod dimensions was more apparent for the shorter hydrolysis reaction times [54].

15.4.2.3 Acid Type

Although H_2SO_4 was the most commonly used acid during the hydrolysis process; other strong acids such as nitric or hydrochloric acid (HCl) have also been shown to successfully degrade cellulose fibres. However, H_2SO_4 appears to be the most effective leading, for example, to a more stable aqueous suspension than that prepared using HCl [54, 55].

It was shown that the nanoparticles prepared using H_2SO_4 present a negatively-charged surface while those prepared using HCl are not charged. During acid hydrolysis *via* H_2SO_4, acidic sulfate ester groups are formed on the nanoparticle surface, creating electric double-layer repulsion between the nanoparticles in suspension. This electric repulsion plays an important role in their interaction with a polymer matrix and with each other [56]. The density of charges on the CNC surface can be determined by elemental analysis or conductimetric titration to calculate the exact sulfur content.

The sulfate group content increases with acid concentration, acid-to-cellulose ratio and hydrolysis time. It was shown that even at low levels, the sulfate groups caused a significant decrease in degradation temperature and increase in char fraction, confirming that the sulfate groups act as flame retardants [57]. To obtain crystals with a higher thermostability, low acid concentration, low acid-to-cellulose ratio and short hydrolysis time should be used.

Because aqueous CNC suspensions prepared from hydrolysis with HCl tend to flocculate, Habibi and co-workers [58] introduced negative charges onto the CNC surface. They achieved it by the selective conversion of hydroxymethyl groups into carboxylic groups by performing TEMPO-mediated oxidation of the CNC surface. After this, these oxidised CNC not only did not flocculate when dispersed in water but they also kept their initial morphological integrity and native crystallinity. The introduction of charges by the oxidation of the whisker surface has been investigated before by Araki and co-workers [16] and Isogai and Kato [59]. Charges at the surface were also introduced by the post-sulfation of HCl-prepared CNC [54].

It has also been shown that the hydrolysis of amorphous cellulosic chains can be performed simultaneously with the esterification of accessible hydroxyl groups to produce surface-functionalised CNC in a single step [60]. This reaction was carried out in an acid mixture composed of HCl and an organic acid (acetic or butyric). The resulting nanocrystals were of similar dimensions to those obtained by HCl hydrolysis.

15.5 Sources of Cellulose Nanocrystals: Morphological Characteristics

CNC can be obtained from a variety of botanical sources including microcrystalline cellulose (MCC), sugar beet primary cell wall cellulose, cotton, and softwood pulp (mostly black spruce) and can also be obtained from animal sources such as tunicate cellulose.

In addition to plant and animal origins, cellulose nanofibres are also extracellularly secreted by certain bacteria belonging to the genera *Acetobacter, Agrobacterium, Alcaligenes, Pseudomonas, Rhizobium*, and *Sarcina* [61]. The most efficient producer of bacterial cellulose is *Acetobacter xylinum* (or *Gluconacetobacter xylinus*), a Gram-negative strain of acetic acid-producing bacteria [62, 63].

Figure 15.6 shows the transmission electron micrographs (TEM) of dilute suspensions of CNC from different sources, where it is possible to see that, regardless of the source, they are produced as stiff elongated nanoparticles.

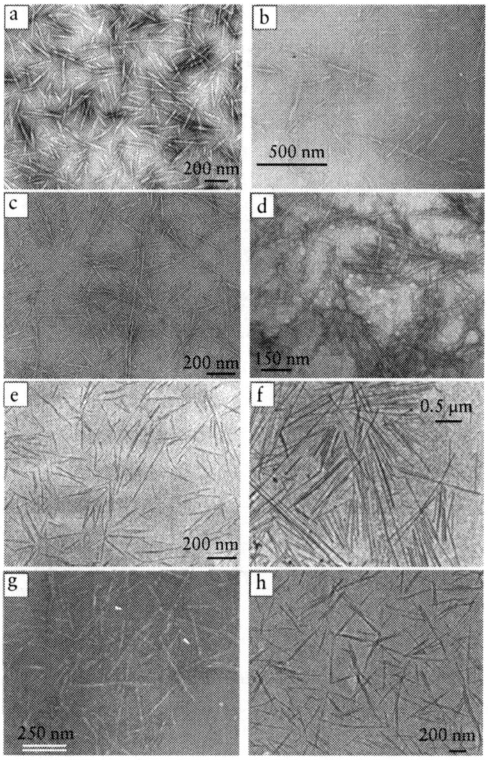

Figure 15.6 TEM of dilute suspensions of CNC from: (a) ramie; (b) *Luffa cylindrica*; (c) sisal; (d) MCC; (e) sugar beet pulp; (f) tunicin; (g) wheat straw; and (h) cotton. (a) Reproduced with permission from Y. Habibi, A.L. Goffin, N. Schiltz, E. Duquesne, P. Dubois and A. Dufresne, *Journal of Materials Chemistry*, 2008, **18**, 5002. ©2008, Royal Society of Chemistry [3]; (b) Adapted from G. Siqueira, J. Bras and A. Dufresne, *Langmuir*, 2010, **26**, 402 [64]; (c) Reproduced with permission from N.L. Garcia de Rodriguez, W. Thielemans and A. Dufresne, *Cellulose*, 2006, **13**, 261. ©2006, Springer [65]; (d) Reproduced with permission from I. Kvien, B.S. Tanem and K. Oksman, *Biomacromolecules*, 2005, **6**, 3160. ©2005, American Chemical Society [66]; (e) Reproduced with permission from M.A.S. Azizi Samir, F. Alloin, M. Paillet and A. Dufresne, *Macromolecules*, 2004, **37**, 4313. ©2004, American Chemical Society [53]; (f) Reproduced with permission from M.N. Anglès and A. Dufresne, *Macromolecules*, 2000, **33**, 8344. ©2000, American Chemical Society [42]; (g) Reproduced with permission from W. Helbert, J.Y. Cavaillé and A. Dufresne, *Polymer Composites*, **17**, 604. ©1996, John Wiley & Sons [1]; and (h) Reproduced with permission from K. Fleming, D.G. Gray, S. Prasannan and S. Matthews, *Journal of the American Chemical Society*, 2000, **122**, 5224. ©2000, American Chemical Society [67]

The persistence of the spot diffractogram when the electron probe is scanned along the rod during transmission electron microscopy observation shows the monocrystalline nature of the cellulosic fragments [68]. Therefore, each fragment can be considered as a cellulosic crystal with no apparent defects. Their dimensions depend on several factors, including the source of cellulose, the exact hydrolysis conditions, and ionic strength. Their diameter and length typically range from 5 to 15 nm and from 200 to 600 nm, respectively. In some cases the higher diameter values of some nanocrystals correspond to aggregates. **Table 15.2** shows the typical geometrical characteristics for nanocrystals derived from different species. Even if the nanocrystal suspensions are often composed of a few laterally bound elementary crystallites that are not separated by conventional acid hydrolysis and the sonication process [69], the length and width of hydrolysed CNC are generally of the order of few hundred nanometres and a few nanometres, respectively. Braun and co-workers [8] observed that the length polydispersity has a constant value, whereas the diameter polydispersity depends on the acid used for isolation. A smaller diameter polydispersity was obtained when using H_2SO_4 instead of HCl because of electrostatic charges resulting from the introduction of sulfate ester groups when using the former.

Another important parameter for CNC is the aspect ratio, defined as the ratio of the length to the width (**Table 15.2**). It determines the anisotropic phase formation and reinforcing properties. The average aspect ratio varies between 10 for cotton and 67 for tunicin.

The shape and dimensions of CNC can be obtained from microscopic observations or by scattering techniques such as small-angle light [96] and neutron [97] scattering. The cross-section of microfibrils observed by transmission electron microscopy is square, whereas their AFM topography shows a rounded profile due to convolution caused by the shape of the AFM tip [73].

15.6 Nanocrystals from Polysaccharides other than Cellulose

Apart from cellulose, nanocrystals can be also prepared from other polysaccharides, such as chitin and starch. Aqueous suspensions of such nanocrystals can also be prepared by acid hydrolysis of the substrate. Chitin nanocrystals are rod-like particles, with typical dimensions of around a few hundred nanometres in length and between 5 and 20 nm in diameter. Starch nanocrystals are platelet-like nanoparticles with lengths between 20 and 40 nm, widths ranging between 15 and 30 nm, and a thickness of between 5 and 7 nm [2]. In the literature it is possible to find several studies on the preparation and characterisation of nanocomposites reinforced with chitin [98-101] and starch [102-105]. Interesting reviews on the preparation of starch nanocrystals and their application in nanocomposites were published by Le Corre and co-workers [106] and Lin and co-workers [107].

Table 15.2 Geometrical characteristics of cellulose nanocrystals from various sources: length (L), cross-section (D) and aspect ratio (L/D)

Source	L (nm)	D (nm)	L/D	Reference
Acacia pulp	100-250	5-15	-	[70]
Alfa	200	10	20	[71]
Algal (*Valonia*)	> 1,000	10-20	∞	[72, 73]
Bacterial	100-several 1,000	5-10 × 30-50 (Nanoparticles had square cross-section)	-	[11, 57, 74]
Banana rachis	500-1,000	5	-	[75]
Capim dourado	300	4.5	67	[64]
Cassava bagasse	360-1,700	2-11	-	[76]
Cladophora	-	20′20 (Nanoparticles had square cross-section)	-	[77]
Cotton	100-300	5-15	10	[51, 78, 79]
Cotton seed linter	170-490	40-60	9	[80]
Curaúa	80-170	6-10	13-17	[55]
Date palm tree (rachis/leaflets)	180-260	6.1	43-30	[81]
Eucalyptus wood pulp	145	6	24	[82]
Flax	100-500	10-30	15	[83]
Grass (*Zoysia*)	200-700	10-60	-	[6, 84]
Hemp	Several 1,000	30-100	-	[85]
Luffa cylindrica	242	5.2	47	[43]
MCC	150-300	3-7	42	[66]
Mulberry	400-500	20-40	-	[86]
Pea hull	240-400	7-12	34	[87]
Ramie	350-700 150-250	70-120 6-8	18-40	[3, 88-90]
Recycled pulp	100-1,800	30-80	-	[13]
Rice husk	_	15-20	10-15	[91]
Sisal	100-500 215	3-5 5	60-43 43	[40, 64, 65]
Sugar beet pulp	210	5	42	[53]
Sugarcane bagasse	200-310	2-6	64	[92]
Tunicin	100-several 1,000	10-20	67	[68, 93]
Wheat straw	150-300	5	45	[1]
Wood	100-300	3-5	50	[52, 94, 95]

- Data is not available from the literature.

15.7 Cellulose Nanocomposite Materials

The main issue for the processing of nanocomposite materials is related to the homogeneous dispersion of cellulose nanoparticles within a polymeric matrix. Water is the initially preferred processing medium because of the high stability of aqueous CNC suspensions, ascribed to the presence of sulfate groups resulting from sulfuric acid hydrolysis, as has been mentioned previously. A high level of dispersion of the filler within the host matrix in the resulting composite film is expected when processing nanocomposites in an aqueous medium.

15.7.1 Effect of Polymeric Matrix on Cellulose Nanocrystal Dispersion and Processing

15.7.1.1 Water-soluble Polymers

Water-soluble polymers are well adapted for the processing of CNC-reinforced nanocomposites [90]. Solid nanocomposites from hydrosoluble matrices are generally obtained by mixing, casting and evaporating the aqueous polymer solution and the aqueous nanocrystal suspension. They can also be obtained by hot pressing or melt extrusion methods from the freeze-dried polymer-nanocrystal mixture.

Investigations for preparing cellulose nanocomposites started in 1995, and materials were prepared by the solvent casting technique. The first publication reporting the preparation of nanocomposites was based on a water-dispersible copolymer such as latex of poly(styrene-*co*-butyl acrylate) (polyS-*co*-BuA) reinforced with tunicin whiskers [68]. Nanocomposite films made with the same copolymer but reinforced with wheat straw [1] and sugar-beet [53] CNC have also been obtained by casting and evaporation of the final aqueous suspension.

Other latexes such as polyhydroxyoctanoate [108-110], polyvinyl chloride [111-114], waterborne epoxy [115], natural rubber [49, 81, 116, 117], and poly(vinyl acetate) [65, 118] have also been used as matrices for the preparation of nanocomposites.

Recently, stable aqueous nanocomposite dispersions containing CNC, poly(styrene-*co*-hexyl acrylate) as the matrix and a reactive silane as stabiliser were prepared *via* miniemulsion polymerisation [71]. Solid nanocomposite films were obtained by solvent casting of the aqueous suspensions.

The preparation of films of cellulosic particle-reinforced starch [42, 97, 119-123], silk fibroin [123], polyoxyethylene (POE) [53, 124-127], poly(vinyl alcohol) (PVA)

[9, 118, 128-130], hydroxypropyl cellulose [128, 129], carboxymethyl cellulose (CMC) [131], or soy protein isolate [132] has also been performed by casting and evaporation of water suspensions.

15.7.1.2 Non-polar Polymers: Alternative Dispersion Methods

In contrast to composites with water-soluble matrices, composites containing non-polar polymer matrices must be prepared in non-aqueous systems. The hydrophilic nature of cellulosic nanoparticles inhibits their homogeneous dispersion in non-polar matrices and limits the compatibility between the reinforcing phase and the matrix.

Several alternatives are available to improve the compatibility between the reinforcing hydrophilic phase and a non-polar matrix in order to obtain a good dispersion. For instance it is possible to disperse the nanoparticles in a suitable organic medium with respect to the polymer matrix. Also, it is possible to coat the surface of the nanoparticles with a surfactant [133]. Chemical modification of the nanoparticle surface is another way to disperse them in organic solvents. It is a classical approach to transform the polar hydroxyl groups located on the surface of cellulosic particles into moieties which are able to enhance interactions with the matrix. The overall objective is to decrease their surface energy in order to improve their dispersion and compatibility with non-polar media.

15.7.1.2.1 Dispersion of Nanoparticles in an Organic Medium

Recently, it was shown that CNC could be dispersed in dimethyl formamide [134, 135] and dimethyl sulfoxide or *N*-methyl pyrrolidine [136] without additives or any chemical modification.

15.7.1.2.2 Coating of Nanoparticle Surface

The coating of the nanoparticle surface with a surfactant was carried out by several authors. For example, cotton and tunicin nanocrystals were coated with a surfactant such as a phosphoric ester of POE nonyl phenyl ether, leading to stable suspensions in toluene and cyclohexane [133] or chloroform [66]. Coated tunicin nanocrystal-reinforced atactic polypropylene [137], isotactic polypropylene [138], and poly(ethylene-*co*-vinyl acetate) [139] were obtained by solvent casting using toluene. The same procedure was used to disperse cellulosic nanoparticles in chloroform with poly(lactic acid) (PLA) [66, 140].

15.7.1.2.3 Chemical Surface Modification

Only a few articles deal with surface grafting; trimethysylilation, ring-opening polymerisation (ROP), cerium-induced grafting and surface acetylation. The chemical grafting has to be mild in order to preserve the integrity of the nanoparticles.

Goussé and co-workers [141] stabilised tunicin nanocrystals in tetrahydrofuran by a partial silylation of their surface. Araki and co-workers [16] prepared sterically stabilised aqueous rod-like CNC suspensions by the combination of HCl hydrolysis, oxidative carboxylation, and the grafting of polyethylene glycol (PEG), having a terminal amino group at one end, using water-soluble carbodiimide. The PEG-grafted CNC displayed drastically enhanced dispersion stability and an ability to redisperse into either water or chloroform from the freeze-dried state.

CNC were surface grafted with poly(ε-caprolactone) (PCL) *via* microwave-assisted ring-opening polymerisation yielding filaceous CNC-*graft*-PCL, which was incorporated into PLA as the matrix [142]. CNC were also functionalised by an esterification reaction with organic acid chloride aliphatic chains of different sizes [90]. CNC-reinforced waterborne polyurethane nanocomposites have been synthesised *via in situ* polymerisation using the casting/evaporation technique [143]. The grafted chains formed a crystalline structure on the surface of the nanoparticles and induced the crystallisation of the matrix. Phenol was enzymatically polymerised in the presence of TEMPO-oxidised cellulosic nanoparticles to prepare nanocomposites under ambient conditions [144].

With the same objective of increasing the apolar character of the nanoparticles, 'long-chain grafting' has also been applied. It consists of grafting agents bearing a reactive end group and a long 'compatibilising' tail. In addition, it has extraordinary possibilities because the surface modification can act as binding sites for active agents in drug delivery systems or for toxins in purification and treatment systems. These surface modifications may also be able to interdiffuse, on heating, to form a polymer matrix phase. The covalent linkage between the reinforcement and the matrix results in near-perfect stress transfer at the interface, giving a composite with excellent mechanical properties. The surface chemical modification by grafting of long-chain isocyanate was recently applied to sisal nanocrystals and *Luffa cylindrica* nanocrystals in separate papers [145, 64] in the context of using these as load-bearing constituents in order to develop new and inexpensive reinforced polymer nanocomposites. Most recently, Follain and co-workers [146] have also applied this chemical modification to *Luffa cylindrica* nanocrystals with the aim of reducing the water affinity of PCL biomaterials.

Nanocomposite materials, through the solvent casting technique with dichloromethane, have been processed from PCL-grafted CNC using the 'grafting onto' [89] and 'grafting from' [3] approaches. Goffin and co-workers [147] used the 'grafting from' approach to graft PCL to CNC by ROP of ε-caprolactone catalysed with tin octoate with the aim of preparing nanocomposites by melt extrusion. The 'grafting onto' approach consists of mixing the nanoparticles with a polymer and a coupling agent to graft the existing polymer onto the surface of the filler. The 'grafting from' approach consists of mixing the nanoparticles with a monomer and an initiator, and the polymer grows directly on the surface of the filler. A co-continuous crystalline phase around the nanoparticles was observed.

A solvent exchange procedure can also be used. For instance, cellulose nanoparticles have been modified with *n*-octadecyl isocyanate using an *in situ* solvent exchange procedure [43].

15.7.2 Processing

Generally, the literature concerning the preparation of CNC-based nanocomposites mainly uses the solvent casting method as the processing approach, especially when water-soluble polymers are used as matrices. In this particular case, the interactions between CNC and the liquid medium are good enough due to the polar nature of both partners.

However, other techniques have recently been gaining ground due to the extensive research on the surface modification of CNC. Those changes, whether chemical or physical ones, improve their dispersability in a wide variety of polymers and through different processing techniques.

15.7.2.1 Melt Extrusion and Impregnation Methods

The melt processing approach is widely used for the processing of thermoplastic polymers, composites reinforced with short fibres and nanocomposites. However, very few studies have reported on its use for nanocomposites reinforced with CNC. The reason for this gap is related to the hydrophilic nature of cellulose. This intrinsic characteristic causes irreversible agglomeration during freeze-drying and aggregation in non-polar matrices due to the formation of additional hydrogen bonds between the amorphous parts of the cellulose nanoparticles.

Therefore, during the past few years, different alternatives have been applied to try to improve the dispersion of CNC. For instance, nanocomposites based on CNC and PLA were obtained by pumping the suspension of nanocrystals into the polymer melt during the extrusion process [7]. An attempt to use PVA as a compatibiliser to promote the dispersion of CNC within the PLA matrix has also been reported [148]. However the CNC dispersion proved rather poor due to the lack of compatibility between the hydrophilic nanofiller and the hydrophobic matrix. Another group has reported a coating approach using polymer- or fatty acid-grafted CNC [87], which leads to successful extrusion in PCL or polyethylene (PE). De Menezes and co-workers [90] functionalised CNC by an esterification reaction to improve their dispersability in low density PE (LDPE). The homogeneity of the ensuing extruded nanocomposite was found to increase with the length of the grafted chains (**Figure 15.7**). Very recently, CNC grafted with PCL (using the 'grafting from' method) were used to reinforce a PCL matrix and the composite was melt extruded [147]. It was evidenced that grafted CNC presented an excellent dispersion within the hydrophobic matrix.

Figure 15.7 Photographs of neat LDPE film and extruded nanocomposite films reinforced with 10 wt% of unmodified and stearoyl acid chloride-grafted cellulose whiskers. Reproduced with permission from A. de Menezes, G. Siqueira, A.A.S. Curvelo and A. Dufresne, *Polymer,* 2009, **50**, 4552. ©2009, Elsevier [90]

The impregnation processing technique involves filtration of the aqueous CNC suspension to obtain a film or dried mat of particles, followed by immersion in a polymer solution. The impregnation of the dried mat was performed under vacuum. Composites have been processed by filling the cavities with transparent thermosetting resins such as phenol-formaldehyde [149-151], epoxy [152], acrylic [153-155], and melamine-formaldehyde [156]. Non-woven mats of cellulose microfibrils have also been used to prepare polyurethane composite materials by the film-stacking method [157].

15.7.2.2 Electrospinning

Electrostatic fibre spinning or 'electrospinning' is a versatile method for preparing fibres with diameters ranging from several microns down to 100 nm, through the action of electrostatic forces [158, 159]. It uses an electrical charge to draw a positively-charged polymer solution from an orifice to a collector. Electrospinning shares the characteristics of both electrospraying and conventional dry spinning of fibres from solution. The process is non-invasive and does not require the use of coagulation chemistry or high temperatures to produce solid threads from solution. This makes the process particularly suitable for the production of fibres from large and complex molecules. Bacterial CNC were incorporated into POE nanofibres with a diameter of less than 1 µm by electrospinning to enhance the mechanical properties of the electrospun fibres [160]. CNC was found to be globally well-embedded and aligned inside the fibres, even though it was partially aggregated. Likewise, electrospun PVA fibre mats loaded with CNC (**Figure 15.8**), with diameters in the nanoscale range and enhanced mechanical properties, were successfully produced [161, 162].

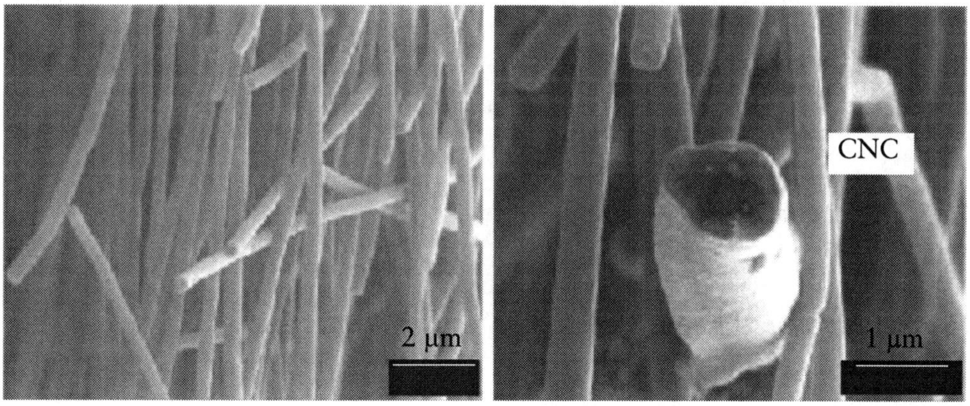

Figure 15.8 Cryo-scanning electron micrographs of electrospun PVA loaded with 15% of CNC. Reproduced with permission from M.S. Peresin, Y. Habibi, J.O. Zoppe, J.J. Pawlak and O.J. Rojas, *Biomacromolecules*, 2010, **11**, 674. ©2010, American Chemical Society [162]

Electrospun polystyrene (PS) [5] and PCL [163] microfibres or nanofibres reinforced with CNC have been obtained by electrospinning. A non-ionic surfactant, sorbitan monostearate, was used to improve the dispersion of the particles in the hydrophobic PS matrix, while surface grafting of the long chains was used for PCL.

15.7.2.3 Multilayer Films

The layer-by-layer (LBL) technique is expected to maximise the interaction between CNC and a polar polymeric matrix, such as chitosan [82]. It also allows the incorporation of high CNC content, giving a dense and homogeneous distribution in each layer.

Podsiadlo and co-workers [164] prepared CNC multilayer composites with a polycation, polydimethyldiallylammonium chloride, using LBL. The authors concluded that the multilayer films had high uniformity and dense packing of the nanocrystals. The preparation of thin films composed of alternating layers of orientated rigid CNC and flexible polycation chains has been reported [165]. Alignment of the rod-like nanocrystals was achieved using anisotropic suspensions of CNC. Green composites based on CNC/xyloglucan multilayers have been prepared using the non-electrostatic cellulose-hemicellulose interaction [166]. More recently, biodegradable nanocomposites were obtained using LBL with highly deacetylated chitosan and CNC [82]. Hydrogen bonds and electrostatic interactions between the negatively-charged sulfate groups on the nanoparticle surfaces and the ammonium groups of the chitosan were the driving forces for the growth of the multilayered films. Self-organised films have also been obtained by the LBL technique using only charge-stabilised dispersions of cellulose nanoparticles with opposite charges [167].

15.7.3 Nanocomposite Properties

15.7.3.1 Mechanical and Thermomechanical Properties

15.7.3.1.1 Cellulose Nanocrystals

CNC have a huge specific surface area and impressive mechanical properties, which make them ideal candidates to improve the mechanical properties of a neat matrix but very few studies reporting the mechanical properties of thin films of nanocrystals have been published. The Young's modulus depends on the preparation procedure and on the nature and strength of interparticle interactions. This modulus can be assumed to be similar, in principle, to that of a paper sheet for which the hydrogen bonding forces provide the basis of its stiffness. It can be experimentally determined from tensile tests performed on films prepared by casting and evaporation of a suspension of nanocrystals. By scanning electron microscopy of the surface of sisal/CNC films obtained by casting, Belbekhouche and co-workers [168] were able to visualise a porous structure similar to that of a paper sheet and nanocrystals with a rod-like morphology. The film consisted of a mat of hydrogen bonded CNC. The porosity and density of this film were found to be around 62% and 0.9 g/cm^3, respectively [168].

A recent study reported a comparison of CNC from different sources and showed the impact of the geometrical characteristics on the specific mechanical properties of the percolating network [117]. It showed that increasing the aspect ratio of the nanoparticles increased the tensile modulus of the films (**Figure 15.9**). It indicated the importance of selecting high aspect ratio whiskers to have an efficient reinforcing effect of a polymeric matrix. It resulted in an obvious reduction of the percolation threshold allowing a reduction of the filler content necessary to reach an effective reinforcing effect, and higher stiffness of the percolating cellulosic network.

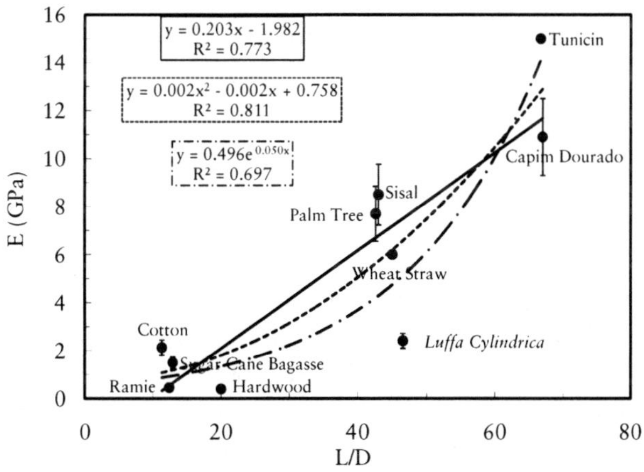

Figure 15.9 Evolution of the Young's modulus of CNC films determined from tensile tests as a function of the aspect ratio of the constituent nanocrystals. E is the tensile modulus of the nanocrystal film, L and D are the length and diameter of the nanocrystals respectively, x and y correspond to the x and y axis respectively, R is the regression coefficient and e is the exponential. Reproduced with permission from J. Bras, D. Viet, C. Bruzzese and A. Dufresne, *Carbohydrate Polymers*, 2011, 84, 211. ©2011, Elsevier [41]

15.7.3.1.2 Cellulose Nanocomposites

Dynamic mechanical analysis (DMA) is a powerful tool to investigate the linear mechanical behaviour of nanocomposites at different temperature and frequency ranges. Classical tensile or compressive tests can be used to evaluate the non-linear mechanical properties.

Favier and co-workers [68, 93] reported the first demonstration of the reinforcing effect of CNC by using DMA. In the shear mode, the authors measured a strong improvement in the storage modulus after adding tunicin nanocrystals into a polyS-*co*-BuA matrix in the latex form. This increase was especially significant above the glass-rubber transition temperature (T_g) of the thermoplastic matrix because of its poor mechanical properties in this temperature range. These outstanding mechanical properties were attributed to a mechanical percolation phenomenon [68, 93]. Above the percolation threshold, the cellulosic nanoparticles can connect and form a three-dimensional continuous pathway due to strong interactions through hydrogen bonding forces [41, 169]. This mechanical percolation effect explained both the high reinforcing effect and the thermal stabilisation of the composite modulus.

Thus, the mechanical performance of the composites is highly sensitive to any factor related to the formation of the percolating nano-network [170]. In this context, three main parameters have been reported: (i) the morphology and dimensions of the nanoparticles; (ii) the processing method; and (iii) the microstructure of the matrix and matrix–filler interactions.

15.7.3.1.3 Morphology and Dimensions of the Nanoparticles

With regard to the morphology and dimensions, the geometrical aspect ratio is an important factor for rod-like particles since it determines the percolation threshold value. This factor is linked to the source of cellulose and preparation conditions for the nanocrystals. A higher reinforcing effect is obtained for nanocrystals with a high aspect ratio. For instance, the rubbery storage tensile modulus was systematically lower for wheat straw whisker/polyS-*co*-BuA composites than for tunicin whisker-based materials [170]. The following relationship was found between the percolation threshold (V_{Rc}) and the aspect ratio (L/D) of rod-like particles [2]:

$$V_{Rc} = 0.7/(L/D) \quad (15.1)$$

Also, flexibility and the possibility of entanglements between nanofibres play an important role in the mechanical properties [40, 53, 81]. Azizi Samir and co-workers [53] used polyS-*co*-BuA reinforced with cellulose rod-like nanoparticles extracted from sugar beet. While linear mechanical measurements performed on the nanocomposites did not show significant differences when varying the strength of the hydrolysis step or the length and flexibility of the nanoparticles, non-linear mechanical tensile tests showed that, as the hydrolysis strength increased, both the modulus and the strength of the nanocomposite decreased, whereas the elongation at break increased. These results showed the strong influence of entanglement on the mechanical behaviour of nanocomposites. When using unhydrolysed cellulose microfibrils extracted from potato pulp, rather than CNC, to reinforce glycerol-plasticised thermoplastic starch, a significant reinforcing effect was observed [110, 171] which was attributed to the tangling effect [172]. Bendahou and co-workers [116] also observed the same reinforcing effect when comparing microfibrillated cellulose with cellulose whiskers in natural rubber nanocomposites. It was ascribed to the tangling effect that provides greater stiffness.

15.7.3.1.4 Effect of Processing Method

Regarding the processing method, slow processes such as casting/evaporation have been reported as producing materials with higher mechanical performance than freeze-drying/moulding and freeze-drying/extruding/moulding. This effect was ascribed to the probable orientation of the rod-like nanoparticles during film processing, resulting from shear stress induced by freeze-drying/moulding and freeze-drying/extrusion/moulding [2].

A rearrangement of the nanoparticles is possible during slow water evaporation because of Brownian motion in the suspension or solution (whose viscosity remains low up to the end of the process, when the latex particle or polymer concentration becomes very high). The nanoparticles have time to interact and to connect each other to form a continuous network. This phenomenon represents the basis of their reinforcing effect. On the other hand, during the freeze-drying/hot pressing process, the nanoparticle arrangement in the suspension is first frozen, and then during the hot pressing stage, the particle rearrangements are greatly limited because of the melt viscosity of the polymer.

15.7.3.1.5 Effect of Matrix Microstructure and Matrix-filler Interactions

The microstructure of the matrix and the resulting competition between matrix-filler and filler-filler interactions are other parameters which affect the mechanical behaviour of cellulose nanocomposites. Classical composite science tends to favour matrix-filler interactions as a fundamental condition for optimal performance but the opposite trend is generally observed when the material is processed *via* casting/evaporation. An unusual behaviour consisting of a diminution in mechanical performance is obtained when increasing the affinity between the filler and the host matrix. This is ascribed to the uniqueness of the reinforcing phenomenon of the cellulosic nanoparticles, resulting from the formation of a hydrogen bonded percolating network. Grunert and Winter [11] prepared nanocomposites with a cellulose acetate butyrate (CAB) matrix and reported a higher reinforcing effect when using unmodified bacterial cellulose nanocrystals compared to trimethylsilylated nanocrystals. They attributed this difference to the restricted filler-filler interaction in the trimethylsilylated nanocrystal nanocomposites.

The transcrystallisation phenomenon reported for polyhydroxyoctanoate [109] and plasticised starch [42] reinforced with CNC also resulted in a decrease of the mechanical properties [172] because the nanoparticles were coated with crystalline matrix domains. In a recent work [145] linear and non-linear mechanical tests showed that better filler-matrix adhesion led to improved mechanical properties. This was ascribed to the chain tangling effect between surface-grafted CNC and the PCL matrix, allowing a higher homogeneity and the formation of hard domains.

When using melt extrusion, the rubbery modulus increased for non-modified CNC while no effect was observed for chemically grafted CNC. There are two probable antagonistic effects: chemical grafting should have improved the dispersion of the filler within the polymeric matrix, but at the same time it decreased the possibility of inter-CNC interactions. Non-linear mechanical tensile tests showed a slight increase of the tensile modulus and decrease of the tensile strength upon addition of CNC [90]. Improved elastic response and tan delta shift and broadening with respect to the matrix was reported for PCL-grafted CNC nanocomposites prepared by melt extrusion. This effect was more pronounced when the grafted CNC content increased [147]. These observations suggested restricted molecular motions due to specific interactions between the nanofiller and polymer chains. In non-percolating systems, for instance for materials processed from freeze-dried CNC, strong matrix-filler interactions enhanced the reinforcing effect of the filler [139].

15.7.3.2 Thermal Properties

Analysis of the thermal properties of materials is essential to determine the temperature range for their processing and use. Important characteristics of polymeric systems, such as the T_g, the melting temperature (T_m) and the degree of crystallinity can be obtained from differential scanning calorimetry (DSC) traces.

On the other hand, the thermal stability can be determined using thermogravimetric analysis (TGA).

15.7.3.2.1 Glass Transition Temperature

As it was reported by several researchers, there is generally no significant variation of T_g when changing the CNC content, regardless of the nature of the polymeric matrix. Studies of POE reinforced with tunicin nanocrystals [125] and work with CAB reinforced with bacterial CNC [11] have confirmed this observation.

In glycerol-plasticised starch-based nanocomposites, peculiar effects of tunicin nanocrystals on the T_g of the starch-rich fraction were reported [42]. The T_g value of the nanocomposites was mostly moisture-sensitive and a plasticisation effect of water was reported for low loading levels (up to 3.2 wt%). However, an anti-plasticisation phenomenon was observed for higher CNC contents (6.2 wt% and over). These results were attributed to the possible interactions between hydroxyl groups on the cellulosic surface and starch, the selective partitioning of glycerol and water in the bulk starch matrix or at the CNC surface, and the restriction of mobility of amorphous starch chains in the vicinity of the starch-crystallite-coated filler surface. This water-plasticising effect, which occurred with increasing humidity, regardless of the CNC content, was also responsible for the decrease of T_g in PVA-based nanocomposites reinforced with sisal [65] and cotton [118] nanocrystals. However, in a moist atmosphere, T_g was found to increase with CNC addition, and this was ascribed to the competitive interactions between PVA, water and cellulose whisker surface [118]. An increase in T_g attributed to cellulose-starch interactions was reported for glycerol-plasticised starch reinforced with CNC prepared from cottonseed linter [80]. However, for tunicin nanocrystal/sorbitol-plasticised starch [119], T_g was found to increase slightly up to about 15 wt% nanocrystals and then to decrease for higher contents. The crystallisation of amylopectin chains upon addition of nanocrystals and the migration of sorbitol molecules to the amorphous domains were proposed to explain these observations. The T_g of electrospun PS fibre with added CNC (and surfactant) tends to decrease with nanocrystal load [5] because of the plasticising effect of the surfactant.

15.7.3.2.2 Melting Temperature

The T_m was reported to be nearly independent of the non-modified filler content. In contrast, T_m was found to increase with increasing modified nanocrystal amount because of the stronger filler-matrix interaction for chemically modified nanocrystals. For instance, in plasticised starch [42, 119], and in POE-based materials [124,

125, 134] filled with tunicin nanocrystals, in CAB reinforced with native bacterial CNC [11], in cotton nanocrystal-reinforced PVA [118]; and also in LDPE extruded nanocomposites [90], the T_m remained roughly constant upon nanocrystal addition.

However, for bacterial nanocrystals, T_m was found to increase when the number of trimethylsilylated nanocrystals increased. In contrast, as the CNC content increased T_m was found to decrease for fully hydrolysed PVA matrices [118]. This was ascribed to stronger interactions between the cellulosic surface and the polymeric matrix.

15.7.3.2.3 Degree of Crystallinity

Regarding to the degree of crystallinity, the nucleating effect of cellulosic nanocrystals is mainly governed by surface chemical considerations [138]. It was shown from both X-ray diffraction and DSC analysis that the crystallisation behaviour of films containing unmodified and surfactant-modified nanocrystals displayed two crystalline forms (α and β), whereas the neat matrix and the nanocomposite reinforced with nanocrystals grafted with, for instance, maleated polypropylene only crystallised in the α form. It was suspected that the more hydrophilic the nanocrystal surface, the more it favours the appearance of the β phase. From DSC measurements Grunert and Winter [11] observed that native bacterial fillers impede the crystallisation of the CAB matrix whereas silylated ones help to nucleate the crystallisation.

Crystallinity was reduced by increasing the filler content for electrospun CNC-reinforced PVA nanofibers [162] but for sorbitol-plasticised starch [119] and for LDPE extruded nanocomposites [90], an increase of CNC content was translated into a significant increase in crystallinity. This phenomenon was ascribed to an anchoring effect of the cellulosic filler, which probably acted as a nucleating agent. For POE-based composites the degree of crystallinity of the matrix was found to be roughly constant up to 10 wt% tunicin nanocrystals [124, 125, 134] and to decrease for higher loading levels [125]. The crystallinity of *N*-octadecyl isocyanate-grafted sisal nanocrystal- reinforced PCL was found to increase by increasing the filler content, whereas no influence with *N*-octadecyl isocyanate-grafted sisal NFC was reported [40]. This difference was ascribed to the possibility of entanglement of the NFC, which tends to confine the polymeric matrix and restrict its crystallisation. A rather limited influence of CNC was evidenced on the crystallinity and melting transition of PCL composites [147]. For nanocomposites based on PCL [145], nanofiller first led to an increase of crystallinity followed by a progressive decrease for higher nanofiller contents. So, a nucleation effect was observed for low nanofiller amounts. In addition, some authors have pointed out the occurrence of a transcrystallisation phenomenon corresponding to a preferential crystallisation of the matrix amorphous chains at the surface of nanocrystals during the preparation process, as was observed for tunicin nanocrystal-filled glycerol-plasticised starch [42]. On the other hand, a higher nucleating effect was observed with surface-modified nanocrystals, probably due to the improvement in the compatibility between both material components.

15.7.3.2.4 Thermal Stability

TGA is performed to investigate the thermal stability of materials. The presence of sulfate groups introduced at the surface of the nanocrystals during hydrolysis with H_2SO_4 reduces the degradation temperature [57, 86]. This behaviour was observed by de Menezes and co-workers [90], who found a reduction in the degradation temperature for modified nanoparticles and this reduction was even greater as the length of the grafted chains increased. Other authors reported no significant influence of tunicin nanocrystals on the degradation temperature of the POE matrix [124, 125]. Choi and Simonsen [131] showed that CNC content appeared to have an effect on the thermal behaviour of CMC plasticised with glycerol, suggesting a close association between the filler and the matrix because the degradation of the nanocomposite was observed in a single step and at a temperature between the degradation temperatures of CMC matrix and CNC.

15.7.3.3 Swelling Properties

The swelling, or kinetics of solvent absorption, can highlight interactions between the filler and the matrix. Swelling measurement involves the drying and weighing of samples, followed by immersing them in a liquid solvent or exposing them to a humid environment. The samples are then removed at specific intervals and weighed until equilibrium is reached. The swelling rate of the sample can be calculated by dividing the gain in weight by the initial weight. The diffusion coefficient can be determined from the initial slope of the solvent uptake curve as a function of time [2]. The study of the kinetics of water absorption is of great interest for nanocomposites reinforced with cellulosic nanoparticles, because of the hydrophilic nature of the reinforcing phase. However, for non-polar polymeric matrices, the absorption of a non-aqueous liquid can also be investigated.

The swelling behaviour in water for cellulose nanoparticle-reinforced thermoplastic starch was investigated and the results showed a higher resistance and a decrease of both water uptake and diffusion coefficient when the CNC content was increased [42, 80, 122]. These phenomena were ascribed to the presence of strong hydrogen bonding interactions between the nanoparticles and also between the nanofiller and the matrix that tend to stabilise the starch matrix when it is in a highly moist atmosphere. Moreover, the crystalline character of cellulose might be also responsible for the decreased water uptake and diffusion coefficient of the material. Other authors [65] found that the addition of sisal nanocrystals stabilised PVA-based nanocomposites with no benefit seen when the nanocrystal content was increased beyond the percolation threshold.

The swelling behaviour of natural rubber reinforced with CNC was analysed in both toluene and water [116]. The toluene uptake decreased strongly when cellulose nanoparticles were added, which was ascribed to the presence of a three-dimensional entangled cellulosic network that strongly restricted the swelling capability and dissolution of the matrix. On the other hand, water uptake increased continuously as the CNC content increased. This was attributed to the increasing amount of hydrophilic material in the nanocomposite film and also to the possible infiltration of water molecules in the interfacial filler/matrix zone [116].

Very recently, Follain and co-workers [146] investigated the sorption behaviour for nanocrystal-reinforced PCL nanocomposites. Under the percolation threshold, an 'anti-plasticising effect' of water was observed, due to the strong intermolecular hydrogen bonding. Above the percolation threshold, the hydrophilic character of nanocrystals dominated, causing a large water mass gain for the whole water activity range. When incorporating surface-modified nanocrystals, a higher water uptake was observed. On the other hand, larger diffusion coefficients reported for nanocomposites indicated an increase of water diffusivity due to the presence of the hydrophilic nanocrystals, whereas a reduction of water diffusivity was found when nanocrystals were surface-modified, owing to the reduction in polymer chain mobility and the water accumulation at the nanofiller/PCL interfaces.

15.7.3.4 Barrier Properties

The origin of the growing interest in the use of nanocellulose-reinforced biobased nanocomposites in barrier applications is the promising performance of nanocellulose with regard to (i) its crystalline structure; (ii) its ability to form a dense three-dimensional percolating network; and (iii) its nucleating effect for semicrystalline matrices [18, 145]. The diffusion pathway is increased by the effect of tortuosity generating changes in the permeability.

Generally, the barrier properties obtained when a nanofiller is dispersed within a polymer matrix are dependent on parameters such as the free volume in the material, the polymer chain segmental mobility, size and dispersion level of the diffusing molecules, the orientation, the amount of dispersed nanoparticles, the formation of a dense cellulose network, the interfacial compatibility between the nanocomposite elements, or the presence of additional additives.

Several authors mentioned a reduction of water vapour permeability (WVP) resulting from the addition of cellulose nanoparticles. This was the case for glycerol-plasticised starch films reinforced with cellulose nanoparticles [100], mango puree-based edible films [173] and chitosan films [174] with nanofibrillated cellulose and also cotton nanocrystal-reinforced CMC films [131].

The reduction of water vapour transmission rate (WVTR) was also observed for PVA/TEMPO-oxidised CNC obtained from Whatman paper [9]. Results were even more interesting in nanocomposites prepared by mixing the PVA matrix with low quantities (10 or 20 wt%) of poly(acrylic acid) (PAA) and nanocrystals, due to the crosslinking with PAA, which reduces the number of hydroxyl groups in the composites and thus the hydrophilicity. The addition of 10 wt% CNC was more effective than 10 wt% PAA, lowering the WVTR to less than 50%, but the combination of CNC and PAA showed the best performance. Also, a very recent work [146] reported a reduction of WVP for nanocomposites of PCL reinforced with unmodified and chemically modified *Luffa cylindrica* nanocrystals.

Contrary to those results, some authors showed an increase of the water barrier properties upon nanofiller loading. These are some examples: CNC/plasticised CMC films [131], surface-carboxylated CNC/PVA films [9] and modified MCC/plasticised starch films [175].

Syverud and Stenius [176] studied the oxygen barrier permeability of pure NFC films, obtaining quite a low value for the oxygen transmission rate (OTR) ranging from 17.0 to 17.8 ml/m^2/day. These values are in the range of the recommended OTR for modified atmosphere packaging (below 10-20 ml/m^2/day). These promising results were ascribed to the low permeability of cellulose enhanced by the crystalline structure of the fibrils. A reduction in dioxygen permeability was also exhibited by MCC cellulose/PLA films [140].

15.8 Conclusions and Future Trends

CNC-based composites have been extensively studied because of the numerous advantages of these nanoparticles. Besides their biodegradability and renewability, these also include: natural abundance and availability, non-food agricultural based economy, low cost, light weight, anisotropy, high strength and stiffness and low energy consumption during processing. Moreover, with their highly crystalline structure (combined with high aspect ratio and nanoscale dimensions) they possess all the qualities required to be ideal candidates for the production of green nanocomposite materials.

Because of its interesting properties such as transparency, high strength, improved oxygen and water vapour permeability, and low thermal expansion, nanocellulose is considered as a potential reinforcing material for different applications. There is a broad range of nanocellulose applications from the food and pharmaceutical industries and packaging, up to electronic devices such as flexible displays, solar cells, electronic paper, and panel sensors.

The potential of nanocomposites in various sectors of research and application is promising; thus, it keeps attracting significant interest from the scientific community and increasing investment.

Acknowledgements

The authors acknowledge The PolyNat Carnot Institute for financial support.

References

1. W. Helbert, J.Y. Cavaillé and A. Dufresne, *Polymer Composites*, 1996, **17**, 604.

2. A. Dufresne in *Monomers, Polymers and Composites from Renewable Resources*, 1st Edition, Eds., A. Gandini and M.N. Belgacem, Elsevier, UK, 2008, p.401.

3. Y. Habibi, A.L. Goffin, N. Schiltz, E. Duquesne, P. Dubois and A. Dufresne, *Journal of Materials Chemistry*, 2008, **18**, 5002.

4. L. Petersson, I. Kvien and K. Oksman, *Composites Science and Technology*, 2006, **67**, 2535.

5. O.J. Rojas, G.A. Montero and Y. Habibi, *Journal of Applied Polymer Science*, 2009, **113**, 927.

6. J.K. Pandey, W.S. Chu, C.S. Kim, C.S. Lee and S.H. Ahn, *Composites Part B: Engineering*, 2009, **40**, 676.

7. K. Oksman, A.P. Mathew, D. Bondeson and I Kvien, *Composites Science and Technology*, 2006, **66**, 2776.

8. B. Braun, J.R. Dorgan and J.P. Chandler, *Biomacromolecules*, 2008, **9**, 1255.

9. S.A. Paralikar, J. Simonsen and J. Lombardi, *Journal of Membrane Science*, 2008, **320**, 248.

10. A.P. Mangalam, J. Simonsen and A. Benight, *Biomacromolecules*, 2009, **10**, 497.

11. M. Grunert and W.T. Winter, *Journal of Polymers and the Environment*, 2002, **10**, 27.

12. G. Morandi, L. Heath and W. Thielemans, *Langmuir*, 2009, **25**, 8280.

13. P.B. Filson and B.E. Dawson-Andoh, *Bioresource Technology*, 2009, **100**, 2259.

14. W. Bai, J. Holbery and K. Li, *Cellulose*, 2009, **16**, 455.

15. X.M. Dong, T. Kimura, J.F. Revol and D.G. Gray, *Langmuir*, 1996, **12**, 2076.

16. J. Araki, M. Wada and S. Kuga, *Langmuir*, 2001, **17**, 21.

17. S.J. Eichhorn, A. Dufresne, M.I. Aranguren, N.E. Marcovich, J.R. Capadona, S.J. Rowan, C. Weder, W. Thielemans, M. Roman, S. Renneckar, W. Gindl, S. Veigel, J. Keckes, H. Yano, K. Abe, M. Nogi, A.N. Nakagaito, A. Mangalam, J. Simonsen, A. Benight, A. Bismarck, L.A. Berglund and T. Peijs, *Journal of Materials Science*, 2010, **45**, 1.

18. G. Siqueira, C. Fraschini, J. Bras, A. Dufresne, R. Prud'homme and M.P. Laborie, *European Polymer Journal*, 2011, **47**, 2216.

19. M.T. Postek, A. Vladár, J. Dagata, N. Farkas, B. Ming, R. Wagner, A. Raman, R.J. Moon, R. Sabo, T.H. Wegner and J. Beecher, *Measurement Science and Technology*, 2011, **22**, 024005.

20. D. Fengel, *Journal of Polymer Science: Part C*, 1971, **36**, 383.

21. D. Fengel and G. Wegener in *Wood: Chemistry, Ultrastructure, Reactions*, Walter de Gruyter, Berlin, Germany, 1984.

22. R.L. Whistler and E.L. Richards in *The Carbohydrates*, Academic Press, New York, NY, USA, 1970.

23. J. Lu, P. Askeland and L.T. Drzal, *Polymer*, 2008, **49**, 1285.

24. A. Dufresne in *Nanocellulose: From Nature to High Performance Tailored Materials*, Walter de Gruyter GmbH & Co. KG, Berlin, Germany and Boston, MA, USA, 2012. Chapter 1.

25. D. Klemm, B. Heublein, H-P. Fink and A. Bohn, *Angewandte Chemie International Edition*, 2005, **44**, 3358.

26. A. Sarko and R. Muggli, *Macromolecules*, 1974, **7**, 486.

27. C. Woodcock and A. Sarko, *Macromolecules*, 1980, **13**, 1183.

28. A. O'Sullivan, *Cellulose*, 1997, **4**, 173.

29. I. Sakurada, Y. Nukushina and T. Ito, *Journal of Polymer Science*, 1962, **57**, 651.

30. K. Tashiro and M. Kobayashi, *Polymer Bulletin*, 1985, **14**, 213.

31. T. Nishino, K. Takano and K. Nakamae, *Journal of Polymer Science Part B: Polymer Physics Edition*, 1995, **33**, 1647.

32. K. Tashiro and M. Kobayashi, *Polymer*, 1991, **32**, 1516.

33. F. Tanaka and T. Iwata, *Cellulose*, 2006, **13**, 509.

34. A. Šturcova, G.R. Davies and S.J. Eichhorn, *Biomacromolecules*, 2005, **6**, 1055.

35. S. Iwamoto, W. Kai, A. Isogai and T. Iwata, *Biomacromolecules*, 2009, **10**, 2571.

36. R.R. Lahiji, X. Xu, R. Reifenberger, A. Raman, A. Rudie and R.J. Moon, *Langmuir*, 2010, **26**, 4480.

37. K. Tashiro, M. Kobayashi and H. Tadokoro, *Macromolecules*, 1977, **10**, 413.

38. K. Nakamae, T. Nishino and H. Ohkubo, *Journal of Macromolecular Science, Part B: Physics*, 1991, **B30**, 1.

39. M. Matsuo, C. Sawatari, Y. Iwai and F. Ozaki, *Macromolecules*, 1990, **23**, 3266.

40. G. Siqueira, J. Bras and A. Dufresne, *Biomacromolecules*, 2009, **10**, 425.

41. J. Bras, D. Viet, C. Bruzzese and A. Dufresne, *Carbohydrate Polymers*, 2011, **84**, 211.

42. M.N. Anglès and A. Dufresne, *Macromolecules*, 2000, **33**, 8344.

43. G. Siqueira, J. Bras and A. Dufresne, *BioResources*, 2010, **5**, 727.

44. B.G. Rånby, *Discussions of the Faraday Society*, 1951, **11**, 158.

45. O.A. Battista, S. Coppick, J.A. Howsmon, F.F. Morehead and W.A. Sisson, *Industrial and Engineering Chemistry*, 1956, **48**, 333.

46. D. Bondenson, A. Mathew and K. Oksman, *Cellulose*, 2006, **13**, 17.

47. W.Y. Hamad and T.Q. Hu, *Canadian Journal of Chemical Engineering*, 2010, **88**, 392.

48. R.H. Marchessault, F.F. Morehead and N.M. Walter, *Nature*, 1959, **184**, 632.

49. G. Siqueira, H. Abdillahi, J. Bras and A. Dufresne, *Cellulose*, 2010, **17**, 289.

50. K. Fleming, D.G. Gray and S. Matthews, *Chemistry - A European Journal*, 2001, 9, 1831.

51. X.M. Dong, J.F. Revol and D.G. Gray, *Cellulose*, 1998, 5, 19.

52. S. Beck-Candanedo, M. Roman and D.G. Gray, *Biomacromolecules*, 2005, **6**, 1048.

53. M.A.S. Azizi Samir, F. Alloin, M. Paillet and A. Dufresne, *Macromolecules*, 2004, **37**, 4313.

54. J. Araki, M. Wada, S. Kuga and T. Okano, *Colloids and Surfaces A: Physicochemical and Engineering Aspects*, 1998, **142**, 75.

55. A.C. Corrêa, E.M. Teixeira, L.A. Pessan and L.H.C. Mattoso, *Cellulose*, 2010, **17**, 1183.

56. Y. Boluk, R. Lahiji, L. Zhao and M.T. McDermott, *Colloids and Surfaces A: Physicochemical and Engineering Aspects* 2011, **377**, 297.

57. M. Roman and W.T. Winter, *Biomacromolecules*, 2004, 5, 1671.

58. Y. Habibi, H. Chanzy and M.R. Vignon, *Cellulose*, 2006, **13**, 679.

59. A. Isogai and Y. Kato, *Cellulose*, 1998, **5**, 153.

60. B. Braun and J.R. Dorgan, *Biomacromolecules*, 2009, **10**, 334.

61. H. El-Saied, A.H. Basta and R.H. Gobran, *Polymer-Plastics Technology and Engineering*, 2004, **43**, 797.

62. R.M. Brown, *Abstracts of Papers of the American Chemical Society*, 2004, **227**, U303.

63. D. Klemm, D. Schumann, F. Kramer, N. Hessler, M. Hornung, H-P. Schmauder and S. Marsch, *Advances in Polymer Science*, 2006, **205**, 49.

64. G. Siqueira, J. Bras and A. Dufresne, *Langmuir*, 2010, **26**, 402.

65. N.L. Garcia de Rodriguez, W. Thielemans and A. Dufresne, *Cellulose*, 2006, **13**, 261.

66. I. Kvien, B.S. Tanem and K. Oksman, *Biomacromolecules*, 2005, **6**, 3160.

67. K. Fleming, D.G. Gray, S. Prasannan and S. Matthews, *Journal of the American Chemical Society,* 2000, **122**, 224.

68. V. Favier, G.R. Canova, J.Y. Cavaillé, H. Chanzy, A. Dufresne and C. Gauthier, *Polymers for Advanced Technologies*, 1995, **6**, 351.

69. S. Elazzouzi-Hafraoui, Y. Nishiyama, J.L. Putaux, L. Heux, F. Dubreuil and C. Rochas, *Biomacromolecules*, 2008, **9**, 57.

70. Y. Pu, J. Zhang, T. Elder, Y. Deng, P. Gatenholm and A.J. Ragauskas, *Composites Part B: Engineering*, 2007, **38**, 360.

71. A. Ben Elmabrouk, W. Thielemans, A. Dufresne and S. Boufi, *Journal of Applied Polymer Science*, 2009, **114**, 2946.

72. J.F. Revol, *Carbohydrate Polymers*, 1982, **2**, 123.

73. S.J. Hanley, J. Giasson, J.F. Revol and D.G. Gray, *Polymer*, 1992, **33**, 4639.

74. C. Tokoh, K. Takabe, M. Fujita and H. Saiki, *Cellulose*, 1998, **5**, 249.

75. R. Zuluaga, J.L. Putaux, A. Restrepo, I. Mondragón and P. Ganan, *Cellulose*, 2007, **14**, 585.

76. E.M. Teixeira, D. Pasquini, A.A.S. Curvelo, E. Corradini, M.N. Belgacem and A. Dufresne, *Carbohydrate Polymers*, 2009, **78**, 422.

77. U.J. Kim, S. Kuga, M. Wada, T. Okano and T. Kondo, *Biomacromolecules*, 2000, **1**, 488.

78. T. Ebeling, M. Paillet, R. Borsali, O. Diat, A. Dufresne, J.Y. Cavaillé and H. Chanzy, *Langmuir*, 1999, **15**, 6123.

79. E.M. Teixeira, A.C. Corrêa, A. Manzoli, F.L. Leite, C.R. de Oliveira and L.H.C. Mattoso, *Cellulose*, 2010, **17**, 595.

80. Y. Lu, L. Weng and X. Cao, *Macromolecular Bioscience*, 2005, **5**, 1101.

81. A. Bendahou, Y. Habibi, H. Kaddami and A. Dufresne, *Journal of Biobased Materials and Bioenergy*, 2009, **3**, 81.

82. J.P. de Mesquita, C.L. Donnici and F.V. Pereira, *Biomacromolecules*, 2010, **11**, 473.

83. X. Cao, H. Dong and C.M. Li, *Biomacromolecules*, 2007, **8**, 899.

84. J.K. Pandey, J.W. Lee, W.S. Chu, C.S. Kim, C.S. Lee and S.H. Ahn, *Macromolecular Research*, 2008, **16**, 396.

85. B. Wang, M. Sain and K. Oksman, *Applied Composite Materials*, 2007, **14**, 89.

86. R. Li, J. Fei, Y. Cai, Y. Li, J. Feng and J. Yao, *Carbohydrate Polymers*, 2009, **76**, 94.

87. G. Chen, A. Dufresne, J. Huang and P.R. Chang, *Macromolecular Materials and Engineering*, 2009, **294**, 59.

88. Y. Lu, L. Weng and X. Cao, *Carbohydrate Polymers*, 2006, **63**, 198.

89. Y. Habibi and A. Dufresne, *Biomacromolecules*, 2008, **9**, 1974.

90. A.de Menezes, G. Siqueira, A.A.S. Curvelo and A. Dufresne, *Polymer*, 2009, **50**, 4552.

91. N. Johar, I. Ahmad and A. Dufresne, *Industrial Crops and Products*, 2012, **37**, 93.

92. E.M. Teixeira, T.J. Bondancia, K.B.R. Teodoro, A.C. Corrêa, J.M. Marconcini and L.H.C. Mattoso, *Industrial Crops and Products*, 2011, **33**, 63.

93. V. Favier, H. Chanzy and J.Y. Cavaillé, *Macromolecules*, 1995, **28**, 6365.

94. J. Araki, M. Wada, S. Kuga and T. Okano, *Journal of Wood Science*, 1999, **45**, 258.

95. U.J. Kim, S. Kuga, M. Wada, T. Okano and T. Kondo, *Biomacromolecules*, 2000, **1**, 488.

96. M.de Souza Lima, J. Wong, M. Paillet, R. Borsali and R. Pecora, *Langmuir*, 2003, **19**, 24.

97. W.J. Orts, L. Godbout, R.H. Marchessault and J.F. Revol, *Macromolecules*, 1998, **31**, 5717.

98. K. Gopalan Nair and A. Dufresne, *Biomacromolecules*, 2003, **4**, 1835.

99. L. Feng, Z. Zhou, A. Dufresne, J. Huang, M. Wei and L. An, *Journal of Applied Polymer Science*, 2009, **112**, 2830.

100. P.R. Chang, R. Jian, J. Yu and X. Maa, *Carbohydrate Polymers*, 2010, **80**, 420.

101. J. Huang, J.W. Zou, P.R. Chang, J.H. Yu and A. Dufresne, *eXPRESS Polymer Letters*, 2011, **5**, 362.

102. H. Angellier, S. Molina-Boisseau, P. Dole and A. Dufresne, *Biomacromolecules*, 2006, **7**, 531.

103. N.L. García, L. Ribba, A. Dufresne, M.I. Aranguren and S. Goyanes, *Macromolecular Materials and Engineering*, 2009, **294**, 169.

104. N.L. García, L. Ribba, A. Dufresne, M.I. Aranguren and S. Goyanes, *Carbohydrate Polymers*, 2011, **84**, 203.

105. P. Mélé, H. Angellier-Coussy, S. Molina-Boisseau and A. Dufresne, *Biomacromolecules*, 2011, **12**, 1487.

106. D. Le Corre, J. Bras and A. Dufresne, *Biomacromolecules*, 2010, **11**, 1139.

107. N. Lin, J. Huang, P.R. Chang, D.P. Anderson and J. Yu, *Journal of Nanomaterials*, 2011, Article ID 573687.

108. D. Dubief, E. Samain and A. Dufresne, *Macromolecules*, 1999, **32**, 5765.

109. A. Dufresne, M.B. Kellerhals and B. Witholt, *Macromolecules*, 1999, **32**, 7396.

110. A. Dufresne, D. Dupeyre and M.R. Vignon, *Journal of Applied Polymer Science*, 2000, **76**, 2080.

111. L. Chazeau, J.Y. Cavaillé, G.R .Canova, R. Dendievel and B. Boutherin, *Journal of Applied Polymer Science*, 1999, **71**, 1797.

112. L. Chazeau, J.Y. Cavaillé and P. Terech, *Polymer*, 1999, **40**, 5333.

113. L. Chazeau, M. Paillet and J.Y. Cavaillé, *Journal of Polymer Science, Part B: Polymer Physics Edition*, 1999, **37**, 2151.

114. L. Chazeau, J.Y. Cavaillé and J. Perez, *Journal of Polymer Science, Part B: Polymer Physics Edition*, 2000, **38**, 383.

115. M. Matos Ruiz, J.Y. Cavaillé, A. Dufresne, C. Graillat and J.F. Gérard, *Macromolecular Symposia*, 2001, **169**, 211.

116. A. Bendahou, H. Kaddami and A. Dufresne, *European Polymer Journal*, 2010, **46**, 609.

117. J. Bras, M.L. Hassan, C. Bruzesse, E.A. Hassan, N.A. El-Wakik and A. Dufresne, *Industrial Crops and Products*, 2010, **32**, 627.

118. M. Roohani, Y. Habibi, N.M. Belgacem, G.Ebrahim, A.N. Karimi and A. Dufresne, *European Polymer Journal*, 2008, **44**, 2489.

119. A.P. Mathew and A. Dufresne, *Biomacromolecules*, 2002, **3**, 609.

120. A.P. Mathew, W. Thielemans and A. Dufresne, *Journal of Applied Polymer Science*, 2008, **109**, 4065.

121. I. Kvien, J. Sugiyama, M. Votrubec and K. Oksman, *Journal of Materiasl Science*, 2007, **42**, 8163.

122. A.J. Svagan, M.S. Hedenqvist and L. Berglund, *Composites Science and Technology*, 2009, **69**, 500.

123. Y. Noishiki, Y. Nishiyama, M. Wada, S. Kuga and J. Magoshi, *Journal of Applied Polymer Science*, 2002, **86**, 3425.

124. M.A.S. Azizi Samir, F. Alloin, W. Gorecki, J-Y. Sanchez and A. Dufresne, *Journal of Physical Chemistry B*, 2004, **108**, 10845.

125. M.A.S. Azizi Samir, F. Alloin, J-Y. Sanchez and A. Dufresne, *Polymer*, 2004, **45**, 4149.

126. M.A.S. Azizi Samir, A.M. Mateos, F. Alloin, J-Y. Sanchez and A. Dufresne, *Electrochimica Acta*, 2004, **49**, 4667.

127. M.A.S. Azizi Samir, F. Alloin and A. Dufresne, *Composite Interfaces*, 2006, **13**, 545.

128. T. Zimmermann, E. Pohler and T. Geiger, *Advanced Engineering Materials*, 2004, **6**, 754.

129. T. Zimmermann, E. Pohler and T. Geiger, *Advanced Engineering Materials*, 2005, **7**, 1156.

130. J. Lu, T. Wang and L.T. Drzal, *Composites Part A: Applied Science and Manufacturing*, 2008, **39**, 738.

131. Y.J. Choi and J. Simonsen, *Journal of Nanoscience and Nanotechnology*, 2006, **6**, 633.

132. Y. Wang, X. Cao and L. Zhang, *Macromolecular Bioscience*, 2006, **6**, 524.

133. L. Heux, G. Chauve and C.Bonini, *Langmuir*, 2000, **16**, 8210.

134. M.A.S. Azizi Samir, F. Alloin, J-Y. Sanchez, N. El Kissi and A. Dufresne, *Macromolecules*, 2004, **37**, 1386.

135. N.E. Marcovich, M.L. Auad, N.E. Belessi, S.R. Nutt and M.I. Aranguren, *Journal of Materials Research*, 2006, **21**, 870.

136. O. van den Berg, J.R. Capadona and C. Weder, *Biomacromolecules*, 2007, **8**, 1353.

137. N. Ljungberg, C. Bonini, F. Bortolussi, C. Boisson, L. Heux and J.Y. Cavaillé, *Biomacromolecules*, 2005, **6**, 2732.

138. N. Ljungberg, J.Y. Cavaillé and L. Heux, *Polymer*, 2006, **47**, 6285.

139. G. Chauve, L. Heux, R. Arouini and K. Mazeau, *Biomacromolecules*, 2005, **6**, 2025.

140. L. Petersson and K. Oksman, *Composites Science and Technology*, 2006, **66**, 2187.

141. C. Goussé, H. Chanzy, G. Exoffier, L. Soubeyrand and E. Fleury, *Polymer*, 2002, **43**, 2645.

142. N. Lin, G. Chen , J. Huang, A. Dufresne and P.R. Chang, *Journal of Applied Polymer Science*, 2009, **113**, 3417.

143. X. Cao, Y. Habibi and L.A. Lucia, *Journal of Materials Chemistry*, 2009, **19**, 7137.

144. Z. Li, S. Renneckar and J.R. Barone, *Cellulose*, 2010, **17**, 57.

145. G. Siqueira, J. Bras, N. Follain, S. Belbekhouche, S. Marais and A. Dufresne, *Carbohydrate Polymers*, 2013, **91**, 711.

146. N. Follain, S. Belbekhouche, J. Bras, G. Siqueira, S. Marais and A. Dufresne, *Journal of Membrane Science*, 2013, **427**, 218.

147. A-L. Goffin, J-M. Raquez, E. Duquesne, G. Siqueira, Y. Habibi, A. Dufresne and Ph. Dubois, *Polymer*, 2011, **52**, 1532.

148. D. Bondeson and K. Oksman, *Composites Part A: Applied Science and Manufacturing*, 2007, **38**, 2486.

149. A.N. Nakagaito, S. Iwamoto and H. Yano, *Applied Physics A*, 2005, **80**, 93.

150. A.N. Nakagaito and H. Yano, *Applied Physics A*, 2004, **78**, 547.

151. A.N. Nakagaito and H. Yano, *Cellulose*, 2008, **15**, 323.

152. Y. Shimazaki, Y. Miyazaki, Y. Takezawa, M. Nogi, K. Abe, S. Ifuku and H. Yano *Biomacromolecules*, 2007, **8**, 2976.

153. H. Yano, J. Sugiyama, A.N. Nakagaito, M. Nogi, T. Matsuura, M. Hikita and K. Handa, *Advanced Materials*, 2005, **17**, 153.

154. M. Nogi, K. Handa, A.N. Nakagaito and H. Yano, *Applied Physical Letters*, 2005, **87**, 243110.

155. S. Iwamoto, K. Abe and H. Yano, *Biomacromolecules*, 2008, **9**, 1022.

156. M. Henriksson and L.A. Berglund, *Journal of Applied Polymer Science*, 2007, **106**, 2817.

157. Ö.M. Seydibeyoglu and K. Oksman, *Composite Science and Technology*, 2008, **68**, 908.

158. Z.M. Huang, Y.Z. Zhang, M. Kotaki and S. Ramakrishna, *Composites Science and Technology*, 2003, **63**, 2223.

159. A. Greiner and J.H. Wendorff, *Angewandte Chemie International Edition*, 2007, **46**, 5670.

160. I. Park, M. Kang, H.S. Kim and H.J. Jin, *Macromolecular Symposia*, 2007, **249-250**, 289.

161. E.S. Medeiros, L.H.C. Mattoso, E. Ito, K.Gregorski, G.H. Robertson, R.D. Offeman, D.F. Wood, W.J. Orts and S.H. Imam, *Journal of Biobased Materials and Bioenergy*, 2008, **2**, 1.

162. M.S. Peresin, Y. Habibi, J.O. Zoppe, J.J. Pawlak and O.J. Rojas, *Biomacromolecules*, 2010, **11**, 674.

163. J.O. Zoppe, M.S. Peresin, Y. Habibi, R.A. Venditti and O.J. Rojas, *ACS Applied Materials and Interfaces*, 2009, **1**, 1996.

164. P. Podsiadlo, S.Y. Choi, B. Shim, J. Lee, M. Cuddihy and N.A. Kotov, *Biomacromolecules*, 2005, **6** , 2914.

165. B. Jean, F. Dubreuil, L. Heux and F. Cousin, *Langmuir*, 2008, **24**, 3452.

166. B. Jean, L. Heux, F. Dubreuil, G. Chambat and F. Cousin, *Langmuir*, 2009, **25**, 3920.

167. C. Aulin, I. Varga, P.M. Claesson, L. Wågberg and T. Lindström, *Langmuir*, 2008, **24**, 2509.

168. S. Belbekhouche, J. Bras, G. Siqueira, C. Chappey, L. Lebrun, B. Khelifi, S. Marais and A. Dufresne, *Carbohydrate Polymers*, 2011, **83**, 1740.

169. V. Favier, G.R. Canova, S.C. Shrivastava and J.Y. Cavaillé, *Polymer Engineering and Science*, 1997, **37**, 1732.

170. A. Dufresne, *Journal of Nanoscience and Nanotechnology*, 2006, **6**, 322.

171. A. Dufresne and M.R. Vignon, *Macromolecules*, 1998, **31**, 2693.

172. M.N. Anglès and A. Dufresne, *Macromolecules*, 2001, **34**, 2921.

173. H.M.C. Azeredo, H.L.C. Mattoso, D. Wood, T.G. Willians, R.J. Avena-Bustillos and T.H. McHug, *Journal of Food Science*, 2009, **74**, N31.

174. H.M.C. Azeredo, H.L.C. Mattoso, R.J. Avena-Bustillos, G. Ceotto Filho, M.L. Munford, D. Wood and T.H. McHug, *Journal of Food Science*, 2010, **75**, N1.

175. P.R. Chang, R. Jian, P. Zheng, J. Yu and X. Ma, *Carbohydrate Polymers*, 2010, **79**, 301.

176. K. Syverud and P. Stenius, *Cellulose*, 2009, **16**, 75.

Abbreviations

3D	Three-dimensional
ABS	Acrylonitrile butadiene styrene
AFM	Atomic force microscopy
Ag	Silver
AR	Aerosil A200
ASTM	American Society for Testing and Materials
ATR	Attenuated total reflectance
BC	Bacterial cellulose
BMIMCl	1-Butyl-3-methylimidazolium chloride
BRHA	Black rice husk ash
C_4H_9OH	Isobutanol
CAB	Cellulose acetate butyrate
$CaCO_3$	Calcium carbonate
CCA	Chromated copper arsenate
CCl_4	Carbon tetrachloride
CH_3OH	Methanol
CMC	Carboxymethyl cellulose
CMF	Cellulose microfibre

CNC	Cellulose nanocrystals
CNF	Cellulose nanofibres
D	Diameter of nanocrystals
DCP	Dicumyl peroxide
DMA	Dynamic mechanical analysis
DSC	Differential scanning calorimetry
DTG	Differential thermogravimetric analysis
E′	Storage modulus
ECM	Extracellular matrices
ECN	*Eucalyptus globulus* cellulose nanocrystals
EFB	Oil palm empty fruit bunch
EN ISO	International Organization for Standardization (European Standard)
EPS	Expandable polystyrene
ESO	Epoxidised soybean oil
FTIR	Fourier transform infrared
G′	Shear storage modulus
H_2O	Water
H_2SO_4	Sulfuric acid
HA	Hydroxyapatite
HCl	Hydrochloric acid
HDPE	High density polyethylene
HDPE-*g*-MA	High density polyethylene-*graft*-maleic anhydride
HSS	High speed steel

IJIRA	Indian Jute Industries' Research Association
KBr	Potassium bromide
L	Length of nanocrystals
L/D	Aspect ratio
LBL	Layer-by-layer
LC	Lignocellulose
LDPE	Low density polyethylene
Lst	Lysostaphin
MAPP	Maleic anhydride modified polypropylene
MCC	Microcrystalline cellulose
MF	Melamine-formaldehyde
MFC	Microfibrillated cellulose
MFI	Melt flow index
MMT	Montmorillonite
MUF	Melamine-urea-formaldehyde
N	Number of cycles
NaOH	Sodium hydroxide
NCC	Nanocrystalline cellulose
NFC	Nanofibrillated cellulose
NFRC	Natural fibre composites
NR	Natural rubber
OTR	Oxygen transmission rate
PAA	Polyacrylic acid

PBS	Polybutylene succinate
PC	Polycarbonate
PCL	Poly(ε-caprolactone)
PE	Polyethylene
PEG	Polyethylene glycol
PEO	Polyethylene oxide
PET	Polyethylene terephthalate
PF	Phenol-formaldehyde
PHA	Polyhydroxyalkanoates
PHB	Poly-*R*-3-hydroxybutyrate
PLA	Poly(lactic acid) or polylactide
PLLA	Poly(L-lactic acid) or poly-L-lactide
PMC	Polymer matrix composites
POE	Polyoxyethylene
PolyS-*co*-BuA	Poly(styrene-*co*-butyl acrylate)
PP	Polypropylene
PP-*g*-MA	Propylene-*graft*-maleic anhydride
PS	Polystyrene
PSF	Poplar seed floss
PVA	Polyvinyl alcohol
PVC	Polyvinyl chloride
RH	Relative humidity
RHA	Rice husk ash

r-HDPE	Recycled high density polyethylene
RHK	Rice husk
r-LDPE	Recycled low density polyethylene
ROP	Ring-opening polymerisation
r-PE	Recycled polyethylene
r-PP	Recycled polypropylene
r-PS	Recycled polystyrene
RRH	Raw rice husk
S	Stress
SAN-*g*-MA	Styrene acrylonitrile-*graft*-maleic anhydride
SC	*Saccaharum cilliare*
SEBS-*g*-MA	Styrene-(ethylene-*co*-butylene)-styrene-*graft*-maleic anhydride
SEM	Scanning electron microscopy
SiO_2	Silica
SMA	Styrene-*graft*-maleic anhydride
SMC	Sheet moulding compound
T_c	Crystallisation temperature
TEA	Tensile energy absorption
TEM	Transmission electron micrographs
TEMPO	Tetramethylpiperidine-1-oxyl radical
T_g	Glass transition temperature or glass-rubber transition temperature
TGA	Thermogravimetry or thermogravimetric analysis
T_m	Melting temperature

TS	Tensile strength
UF	Urea-formaldehyde
UV	Ultraviolet
v-HDPE	Virgin high density polyethylene
v-LDPE	Virgin low density polyethylene
v-PP	Virgin polypropylene
WF	Wood flour
WPC	Wood/plastic composite(s)
WPU	Waterborne polyurethane
WRHA	White rice husk ash
WSHCH	Water-soluble high molecular weight chitosan
WVP	Water vapour permeability
WVTR	Water vapour transmission rate
ΔH_c	Crystallisation enthalpy
ΔH_m	Fusion enthalpy

Index

1-Butyl-3-methylimidazolium chloride, 283

A

Abrasion resistance, 26, 36, 240
Absorb, 14, 116, 160, 194, 246
  Absorbance, 92, 201, 231, 264
  Absorption, 13, 16, 29-32, 35, 39-40, 118, 126-133, 147, 162, 173, 175, 183, 194, 199-202, 208-209, 231-234, 245-246, 248, 270, 279, 285, 333
Accelerator, 64, 70
Acid, 18, 28-29, 36, 49, 62, 66, 94, 97, 106, 161, 172, 180-181, 202, 232, 240, 247, 260, 265-266, 269-270, 278, 280-281, 287-288, 311-316, 318, 320-322, 324, 335
  number, 28
Acidic, 231, 286, 315
Acrylic(s), 147, 245, 287, 324, 335
  acid, 287, 335
Acrylonitrile, 125, 130, 134
  butadiene styrene, 125, 134-135
Activation, 247-248
Additives, 26-27, 43, 45, 63, 118, 124-125, 134-135, 138, 143, 151, 194, 246-247, 260, 267, 282, 288, 321, 334
Adhesion, 19, 24-25, 34, 54, 63, 88-89, 98, 106, 131, 151-153, 175, 177-179, 182, 184, 187, 193, 205, 214, 217, 223, 245, 267, 270, 283, 285, 330
  Adhesive, 76, 146-154, 159-161, 163-164, 199, 246, 263, 267
Adsorption, 153, 232, 240, 243-244, 248-251, 255
Aerosil, 242, 246-247
Aerosil A200, 242
Ageing, 183, 198, 248
Agent, 19, 64, 66, 129, 131-133, 179-180, 183-184, 193, 214, 263, 267-268, 323, 332
Agglomeration, 25, 31, 34-35, 108, 207, 323
Aggregate, 281, 308

Aggregation, 88, 323
Agriculture, 42, 81, 120, 224, 236, 240
Aligned, 24, 87, 222, 308, 325
Alignment, 24, 314, 326
Aliphatic, 171, 180-181, 200-201, 263-264, 269, 288, 322
Alkali, 15, 75, 88, 131, 171-172, 178, 185, 240
Alkaline, 15, 88, 151, 240, 260
Aluminium, 17, 56, 153, 199-200, 240
American Society for Testing and Materials, 28-29, 44, 89, 94, 107, 121
Amorphous, 3, 15, 87-89, 169, 171-172, 181, 202, 240-243, 245, 252, 261-262, 280, 306-309, 313, 316, 323, 331-332
Analysis, 3, 5, 7-8, 10, 26, 34, 42, 44, 78-79, 88, 91, 107, 110, 112-113, 117, 120, 122, 134, 139, 165, 180, 196-197, 199-200, 205, 210, 212, 226-227, 231-232, 236-238, 241, 243, 247, 249, 251, 254, 263-264, 269-270, 273, 282, 310, 315, 328, 330-332
Anhydride group, 34
Anisotropy, 24, 311, 314, 335
Antibacterial, 53-55, 286
Antimicrobial, 55, 294
Antioxidant, 268
Appearance, 48, 117, 122, 126, 186, 195, 208, 263, 332
Application, 4, 6, 21, 27, 42, 51, 53-54, 56, 63, 73, 80, 145-147, 149, 151-152, 158, 160-161, 168, 186, 191, 230, 245, 260, 269, 281, 288, 291, 318, 336
Aqueous, 239, 264, 310, 313-316, 318, 320-322, 324, 333
Aramid, 3, 61, 66, 71, 145, 261, 263
Aromatic, 121, 171-172, 245, 260-261, 263-264, 312
Aspect ratio, 13, 23-24, 87, 280-281, 295, 318-319, 327-328, 335
Atactic, 268, 321
Atmosphere, 107, 200, 232, 246, 270, 331, 333, 335
Atomic force microscopy, 311-312, 318
Attenuated total reflectance, 196-198, 203-204

B

Bacteria, 2, 49-50, 55, 279, 291, 294, 316
Bacterial cellulose, 51-54, 175, 177, 291-295, 316, 330
Band, 89, 91, 201, 204
Barrier, 248, 268, 279, 284, 287, 312, 334-335
Bearing, 14, 181, 187, 322
Bending, 148, 155, 181, 311
Binder, 20, 193, 261, 266, 274

Binding, 31, 283, 322
Bioactivity, 47, 51
Biobased, 2, 4, 20, 115, 180, 193, 268, 272, 277-279, 283, 288, 290, 295-296, 299, 302, 304, 334, 341, 346
Biocompatible, 49, 51, 56, 290-291
Biodegradability, 2, 6, 47, 57, 63, 85, 117-118, 120, 132, 145, 229, 244, 277-279, 287, 291, 306, 335
Biodegradable, 1, 17, 61-62, 66-67, 69, 71, 145, 164, 177, 179, 181, 190, 193, 262, 269-270, 278-279, 283, 286-291, 297, 326
Biological, 51-53, 56, 119, 121, 139, 278, 296, 306
Biomass, 1-10, 12, 14, 16, 18, 20, 22, 24, 26, 28, 30, 32, 34, 36, 38, 40, 42, 44, 46, 48-50, 52, 54, 56, 58, 60, 62, 64, 66, 68, 70, 72, 74, 76, 78, 80, 82, 84, 86, 88, 90, 92, 94, 96-98, 100, 102, 104-114, 116, 118, 120, 122, 124, 126, 128, 130, 132, 134, 136, 138, 140, 142, 144, 146, 148, 150, 152, 154, 156, 158, 160, 162, 164, 166, 168, 170, 172, 174, 176, 178, 180, 182, 184, 186, 188, 190, 192-196, 198, 200, 202, 204, 206, 208, 210, 212, 214, 216, 218, 220, 222, 224, 226, 228-238, 240, 242, 244, 246, 248, 250-252, 254-256, 258-260, 262, 264-266, 268, 270-272, 274, 276, 278-280, 282, 284, 286-288, 290, 292, 294, 296, 298, 300, 302, 304, 306, 308, 310, 312, 314, 316, 318, 320, 322, 324, 326, 328, 330, 332, 334, 336, 338, 340, 342, 344, 346, 348
Biomaterial, 53-54, 181
Biomedical applications, 47, 49, 51, 53-55, 57, 59, 287, 291-293
Blend, 68, 129, 133, 197, 216, 266, 268-269
Blending, 63, 131, 197, 266, 268-269, 281-282, 288
Board, 19, 38-39, 125-126, 130-131, 133, 241
Boiling, 29-30, 39-40, 196, 250-251
Bond, 20, 33, 50-51, 74, 105, 147-148, 151-152, 154, 161, 175
Bonding, 22, 31, 50-51, 76, 86, 94, 105, 108, 127, 146-147, 149-154, 160-161, 163-164, 171, 177, 183, 202, 245, 263, 269, 280, 285, 290, 309-310, 327-328, 333-334
Bone, 51-52
Bovine serum albumin, 55, 284
Branched, 3, 87, 171, 285
Breaking, 19, 108, 214
Brittle, 80, 242
Brittleness, 244, 288
Brownian motion, 329
Building, 20, 48, 51, 58, 73, 107, 111, 118, 120, 136, 138, 306
Bulk, 12, 39, 239, 243, 309, 331
Butadiene, 125, 130, 134

Butanol, 233
Butyl acrylate, 320
By-product, 12, 18, 240, 251, 260, 266

C

Calcium carbonate, 25, 29, 35
Calcium oxide, 17
Calorimetry, 197, 200, 212, 241, 243, 247, 249, 251, 254-255, 330
Capacity, 26, 38, 152, 155, 161-162, 193, 241, 243, 249-251, 279, 287, 295
Capillary, 216, 244
  rheometry, 216
Carbon dioxide, 2, 14
Carbon fibre, 66, 145, 260
Carbon tetrachloride, 231, 233,
Carboxyl, 35, 261, 265
Carboxymethyl cellulose, 52, 293, 321, 333, 335
Catalyst, 18, 63, 70, 194
Cavity, 28, 65, 68-69, 71, 87, 162
Cell, 3, 18-19, 48, 86-87, 133, 168-169, 171, 173, 245, 294, 307-308, 310, 316
  Cellular, 55, 306-307
Cellulose, 2-5, 7-8, 11-12, 15-17, 25, 29, 33-34, 47-59, 69, 74, 86-87, 89, 97-98, 106, 112, 116-117, 129, 139, 145-146, 167-173, 178, 183, 188, 193-195, 197, 199, 201, 203, 205, 207, 209, 211, 213, 215, 217, 219, 221, 223, 225, 227, 230, 237, 240, 244-245, 259-260, 262-263, 277-285, 287, 289-299, 301-303, 305-321, 323-331, 333-335, 337-341, 343, 345, 347
  acetate, 53, 55-56, 282, 330
  acetate butyrate, 282, 330-332
  microfibre, 283
  nanocrystals, 117, 277, 279-282, 284-285, 287-293, 295-296, 305-307, 309, 311-335, 337, 339, 341, 343, 345, 347
  nanofibres, 280, 284, 288, 292, 295, 316
Ceramic, 85, 160, 254, 261
Chain, 3, 30, 50, 86, 106, 124, 170-171, 194, 202-204, 218, 220, 223, 246, 291, 295, 309-312, 322, 330, 334
Chamber, 197, 231
Char, 211, 315
Characterisation, 98, 106-107, 110, 194, 197, 270, 293, 318
Chemical, 2, 5, 8, 12, 15-17, 19, 21-23, 35, 37, 48-49, 51, 63, 74, 85-87, 89, 94, 97-98, 105, 116, 121-122, 125, 127, 135, 140, 145, 149, 151-153, 163, 169-172, 175, 179, 195-196, 203, 225, 231-236, 240, 244-248, 252-254, 257, 259-

260, 265-267, 272, 277, 279, 282, 285, 288-289, 296-298, 302-303, 305, 309, 317, 321-323, 325, 330, 332, 339-340
bonding, 153, 163
composition, 12, 15, 22, 48, 86, 89, 94, 97-98, 169-172, 196, 236, 240, 244, 259
modification, 16-17, 19, 37, 94, 288, 321-322
properties, 23, 51, 63, 172, 195
reaction, 19
resistance, 21, 152, 231-235
structure, 116, 122, 171, 309
Chemistry, 7-8, 10, 45, 81, 102, 113, 149, 171, 179, 189, 225-227, 244, 252, 255, 297-298, 301, 317, 325, 336-337, 339, 343, 345
Chiral, 181
Chitosan, 52, 55, 216, 278, 286-287, 326, 335
Chromated copper arsenate, 330-332
Circular, 27, 38, 293
Clamping, 32, 157, 162, 176
Classification, 11, 119, 122, 167, 244
Clay, 25, 52-54, 115, 126, 129
Clean, 151, 175, 239-241, 243, 245, 247, 249, 251, 253, 255, 257
Cleaning, 54, 151, 240, 249-251
Cleavage, 148, 152, 265, 313
Coating(s), 1, 63, 106, 195, 292, 321, 324
Coefficient, 67, 126-127, 223, 247, 284, 327, 333-334
Collagen, 53, 55, 287
Colloid, 255, 272
Colour, 27, 122, 194-195
Combustion, 121, 242-243
Co-monomer, 291
Compatibiliser, 19, 26-35, 195, 216, 267-269, 324
Compatibility, 86, 108, 117, 160, 261, 267-268, 321, 324, 333-334
Compatible, 56, 125, 135, 152, 193, 283
Component, 2-3, 20, 86, 107, 116-117, 146, 149, 155, 169-170, 193-195, 230, 240, 261, 265, 279, 283
Composite, 4-6, 12, 15-17, 19-20, 22-24, 26, 31, 33, 35-36, 38-39, 42, 44, 47-49, 51-57, 59, 61, 63-72, 74-76, 78-82, 85, 87, 93, 95, 97, 100-101, 106-107, 110-111, 113, 115, 117-118, 124-126, 128, 130-133, 140, 142, 145-148, 151, 154-164, 167-168, 173, 175, 177-179, 181-183, 185, 188-191, 193, 197-200, 207, 214-215, 217, 223, 230-231, 234, 244, 246-247, 251, 253, 255, 261-263,

267-273, 290, 305, 310, 320, 322, 324-325, 328, 330, 341, 344-345
Composition, 12, 15, 17, 22, 48-49, 51, 54, 86, 89, 94, 97-98, 116, 120, 134, 146, 159, 169-174, 196, 218, 222, 236, 240, 244, 259-260, 279, 291, 308
Compound, 66, 73-74, 86-87, 115, 117, 122, 135, 138, 178, 268
Compounding, 27-28, 33, 36-37, 75, 117, 122, 194, 197, 244, 295
Compression, 5, 64-68, 73, 75, 87, 89, 94, 97, 106-108, 125, 128, 132, 181, 184, 197, 231
strength tests, 108
Compressive stress, 109
Concentration, 89, 97, 122, 149, 159-160, 198, 214, 235, 245, 290, 314-316, 329
Condensation polymerisation, 108
Conductivity, 39, 107
Construction, 5, 22, 58, 73, 118, 120, 136, 138, 145, 167, 176-177, 245
Consumption, 5, 56, 73, 75, 77, 120, 176, 186, 230, 295, 306, 335
Contact angle, 199, 207-208
Contaminated, 122, 151
Contamination, 239, 248
Contraction, 54, 310
Conversion, 15, 29, 259-260, 264, 281, 316
Converting, 121
Cooling, 21, 38, 160, 162, 200
Copolymerisation, 175
Copper, 126
Core, 177, 308
Correlation, 34, 217, 223
Corrosion resistance, 63
Cosmetics, 185, 287
Cotton, 11-13, 23, 48, 50, 53, 62, 74-75, 86, 115, 145, 167, 170, 172-173, 186, 229, 262, 270, 285, 287-288, 316-319, 321, 327, 331-332, 335
Coupling, 19, 116, 118, 125-127, 129-133, 135, 153, 175, 183-184, 194, 245, 323
Crack, 31, 184-185, 214-215
Creep, 21-22, 36, 129
resistance, 22, 36
Cross-section, 13, 306, 318-319
Crosslinked, 53, 55, 88, 131, 216, 245, 279
Crosslinking, 55, 108, 131, 194, 202, 212, 218, 223, 267, 284, 335
Crude oil, 240, 248, 250-251
Crystal, 15, 171, 198, 268, 310-312, 318
Crystalline, 3, 15, 95, 169, 171, 181, 200, 202, 242, 262, 268, 279-282, 291,

293, 305, 307-313, 322-323, 330, 332-335
Crystallinity, 15, 50-51, 53, 86, 200, 202, 212, 215, 223, 247, 268-269, 282, 291-292, 310, 316, 330, 332
Crystallisation, 15, 212, 322, 331-332
enthalpy, 212
temperature, 212
Cure, 64, 88, 94, 267
rate, 267
time, 64
Curing agent, 64, 267
Curing kinetics, 267
Curing reaction, 21, 37
Cutting, 16, 27, 76-79, 151
Cyclic, 180, 184-185

D

Damage, 63, 68, 76-77, 79-80, 150, 159, 179, 184, 246, 294
Damping, 26, 150
Debonding, 80, 108, 184
Decomposition, 91, 111, 184, 211, 232, 236, 243, 267
Defect, 305, 309
Deform, 158, 215, 262
Deformation, 76, 200, 202-203, 244, 264, 310, 312
Degradability, 179, 187, 195, 295-296
Degradation, 9, 14, 99, 135, 173, 181, 193, 202, 210-211, 218, 220, 223-225, 227, 230, 234, 239-242, 244-248, 268, 275, 286-287, 289, 295-296, 298, 315, 333
temperature, 247, 268, 286, 315, 333
Dense, 326, 334
Density, 13-15, 20, 25-26, 39, 47, 56, 72, 75, 85, 117, 121-122, 125, 127, 130-131, 133, 145, 167, 173-174, 176, 178, 182, 187, 193-197, 199-201, 203, 205, 207, 209, 211, 213, 215-217, 219, 221, 223, 225, 227, 229-230, 242, 244-245, 251, 268, 279, 305, 315, 324, 327
Derivative(s), 51, 55, 195, 270, 282, 286, 290
Detector, 110, 199
Diameter, 13, 15, 23-24, 69, 72, 77-79, 87, 156, 158, 168-169, 173, 178, 195, 241, 283, 318, 325, 327
of nanocrystals, 327-328
Dibutyl phthalate, 267
Dichloromethane, 323
Dicumyl peroxide, 132

Die(s), 16, 27, 72, 80-81, 87, 132, 216
Dielectric, 160, 246
Differential scanning calorimetry, 197, 200, 212-213, 247, 330, 332
Differential thermogravimetric analysis, 210
Diffraction, 15, 95, 243, 310-311, 332
Diffusion, 20, 126-127, 153, 194, 199, 223, 243, 246, 284, 312-313, 333-334
  Diffusivity, 127, 334
Diltute, 97, 265, 316-317
Dimensional stability, 19, 118, 284
Dimethyl sulfoxide, 321
Dipped, 37, 72
Discolouration, 194-195
Discontinous, 87, 261
Dispersing, 107, 194, 260, 282, 290-291, 295
  Dispersion, 19, 25, 34, 93, 126, 178, 195, 245, 268, 281-282, 285, 290, 292, 295, 312-313, 320-324, 326, 330, 334
Dissipation, 33, 87
Dissolution, 54, 334
Distribution, 23-24, 27, 34, 88, 149, 155, 157, 161, 176, 212, 214, 243-244, 246, 293, 313, 326
Dried, 27-28, 37, 66, 68-69, 71, 88, 97, 106, 117, 197, 199, 230-232, 294, 320, 322, 324, 330
  Drying, 37, 72, 151, 281, 323, 329, 333
    time, 151
Drug, 52-55, 279, 322
Ductile, 48, 80
DuPont, 32
Durability, 48, 55, 126, 232
Dynamic, 26, 32, 44, 155, 180, 200, 216, 218-223, 241, 310, 328
  mechanical analysis, 32, 44, 180, 328
  mechanical properties, 26, 32

E

Edible, 285, 287, 295, 335
Efficiency, 5, 34-35, 61, 63, 75, 105, 147, 156, 305
Elastic modulus, 53, 293, 310-312
Elasticity, 53, 105, 216, 309
Elastomer, 126, 248
Electric, 149, 186, 246, 315
  Electrical, 5, 14, 26, 35, 47, 52-53, 55-56, 73, 107, 153, 179, 228, 246, 267,

325
Electron, 57, 88, 107, 110, 196-197, 199, 205, 242, 282, 316, 318, 325, 327
Electronic, 88, 134, 136-137, 336
Electrostatic, 55, 153, 318, 325-326
Elongation, 13-17, 19, 30, 130, 173, 180, 246, 248, 268, 270, 292, 329
at break, 16-17, 19, 30, 130, 246, 248, 329
Embedded, 47, 85, 138, 181, 325
Emulsion, 133, 181, 250
End group, 116, 322
Energy, 19, 31-33, 35, 56, 105, 107, 116-117, 120-121, 138, 145, 160-162, 164, 176, 183-184, 198-199, 207-208, 215, 229, 246-247, 253, 259-260, 263-265, 271, 281, 291, 305-306, 312, 321, 335
consumption, 120, 306, 335
Engineering, 7-9, 22, 42, 45, 51-52, 57-59, 63, 80-81, 83, 85, 102-103, 112, 134, 137, 139-141, 143, 145, 147, 158, 165, 173, 188-189, 191, 202, 224-225, 227-228, 236, 238, 252-254, 256-257, 262, 272-273, 279, 294-295, 298, 300-302, 336, 339-342, 344, 346
Enthalpy, 200, 212, 247
Environment, 1-2, 6-7, 56, 80-82, 105, 111-112, 120, 139, 188, 190, 193, 229, 237, 240, 249, 255, 260, 262, 271, 299, 306, 333, 337
Environmental, 1-2, 5-6, 11, 34, 47-48, 57, 61, 80, 105, 111, 118, 120, 145, 150-152, 174-175, 177, 179, 186, 193, 199, 226, 229, 249, 251, 253-254, 257, 262, 265-266, 269, 275, 277, 290, 296, 305
protection, 262, 290
Enzyme, 53, 55
Epoxidised soybean oil, 180, 290-291
Epoxy, 36, 47, 62, 64, 66, 71, 73, 77-78, 80, 153, 163, 176, 179-180, 184-185, 189, 263, 266-267, 290-291, 320, 324
Equipment, 20, 27, 69, 87, 145, 162, 186, 193, 241, 285
Esterification, 94, 175, 269, 316, 322, 324
Ethanol, 86, 266
Ethylene, 126, 245, 248, 321
-propylene-diene terpolymer, 248
Eucalyptus globulus cellulose nanocrystals, 290
Evaporation, 56, 281, 320-322, 327, 329-330
Expandable polystyrene, 125
Expansion, 156, 247, 336
coefficient, 247
Exposure, 65, 68, 97, 126, 132, 161, 194-195, 197, 199, 202, 204-207, 209-210, 214-216, 218, 221-222, 268

Extension, 77, 181
Extracellular matrices, 51, 55
Extraction, 86, 105, 168, 178, 265, 312
Extruded, 68, 124-125, 282, 324, 332
Extrusion, 5, 68-69, 75, 87-88, 107, 120-122, 125-128, 130-133, 181, 320, 323-324, 329-330

F

Fabric, 24, 177, 181
Fabrication, 6, 22, 67-68, 70, 72-73, 106, 111, 267-268, 271, 283
Failure, 13, 22, 25, 30-32, 63, 79, 93, 108, 154, 156-157, 160, 162-165, 184-185, 246
Feedstock, 47, 57, 120, 259, 265, 289
Fibre, 2-4, 11-37, 39, 41, 48-49, 54-55, 61-64, 66-68, 70-81, 85, 87-89, 91, 93-95, 97-98, 100, 105-111, 115, 117, 119-121, 123, 125, 127-131, 133-139, 141, 143, 145-147, 149, 151, 153-155, 157, 159-165, 167-171, 173-179, 181-187, 189, 191, 207, 214-215, 229-234, 236, 244-246, 260-262, 267-268, 270, 292, 308, 310, 325, 331
-reinforced plastics, 145, 167, 182, 184, 186
Fibril, 183, 280-281, 307-308
Fibrous, 54, 102, 167, 177, 195, 240
Fickian diffusion, 284
Filament, 13, 64
Filler, 4, 19-20, 23, 25-26, 28, 35, 88, 130, 193-195, 197, 214, 239-241, 243-249, 251, 253, 255, 257, 260-261, 266, 268-270, 282, 285, 312, 320, 323, 327-334
Filling, 245, 247, 324
Film(s), 56, 67-68, 74, 149, 199, 207, 267, 269-270, 282, 284, 286-287, 293, 320, 324-327, 327, 329, 332, 334-335
Filtration, 313, 324
Fire, 27, 36
Flame retardancy, 126
Flax, 3-4, 11-13, 48-49, 62, 66, 68, 71, 74-75, 86, 99, 115, 128, 145, 167-169, 172-177, 182-186, 188, 229-230, 262, 311, 319
Flexibility, 15-16, 56, 66, 86, 117, 306, 309, 329
Flexible, 22, 56-57, 66, 286, 326, 336
Flexural, 28, 30-31, 33, 39-40, 67, 107-109, 149, 176, 180-182, 245, 248, 267, 269
modulus, 30-31, 39, 180, 245, 248
storage modulus, 33
Flow, 21, 28, 35, 37, 70, 120, 122, 127, 194, 196, 200, 213, 216, 222, 228, 282,

291, 314
Fluid, 56, 224
Fluoropolymers, 151
Foam, 120, 125, 133
Food, 9-10, 40-41, 81, 124, 177, 181, 239-240, 245, 259, 262, 279, 287, 297, 301, 335-336, 346
packaging, 245, 279, 287
Force, 43, 54, 65, 77-79, 116, 148, 157, 182, 310-311
Formation, 18, 20, 27, 33-34, 51, 55, 153, 158, 175, 195, 202-204, 207, 215, 223, 226, 245, 247, 284-285, 292-293, 314, 318, 323, 328, 330, 334
Forming, 18, 20-21, 34, 168, 282
Fossil fuel, 120
Fourier-Transform, 35, 122, 196, 198, 200, 264
infrared, 35, 88, 92, 94, 196-198, 200-204, 223, 264, 269-270, 273
infrared spectra, 92, 198
infrared spectroscopy, 94, 197-198, 203-204, 273
Fraction, 33-34, 64, 66, 78, 97, 176-177, 182-183, 200, 315, 331
Fractionation, 266
Fracture, 14, 22, 93, 128, 131, 183
Frame, 137-138
Free energy, 199, 207, 247
Free radical polymerisation, 261
Frequency, 32, 89, 160-161, 200, 217-221, 223, 247, 290, 328
Friction, 26, 33, 67, 107, 184
Fuel, 1, 5, 22, 61, 73, 75, 120, 176-177, 186, 230, 239, 248-251, 254, 260, 264, 271, 277, 305
Functional group, 263, 290
Fungi, 2, 50
Fusion, 162, 212
Fusion enthalpy, 212

G

Galactose, 86, 171, 240
Gas(es), 6, 150-151, 161, 241-242, 305, 312
Gel(s), 50, 54-55, 64, 67, 70-71, 281-282, 284-285, 287, 290, 292, 295
Geometry, 76-78, 80, 85, 87, 157, 171, 200, 310
Glass, 3, 16-17, 20, 22-23, 25-26, 29, 33, 35, 47-48, 61, 66, 71-73, 75, 79-80, 89, 97, 115, 117, 120, 123-124, 135-136, 138, 145, 173-174, 176-179, 181-185, 244, 261, 263, 265, 286, 305, 328, 331
fibre, 16-17, 23, 26, 29, 48, 66, 73, 79, 115, 117, 135-136, 138, 178, 182, 184-

185, 244
Glass/glass-rubber transition temperature, 265, 286, 290, 295, 328, 330-331
Gold, 107
Government, 116, 123
Graft, 125-126, 132, 175, 322-323
Grafted, 34, 126, 131-132, 177, 182, 184, 245, 269, 294, 322-324, 330, 332-333
Grafting, 26, 126, 287, 294, 322-324, 326, 330
Green chemistry, 7, 297
Green polymer, 10, 57, 107, 113, 236
Growth, 20, 54, 115, 138, 174, 176, 184-186, 278, 306-307, 326

H

Handling, 22, 294
Hardener, 64, 179
Hardness, 16, 26, 180-181
Healing, 51, 53-54, 294
Health, 14, 61, 177, 290
Heart, 53, 294
Heat, 21, 27, 33, 38, 65, 72, 147, 160, 162, 179, 181, 194, 213, 246, 254, 268, 288
Heating, 32, 38, 67, 69, 87, 107, 147, 161-163, 200, 213, 232, 260, 322
Hemicellulose, 2, 17, 49, 86-87, 97-98, 116, 168-171, 173, 201, 203, 240, 244-245, 259, 262-263, 266, 280, 287, 306-307, 326
High density, 26, 75, 121, 193-195, 197, 199-201, 203, 205, 207, 209, 211, 213, 215, 217, 219, 221, 223, 225, 227, 268
polyethylene, 26, 121, 123-124, 126-133, 135, 193-223, 225, 227, 268
polyethylene-graft-maleic anhydride, 132-133
High molecular weight, 171, 181, 286, 309
High pressure, 67-68, 193, 341
High speed steel, 77
High temperature, 16, 21, 27, 132, 137
High viscosity, 70
Hole, 76-78, 80, 156-157, 159
Homogeneity, 19, 88, 173, 266, 324, 330
Homogeneous, 54, 194, 282, 292, 312, 320-321, 326
Homopolymer, 28, 50, 86, 196
Horizontal, 37
Humidity, 18, 149, 197-198, 202, 208, 231, 270, 312, 331
Hybrid, 22, 62, 73, 79, 186, 239, 248, 251, 284, 287

Hydraulic, 27-28, 38, 69
Hydrocarbon, 245, 250
Hydrochloric acid, 232, 235, 315-316, 318, 322
Hydrogel, 287
Hydrogen bonding, 50-51, 86, 171, 269, 280, 285, 290, 309-310, 327-328, 333-334
Hydrolysis, 117, 263, 280-281, 288-290, 311-316, 318, 320, 322, 329, 333
Hydrophilic, 29, 35, 116-118, 171, 173, 175, 193, 207-208, 230, 234, 236, 244-246, 288, 321, 323-324, 332-334
Hydrophilicity, 85, 292, 335
Hydrophobic, 3, 19, 85, 87, 117-118, 125, 172, 175, 181, 193, 195, 207, 244, 246, 250, 266, 281, 291, 324, 326
Hydroxyapatite, 51-53, 141
Hydroxyl group, 18, 34
Hygroscopic, 13, 17, 86
Hysteresis, 244

I

Immersion, 208-209, 248, 324
Immiscible, 269
Immobilisation, 55
Impact, 1, 16, 19, 22, 26-28, 31, 35, 37, 67, 73, 76, 118, 125, 128, 130, 133-134, 137, 149, 169, 177, 179-184, 186-187, 197-198, 215-216, 246, 260, 267, 269, 305, 312, 327
Impact resistance, 149, 179, 312
Impact strength, 16, 19, 22, 31, 35, 67, 128, 130, 133, 177, 179, 181, 183-184, 198, 215-216, 246, 269
Impregnated, 20, 26, 71, 73, 87
  Impregnating, 294
  Impregnation, 37, 54, 72, 75, 323-324
Impurities, 122, 127, 134
In situ, 28, 37, 126, 281-282, 287, 292-293, 322-323
In vitro, 284, 289
Indian Jute Industries' Research Association, 22, 38-40, 42, 46
Induced stress, 159
Industry, 5, 16, 61, 73, 75, 83, 87-88, 119, 122, 145, 154, 177-178, 185-187, 196, 230, 240, 245, 249, 251, 255, 260, 263, 267, 270, 288, 306
Infrared, 34-35, 88, 91, 122, 196, 198, 200, 264, 269
  spectroscopy, 198, 200
Infrastructure, 123

Initiation, 31, 184
  Initiator, 71, 132, 268, 323
Injection, 5, 28, 64, 68-70, 75, 87-88, 107, 128, 133, 136-137, 165, 178, 181
  moulding, 5, 28, 64, 68-69, 75, 87-88, 107, 128, 136, 181
Inlet, 70-71
Inner diameter, 241
Inorganic, 2, 5, 148, 167, 239-240, 246, 248, 251, 265, 293
Insoluble, 87, 172, 240
Instability, 19, 118, 232
Institute, 41, 191, 253, 255, 257, 296, 336
  Institution, 81
Insulation, 11-12, 14, 73, 119, 125, 177, 241, 246
Intensity, 91, 96, 197-198, 203-204, 264
Interaction, 19, 25-26, 32-34, 54, 93, 95, 100, 245, 268-270, 310, 315, 326, 330-331
Interface, 20, 23-24, 30, 32, 34, 105, 147, 152-154, 160, 162-163, 175, 177, 182, 184, 244-246, 255, 272, 283, 322
International Organization for Standardization (European Standard), 198
Interphase, 20, 24, 184
Intrinsic, 117-118, 306, 323
Ion, 56, 260
  Ionic, 54, 153, 264, 266, 283-284, 287, 318, 326
  Ionisation, 151
Irradiation, 55, 212
Irreversible, 19, 55, 323
Isobutanol, 231, 233
Isolate, 263, 321
  Isolation, 87, 263, 312, 318
Isotactic polypropylene, 321
Isothermal, 247
Isotropic, 24
Izod, 28

J

Joining, 3, 76, 106, 145-147, 149-151, 153-155, 157-165
  Joint, 146-164
Jute, 4, 11-17, 22-26, 28-30, 32-35, 38-43, 46, 48-49, 62, 68, 73, 75, 86, 120, 145, 167, 172-174, 178, 183-184, 186, 229-230, 262

K

Kevlar, 292, 306, 309
Kinetic, 247, 249-250

L

Lamella, 170, 307
Laminate, 40, 71, 78, 80, 261
Latex, 133, 281, 295, 320, 328-329
Lattice, 310
Layer, 20, 57, 67, 72, 74, 89, 94, 97-98, 151-153, 168-170, 307, 315, 326
Layer-by-layer, 326
Length, 13, 15, 23-24, 27, 34, 37, 69, 87, 157, 167-168, 171, 173, 176-177, 182-184, 195, 198, 230, 280, 306, 318-319, 324, 327, 329, 333
  of nanocrystals, 327-328
Light, 5, 56, 61-63, 75, 145, 194, 197, 202, 230, 268, 292, 312, 318, 335
Lignin, 3-4, 6, 12-17, 49, 86-89, 97-98, 106, 116, 145-146, 168-173, 177-178, 193-195, 197, 201, 203-204, 212, 216, 240, 244-245, 259-271, 273, 275, 280, 306-308, 313
Lignocellulose, 12, 18, 33, 106, 175, 263
Linear, 3, 21, 50, 86, 106, 108, 170, 200, 242, 245, 268, 285, 287-288, 310-311, 328-330
Linkage, 3, 18, 30, 35, 322
Liquid, 19, 21, 25, 54, 64, 66, 150, 218, 241, 249, 266, 283-284, 308, 312, 323, 333
Load, 14, 20, 31, 89, 108, 146-149, 152, 155, 158, 160-163, 175, 181, 187, 215, 291, 322, 331
  Loading, 24-27, 29-31, 35, 108-111, 126, 128, 130, 132-133, 146, 152, 160, 162-164, 176, 184, 187, 194, 207-209, 214, 232-236, 246-248, 295, 331-332, 335
Long chain, 86
Long-term, 56, 126-127, 202, 232, 246, 268
Loss, 33, 91, 99-100, 111, 134, 183, 194-195, 211-212, 218-219, 232, 235-236, 246, 291
  modulus, 33, 218-219
Low density, 14, 25, 85, 117, 121, 125, 133, 167, 176, 178, 182, 187, 216, 229, 244, 268, 279, 305, 324
  polyethylene, 121, 124, 130-131, 133, 216-217, 268-269, 324, 332
Low molecular weight, 4, 21, 265-266, 286-287, 292
Low temperature, 53

Lysostaphin, 53, 55

M

Machine, 28, 37, 65, 67, 87, 107, 157, 198
Macromolecular, 10, 87, 139-140, 188, 300, 338, 341-344, 346
  Macromolecule, 170, 230, 264
Macroscopic, 107, 194, 223, 280, 305
Main chain, 202
Maintenance, 147, 155, 159
Maleated, 27-28, 245, 332
Maleic anhydride modified polypropylene, 183-184
Management, 21, 42, 139, 142, 179, 226, 253, 257
Manufacture, 36, 118, 138, 158, 176
  Manufacturing, 20, 28, 41, 44, 59, 73-74, 83, 87, 101, 115, 139-142, 146, 154, 165, 168, 173, 175, 179, 185, 187-191, 194, 227, 251, 262, 272-274, 304, 344-345
Market, 1, 115-116, 134-135, 139, 149, 179, 262, 266, 279, 298
Material(s), 4, 1-6, 8-10, 14, 16-17, 20-24, 26, 28, 33, 41-45, 47-55, 57-63, 65-69, 71, 74-77, 79-83, 85-89, 93-97, 100-102, 106-107, 111-113, 115-117, 119-124, 126-127, 130-136, 138-142, 145-147, 149-165, 167, 169, 173, 176-191, 193-197, 199, 201, 203, 205, 207, 209, 211, 213, 215-219, 221, 223-227, 229-230, 232, 236, 238-239, 243-248, 251-256, 259-263, 265-275, 277-279, 284, 286-288, 290-292, 294, 296-299, 301-304, 306-314, 317, 320, 323, 325, 328-331, 333-338, 336, 341-342, 344-346
Matrix, 3-4, 15-16, 19-25, 29-35, 47-48, 54-56, 61-65, 67-68, 74-76, 80, 85, 87-89, 93-94, 96-97, 99-100, 105-111, 115-117, 125-127, 130-131, 133, 135, 137, 147, 161, 163, 168-169, 175, 177-178, 181-185, 187, 193-195, 214-216, 230, 244-246, 248, 261-263, 267-270, 282-285, 287, 290-292, 294-295, 307, 312, 315, 320-322, 324, 326-335
Measurement, 307, 333, 337
Mechanical properties, 1, 4, 14-16, 19, 23-27, 29, 32, 34-35, 42, 44, 47-49, 51-55, 61-63, 71, 85, 88, 93, 95-96, 100, 105-106, 108, 111, 117-119, 122, 126-135, 145, 168-169, 172-173, 175, 177-179, 181-184, 187, 194-195, 197, 212, 214, 223, 244-245, 247-248, 267, 270, 279-280, 282, 284-288, 290-295, 306, 309-310, 322, 325-330
Mechanical resistance, 80
Mechanical strength, 4, 230, 284, 309
Mechanical testing, 74, 198
Mechanism, 19, 72, 93, 152, 267, 284, 310
Melamine-formaldehyde, 36, 324

Melamine-urea-formaldehyde, 36, 39
Melt, 22, 28, 75, 125, 127, 130-131, 149, 181, 193, 196-197, 200, 216-218, 279, 281-282, 285, 288, 295-296, 320, 323-324, 329-330
  flow index, 28-29, 127, 196
  fracture, 131
  rheology, 200, 216
  temperature, 181
  viscosity, 216, 218, 329
  Melting, 67, 69, 162, 172, 200, 212, 281, 290, 330-332
    point, 67, 212
    temperature, 69, 172, 212, 281, 290, 330-332
Membrane, 53, 55-56, 284, 294-295, 337, 345
Mercerisation, 15, 310
Metallic, 56-57, 66-67, 153-156, 158, 160
Metastable, 171
Methanol, 231, 233, 307
Microcrystalline, 280, 316
  cellulose, 316-317, 319, 335
Microfibrillated cellulose, 280-281, 293, 295, 329
Microscope, 57, 80, 199
  Microscopy, 88, 107, 196-197, 199, 205, 282, 311, 318, 327
Microstructure, 31, 34-35, 159-161, 264, 328-329
Microwave, 146, 159-164
Migration, 31-32, 331
Mill, 88, 259-260, 267
Mineral filler(s), 25, 28, 36
Miscibility, 269
Miscible, 134, 269
Mixed, 15, 25, 27-28, 64, 68-69, 89, 94, 97, 106, 122, 126-127, 129, 134, 138, 147, 179, 181, 186, 197, 230
  Mixer, 28, 34, 68-69, 131, 197
  Mixing, 25, 27-28, 34, 64, 68, 71, 75, 110, 125-126, 132, 175, 193, 268, 279, 320, 323, 335
  Mixture, 106-107, 115, 181, 186, 240, 281, 310, 316, 320
Mobility, 30, 32-33, 212, 331, 334
Model, 55, 79, 222, 283, 308, 310
  Modelling, 76-77, 131, 187
Modification, 16-19, 37, 55, 88, 94, 97, 118, 151, 177-178, 185, 260, 288, 293, 295, 321-323
  Modified, 2, 27, 51, 55, 71, 73, 88, 94-96, 100, 176-181, 264, 266, 270, 279,

282, 287, 323-324, 330-335
Modifier, 26-27, 35, 137
Moduli, 30, 33-34, 117, 200, 293, 311
Modulus, 13, 16-17, 20, 26, 30-34, 39, 47, 49, 53, 71, 95-96, 100, 105, 117-118, 120, 128-131, 134, 173-174, 176, 178, 180-185, 187, 198, 214-216, 218-221, 244-245, 248, 268, 270, 282, 285-286, 290, 293, 306, 309-312, 326-330
Moisture, 13-14, 16, 18, 27, 37, 68-69, 97, 118, 126-133, 146-147, 159, 171, 173, 175, 193-194, 197, 230-234, 242, 244-246, 285, 295, 331
Molecular structure, 170, 227
Molecular weight, 4, 21, 134, 171, 181, 197, 202, 213, 264-266, 286-287, 292, 309
Molten, 27-28, 69, 87, 194, 216, 228
Monitor, 107
Monitoring, 56
Monomer, 194, 282, 288, 323
Montmorillonite, 53-54
Morphology, 3, 15, 89, 97-99, 107, 172, 174, 194-196, 205, 212, 242, 289, 292, 295, 305, 315, 327-328
Motion, 26, 146, 329
Mould, 22, 28, 38, 64-71, 87, 89, 97, 106, 151
Moulded, 22, 28, 36, 38, 40, 66, 72, 94, 132, 136-137, 177-178, 184, 248
Moulding, 5, 16, 27-28, 38, 64-70, 73-75, 87-89, 97, 106-107, 128, 133, 136, 178, 181, 184, 193, 197, 231, 329

N

Nanocomposite, 51, 54-55, 277, 279, 282-284, 286, 289, 306-307, 320, 323-324, 326, 329, 332-335
Nanocrystalline cellulose, 52, 277, 279, 281, 283, 285, 287, 289, 291, 293, 295-297, 299, 301, 303, 306
Nanofibrillated cellulose, 280-281, 286, 305, 332, 335
Nanoparticle, 312, 315, 321, 326, 329, 333
Nanoscopic, 280
Nanostructure, 51, 293
Nanotechnology, 4, 7, 10, 113, 297-298, 344, 346
Natural fibre composites, 20, 22, 115, 117-119, 121, 123-131, 133-139, 141, 143, 146, 168, 178, 182-187, 191, 268
Natural resource, 252, 271
Natural rubber, 248, 279, 295-296, 320, 329, 334
Negatively-charged, 326
Network, 3, 21, 49, 51, 54-55, 245, 282, 287, 291-292, 294-295, 327-330, 334

Neutral, 4, 97
Newtonian fluid, 224
Nitrogen, 27, 107, 200, 232, 243, 265, 286
Non-linear, 328-330
Non-Newtonian fluid, 224
Non-polar, 175, 217, 244, 246, 250, 295, 321, 323, 333
Non-toxic, 5, 286
Notched, 31, 198
Nozzle, 68-69
Nucleation, 212, 214, 332
Number of cycles, 89, 92-93
Nylon, 3, 25, 62, 66, 135, 263

O

Oil palm empty fruit bunch, 267
Optical, 122, 284, 286
Optimisation, 76-77, 79
Organic, 4, 86, 121, 148-149, 239-240, 248, 251, 259, 266, 281-282, 288, 290, 293, 316, 321-322
Orientation, 15, 23, 26, 34, 48, 71, 76, 87, 157, 173, 291, 308, 310, 329, 334
  Oriented, 24, 48, 67-68, 70, 73, 308, 310
Orifice, 325
Oscillation, 32, 200, 217-218, 222
  Oscillatory, 200, 216
Outlet, 71
Output, 78, 161
Oven, 27, 66, 68-69, 197, 230-231
Overlap, 122, 152, 163
Oxidation, 151, 194-195, 204, 242, 254, 268, 280, 311, 316
  Oxidative, 3, 194, 204, 246-247, 322
    degradation, 246
    stability, 247
Oxygen, 50, 263, 335-336
  transmission rate, 279, 295-296, 335

P

Packaging, 1, 12, 22, 48, 51, 120, 124, 134, 136, 175, 181, 185, 244-246, 269-270, 279, 286-288, 291, 335-336
Packing, 38, 309, 326
Panel, 67, 73, 186, 263, 336

Paper, 12, 20, 22, 38, 40-41, 69, 74, 124, 133, 145, 177, 180, 185, 187, 196, 199, 224, 231, 237, 239, 259-260, 263-265, 267, 293, 327, 335-336
Particle(s), 19, 25, 48, 53-54, 75, 85, 88, 107, 126, 130-131, 133, 151, 153, 194, 207, 214, 242, 245, 261, 268, 277, 305, 312-313, 318, 320-321, 324, 326, 328-329
Particulate, 24, 88, 121
Peel strength, 147
Peeling, 152, 154
Penetration, 54, 312
Performance, 1, 16, 24, 42, 52, 54, 62, 71, 73-75, 80, 93, 115, 159, 168, 174, 177, 179-185, 194-195, 229, 246, 262-263, 277, 279, 285, 288, 296, 308-309, 328-330, 334-335, 338
Permanent, 150, 154-156
Permeability, 287, 334-336
Permit, 159, 306
Peroxide, 132
Petrochemical, 290
pH, 97, 264, 287, 345
Phase separation, 55, 284
Phenol-formaldehyde, 36, 39, 106-108, 111, 175, 324
Phenolic resin, 20, 115, 265
Physical properties, 116, 125, 127, 130, 134, 138, 170, 194, 229, 259, 263, 268, 306, 309
Physicochemical properties, 55, 232, 236
Physics, 9, 44, 190, 298-299, 302, 338, 343, 345
Pigments, 239-240, 251
Plant, 4, 13, 15, 25, 38, 48-49, 66, 86-87, 115, 119, 121, 167-179, 181, 183, 185-187, 189, 191, 201, 226, 230, 241, 251, 260, 262-263, 265, 271, 288-289, 293, 305-306, 308-309, 311, 316
Plasma, 75, 151, 175
Plastic, 29, 35, 62, 64, 73, 118, 120-122, 125-127, 134, 151, 155, 165, 176, 193-194, 244-246, 291, 312
  Plasticisation, 87, 246, 331
  Plasticised, 270, 329-333, 335
  Plasticiser, 216, 291-292
  Plasticising, 287, 331, 334
Plate, 38, 64-67, 77, 80, 156-158, 200, 241
Polar, 19, 26, 34, 175, 193, 207, 217, 244, 246, 250, 295, 321, 323, 326, 333
  Polarity, 199, 310
Poly(ε-caprolactone), 322-324, 326, 330, 332, 334-335

Poly(lactic acid) or polylactide, 49, 55, 62, 66-67, 69, 74-75, 161-164, 180-181, 269-270, 288-289, 321-322, 324, 335
Poly(L-lactic acid) or poly-L-lactide, 181, 183, 270
Poly(styrene-*co*-butyl acrylate), 320, 328-329
Poly-L-lactide, 181
Poly-p-phenylene, 312
Poly-*R*-3-hydroxybutyrate, 291-292
Polyacrylic acid, 335
Polybutylene succinate, 269
Polycarbonate, 22, 66, 134, 137
Polycondensation, 181
Polyester, 20, 36-37, 40, 47, 62, 64-66, 70, 72, 75, 77-79, 88-89, 93-97, 100, 115, 135, 178-179, 181, 183, 185-186, 288, 290
Polyethylene, 1, 26, 48, 62, 75, 121, 124, 133, 135, 193-195, 197, 199-203, 205, 207, 209, 211, 213, 215-217, 219, 221, 223, 225, 227, 239, 245, 248, 263, 268-269, 290-291, 312, 322, 324
  glycol, 290-293, 322
  oxide, 269
  terephthalate, 120-121, 123-124, 130, 135, 268-269
Polyglycerol, 180
Polyhydroxyalkanoates, 49, 62, 278, 291, 295
Polyhydroxybutyrate, 75
Polyisoprene, 295
Polymer, 2-10, 15, 19-21, 23-26, 30-35, 41-45, 47-50, 55, 57-58, 60-65, 67-69, 72-81, 83, 85-87, 89, 95, 102-103, 105-113, 115-117, 119-125, 127-131, 133-135, 137-143, 145, 147, 153, 164, 172-173, 175, 179, 181, 188-190, 193-195, 197, 202-204, 212, 214-215, 222, 224-227, 229-231, 234, 236-240, 244-246, 248, 251, 253, 255-256, 260-263, 265-266, 268-269, 271-272, 274-275, 278-279, 281-282, 285, 289, 291-292, 295-307, 309, 311, 313, 315, 317, 319-325, 327, 329-331, 333-347
  matrix composites, 61, 147
  resin, 106, 108, 110-111, 129, 231, 234, 236, 262
  Polymeric, 9, 19, 26, 44-45, 56, 75, 88, 105, 145, 151, 175, 190, 224, 229, 232, 236, 239, 248, 255-256, 261, 287, 290, 312, 320, 326-327, 330-333
  Polymerisation, 3, 18, 21, 86, 108, 133, 171, 181, 261, 281-282, 287, 292, 320, 322
Polymethyl methacrylate, 55
Polyolefins, 26, 34, 120, 124-125, 130, 134, 151, 194, 202
Polyoxyethylene, 320-321, 325, 331, 333

Polypropylene, 1, 25-30, 32-35, 48, 62, 66-69, 75, 78, 121, 124, 126-131, 133, 135-136, 161-163, 176-178, 182-184, 191, 193, 195, 239, 244-248, 263, 268-269, 291, 321, 332
Polypropylene-graft-maleic anhydride, 126, 130-131
Polystyrene, 62, 120-121, 125-126, 129, 133-135, 193, 231, 239, 263, 268, 326, 331
Polytetrafluoroethylene, 151, 161
Polyurethane, 36, 72, 263, 290, 322, 325
Poly(vinyl acetate), 147
Poly(vinyl alcohol), 52-54, 64, 269-270, 320, 324-325, 331-332, 334-335
Poly(vinyl chloride), 26, 66, 120-121, 129, 194, 263, 320
Poplar seed floss, 193, 195-201, 203-223, 225, 227
Pore, 242-244
  size, 243-244
Porosity, 14, 56, 255, 287, 292, 327
Porous, 50, 54-55, 169, 243, 247, 265, 327
Potassium bromide, 198
Potential, 2, 4-6, 20, 25, 31, 47, 54-56, 105-107, 109, 111, 113, 116, 134, 146, 164, 173, 180, 187-188, 193, 229-230, 244-245, 247, 259, 261-263, 265-271, 273, 275, 277, 279, 281, 284, 286-288, 290-292, 294, 309-310, 336
Power, 26, 56, 77, 121, 161, 222, 228, 246, 254
Precursor, 20, 87, 193, 265-266
Preparation, 4, 6, 15, 26, 47, 54-55, 62, 105, 147, 150-151, 154-155, 159, 164, 195, 230, 251, 267, 270, 281, 284, 293, 318, 320, 323, 326, 328, 332
Prepreg, 37-38, 72
Press, 10, 23, 27-28, 38, 42-43, 67, 102, 113, 139, 165, 188, 197, 226, 236, 255, 272-273, 293-294, 337
Pressure, 19, 21, 28, 37-38, 64-65, 67-69, 71-72, 116, 132-133, 147, 149-150, 152, 161, 193, 199, 241
Pretreatment, 88, 97-98, 151, 175, 184
Price, 117-120, 122, 134, 136, 176, 262
Procedure, 56, 147, 311-313, 321, 323, 326
Process, 1, 27, 33, 35, 37-38, 53-54, 64, 66-74, 76-81, 101, 117, 122, 125, 131-132, 151, 159-161, 165, 175-176, 178, 194, 202-203, 206, 208, 211, 214, 218, 223, 230, 243, 248-249, 251, 260, 264-266, 268-271, 273, 281-282, 284, 292, 294, 310, 312-313, 315, 318, 324-325, 329, 332
  Processability, 196, 277, 288, 291, 296
  Processing, 2, 5, 11, 14, 16, 19, 23-25, 27, 36, 49, 54, 61-71, 73, 75-77, 79-81, 83, 85, 87, 107, 117-118, 121-122, 124, 132, 135, 137, 145-147, 150-151,

161-162, 164, 180, 184, 187, 193-194, 197, 230, 253-254, 261, 268, 271, 277, 281-282, 285, 288, 290-293, 295-296, 298, 306, 320, 323-324, 328-330, 335
Producer, 176, 316
Product, 12, 18, 36, 65, 70, 76, 81, 111, 136-137, 146-147, 165, 168, 181, 240, 242, 246, 251, 260, 266, 270, 273, 291
Production, 3, 6, 12, 51, 55, 59, 64, 66, 68, 73, 86-88, 97, 101, 119-120, 168, 176, 178, 180-181, 187, 191, 193, 195, 229-230, 239-240, 245, 251-252, 259-260, 263-265, 267, 279, 288, 290, 295-296, 305, 315, 325, 335
Profile, 72, 242, 293, 318
Profit, 124, 136
Properties, 1-6, 12-17, 19-20, 23-29, 31-32, 34-38, 42, 44, 47-49, 51-56, 61-63, 69, 71, 73, 76, 79-80, 85, 88, 93, 95-96, 100, 102, 105-108, 111, 116-120, 122, 124-135, 138, 145-147, 150, 153, 160, 167-170, 172-185, 187, 194-195, 197, 200, 205, 207-208, 210, 212, 214, 216, 223, 227, 229, 232, 236, 239-240, 244-248, 251, 259-261, 263, 266-271, 277, 279-280, 282, 284-288, 290-296, 305-306, 309-310, 312, 314-315, 318, 322, 325-330, 333-336
Propylene, 245, 248, 290
Protection, 54, 186, 262, 290
Protein, 266, 287, 321
Prototype, 115, 186
Pseudoplastic, 222
Purification, 49, 230, 312, 322
Purified, 106, 311
Purity, 122, 207, 266, 292
Pyrolysis, 241, 251-252, 296

Q

Quality, 20, 56, 65, 71, 76-77, 80, 117-119, 122, 125, 136, 138, 146, 149, 157, 159, 164, 178, 181, 187, 241, 260, 290, 295
Quality control, 80, 159

R

Radiation, 160-161, 194, 202, 208, 212, 268, 270
Radical, 194, 261, 311
Radius, 243-244
Ratio, 13, 23-24, 61, 69, 87, 89, 91, 97, 126-127, 129, 132, 145, 156, 180, 195, 229, 280-281, 295-296, 309, 315-316, 318-319, 327-328, 335
Raw material, 49, 62, 66, 97, 117, 119, 122, 195, 251, 259
Raw rice husk, 239-242, 246-247, 251

Reaction, 18-19, 21, 37, 88, 97-98, 135, 202, 248, 264, 267, 315-316, 322, 324
temperature, 264
time, 18, 315
Reactivity, 50-51, 248, 265
Reactor, 97, 121, 241, 251, 293
Recovery, 121
Rectangular, 199
Recycle, 134, 246
Recycled, 1, 74, 115-117, 119-127, 129-139, 141, 143, 181, 262, 319
high density polyethylene, 126-133
low density polyethylene, 130-131, 133
polyethylene, 133
polypropylene, 126-129, 131, 133
polystyrene, 125, 133
Reduction, 30, 36-37, 77-78, 173, 183, 185, 195, 270, 287, 312, 327, 333-335
Regeneration, 51-52, 55, 284-285
Reinforcement, 2-5, 20, 23-25, 32, 48, 55, 61-65, 68, 70, 72, 85-88, 94, 105, 107-109, 111, 113, 130, 135, 145, 168, 173, 176, 188, 229-230, 261-262, 267, 270, 277, 283, 286, 288, 291, 322
Reinforcing agent, 193, 214
Relative, 18, 21, 23, 33, 48-49, 145-146, 153, 198, 231, 243, 312
humidity, 18, 198-199, 207, 231, 233-234, 312
Reliability, 187
Renewable resource, 1, 21, 260, 278
Research, 1-2, 6, 11, 22, 36, 39-42, 45-46, 48, 50, 52, 56, 59, 76, 83, 111, 116, 118, 125, 134, 140, 179, 193, 229-230, 239, 252-254, 256, 271-272, 277, 279, 281, 295-296, 304, 306, 323, 336, 341, 344
Residue, 111, 196, 211, 236, 240, 243
Resin, 12, 20, 25, 36-37, 43, 64-67, 69-75, 87-89, 93-94, 96-97, 100, 106, 108, 110-111, 115, 129, 135, 179-181, 184, 194, 230-231, 234, 236, 262, 265-267
transfer moulding, 69-70, 74
Resist, 172
Resistance test, 108
Resolution, 196, 198
Resonant frequency, 32
Retaining, 279, 282
Retention, 33-34, 147
Reuse, 179
Review, 118, 266, 277
Rheology, 43, 194, 200, 216, 223-224, 227-228, 298

Rice husk, 6, 128, 130, 136-137, 239-245, 247-249, 251-253, 255, 257, 262, 319
ash, 239, 241-249, 251, 253, 255, 257
Rigid, 21-22, 48, 119, 246, 262, 326
Rigidity, 4, 15, 173, 262
Ring, 50, 171, 181, 261, 263-264, 307, 312, 322
-opening polymerisation, 181, 322-323
Risk, 21, 27
Roll, 37
Rollers, 37
Room temperature, 22, 64-65, 67, 69, 94, 97, 147, 199, 231
Rotation, 68-69, 259, 309
Roughness, 85, 89, 94, 98, 153, 175, 194, 205, 208
rpm, 27, 69, 77-79, 197
Rubber, 3, 5, 7, 43, 124, 130, 239, 248, 266, 279, 295, 301, 320, 328-329, 334
Rubbery, 218, 328, 330
Rule of mixtures, 128

S

Saccaharum cilliare, 3, 106-111, 273
Safety, 47, 57, 134
Sample, 29, 31-32, 34, 89, 111, 129, 163, 196-200, 207, 211-212, 215, 218, 222, 236, 247, 333
Saturated, 72, 250
Saw, 38
Scaffold, 294
Scale, 20, 38, 49, 73, 132, 181, 185, 187, 194, 266-267, 305-306, 312
Scanning electron microscopy, 88-90, 93-95, 97, 107, 110, 196-197, 199, 205-207, 223, 282, 289, 327
Scattering, 174, 318
Seal, 259, 271
Sectional, 72, 283, 308, 312
Segment 290
Segmental mobility, 334
Semi-crystalline, 95, 169, 171, 334
Sensitivity, 16, 222
Separation, 49, 55, 122-123, 178, 265, 284
Serum, 55, 284, 295
Shape, 21, 23, 48, 61, 65, 68, 72, 76-77, 117-118, 146, 199, 212, 242, 292, 318
Shear, 24, 32, 34, 147-148, 152, 175, 182, 216, 222-223, 281, 290-291, 328-329
modulus, 290

rate, 222
storage modulus, 32-33, 216
stress, 24, 148, 329
Sheet, 28, 65, 73-74, 178, 284, 327
moulding compound, 73-74, 178,
Shrinkage, 19, 87
Silica, 17, 67, 239-243, 245, 248, 252, 293
Silicone, 37, 251, 255
Silk fibroin, 55, 320
Silver, 53, 55-56, 242, 293-294
Size, 23, 32, 37, 64, 71, 87, 106, 160, 194, 231, 242-245, 293, 313, 334
Soaking, 29, 39, 71
Sodium hydroxide, 88-89, 97, 106, 178, 185, 232, 235, 264-265, 284, 310
Soft, 53, 158, 167, 177, 216, 290, 297
Softening, 25, 28, 68, 281-282
Software, 199-200
Solid, 21, 72, 97, 117, 120, 122-124, 130, 134, 151, 156, 195, 198, 218, 225, 243, 264, 269, 282, 290, 320, 325
Solubilisation, 266
Solubility, 36, 51, 97, 261, 264, 266
Soluble, 3-4, 19, 86, 88, 116, 145, 171-172, 244, 264, 281, 286-288, 320-323
Solution, 1, 22, 36-37, 55, 88, 97, 106, 134, 137, 151, 260, 262, 264-265, 270, 284, 320, 324-325, 329
Solvent, 18, 22, 147, 151, 199, 266, 281-283, 286-290, 295, 320-321, 323, 333
Sonication, 280-281, 313, 318
Sorption, 19, 244-246, 249-251, 295, 334
Specific gravity, 5, 16-17, 29-30, 39, 125
Specific surface, 55, 243-244, 281, 326
Specification, 36, 94, 111, 236, 270
Spectra, 91-92, 198
Spectrometer, 198, 264
Spectroscopy, 88, 94, 196-198, 200, 203-204, 273, 311
Speed, 27, 37, 76-79, 131, 197-198
Spherical, 243, 305
Spinning, 16, 178, 288, 325
Square, 24, 158, 318-319
Stabilisation, 328
Stabilise, 54, 333
Stabiliser, 268, 271, 320
Stability, 4, 9, 14, 19, 21-22, 52-53, 91, 110, 117-118, 126-131, 133-134, 147,

151, 178, 181, 202, 208, 210, 212, 224-225, 227, 234, 242-243, 246-247, 261, 268-271, 279, 284, 286-288, 292, 298, 313, 320, 322, 331, 333
Stacking, 23, 67-68, 80, 325
Standard, 39, 76, 89, 106-107, 121, 161, 198, 230, 236, 294
Staphylococcus aureus, 55, 294
State-of-the-art, 187
Static, 155, 185, 199
Steps, 26, 64, 76, 115, 125, 147, 162
Sticking, 25, 64, 67, 71
Stiffness, 4, 16, 19, 34, 47-48, 53, 61, 134, 156, 173-174, 176-178, 182-183, 214, 216, 218, 246, 262-263, 270, 279, 291, 295, 312, 327, 329, 335
Stimulation, 56
Stirrer, 106, 231
Storage, 22, 32-33, 66, 119, 194, 216, 218-221, 286, 291, 328
Storage modulus, 32-34, 216, 218-221, 286, 328
Strain, 26, 30-32, 53, 109, 134, 155, 282, 291, 316
Strength, 4-5, 13-17, 19-20, 22-26, 28-32, 34-35, 39-40, 47-49, 53, 61, 66-67, 71-72, 89, 93, 95-96, 100, 105-108, 117-119, 125, 128, 130-133, 145-149, 151-152, 154-157, 160-164, 173-174, 176-185, 187, 198, 214-216, 223, 230, 244-246, 248, 262-265, 267-270, 282, 284-285, 291-292, 306, 308-309, 318, 327, 329-330, 335-336
Stress(es), 3-10, 19-21, 24-26, 29-32, 34, 41-45, 47, 49, 53, 58-60, 79, 81-83, 89, 92, 101-103, 105, 109, 111-113, 115, 117, 122, 139-142, 148-149, 151-152, 154-155, 157-161, 165, 170, 173-176, 178, 181-183, 185-186, 188-191, 194, 214-215, 217-228, 236-238, 241-243, 248-249, 251-257, 259-263, 268, 270-275, 278, 280, 282-286, 297-304, 306, 309-312, 317, 322, 324-327, 329, 336-346
  relaxation, 21
  -strain curve, 30-31
Stretch, 53
  Stretching, 91, 198, 200-203, 264
Structure, 3, 15, 21, 26, 47-51, 53-54, 56, 86-88, 116-117, 122, 146, 152, 155, 159-160, 164, 168-171, 173-176, 181, 195-196, 201, 223, 226-228, 236, 240, 242-243, 247, 263-265, 268, 277, 284, 287, 291-292, 305-310, 322, 327, 334-335
Styrene-(ethylene-*co*-butylene)-styrene-graft-maleic anhydride, 129, 135
Styrene acrylonitrile-graft-maleic anhydride, 135
Styrene-graft-maleic anhydride, 125, 135
Substitution, 51, 56, 180

Substrate, 115, 147-154, 159, 295, 313, 318
Sulfur, 263, 265, 315
Sulfuric acid, 311-312, 315, 318, 320, 333
Supply, 56, 119, 138, 230
Surface, 19-20, 25-26, 32, 34, 48, 54-56, 63-64, 66-67, 70-71, 76, 80, 85-86, 88-89, 93-95, 97-98, 106-107, 116, 118, 131-132, 147, 150-155, 159, 164, 175, 177, 185, 194, 199, 203-205, 207-208, 242-245, 249-251, 255, 270, 279, 281, 288-290, 292, 295, 312-313, 315-316, 321-323, 326-327, 330-335
  free energy, 199, 207
  tension, 153, 251
Surfactant, 295, 321, 326, 331-332
Suspension, 313-315, 320, 324, 327, 329
Sustainable, 21, 49, 138, 140, 162, 229, 259, 271, 277, 288
Swell, 15, 19, 116
Symposium, 41-42, 103
Synergistic, 2, 277
Synthesis, 2, 106, 111, 230, 265-266, 279
Synthetic polymers, 1-2, 4, 25, 61, 119, 187, 263, 269, 292
Synthetic rubber, 266

T

Talc, 25, 35, 267
Tapping, 157-158
Technical, 4, 6, 16, 39-40, 42, 46, 105, 143, 168-169, 176, 178, 182, 187, 196, 224, 236, 239, 264, 266, 273, 306
Temperature, 16, 18, 21-22, 27, 32-33, 35, 37, 53, 64-65, 67, 69-72, 89, 91, 94, 97, 99-100, 111, 118, 124, 132-135, 137, 147, 149-150, 152, 159, 161-162, 172, 181, 183-184, 193, 197, 199-200, 202, 208, 210-213, 217, 231, 236, 247, 251, 264-265, 267-268, 281-282, 286, 288, 290, 314-315, 328, 330-331, 333
  range, 32, 72, 91, 200, 211, 328, 330
Template, 281-282, 295
Tensile energy absorption, 31-32, 35
Tensile properties, 53, 100, 176, 194, 248
Tensile strength, 13, 16-17, 23, 26, 28-29, 39-40, 48-49, 52-53, 72, 89, 93, 95-96, 100, 105, 107-108, 119, 131, 174, 176, 178, 181-183, 198, 214, 223, 244, 246, 248, 268-270, 285, 291-292, 330
Tensile stress, 109, 157
Tensile testing, 97
Tension, 89, 93, 153, 251
Terpolymer, 248

Tertiary, 121
Test, 28, 30, 39, 89, 108, 132-133, 183, 198, 214, 311
Tetrahydrofuran, 322
Tetramethylpiperidine-1-oxyl radical, 311, 316, 322, 335
Theory, 42, 152-153, 222, 247
Thermal analysis, 26, 88, 196-197, 200, 241, 243, 249, 251, 254, 269
Thermal conductivity, 39
Thermal degradation, 99, 135, 173, 181, 211, 234, 239-242, 244-245, 247-248, 268
Thermal properties, 26, 47, 52, 73, 180-181, 210, 247, 268, 279, 288, 290-292, 330
Thermal stability, 21-22, 52-53, 91, 110, 117, 129, 134, 147, 151, 178, 181, 210, 212, 234, 242-243, 246-247, 261, 269-271, 279, 286-288, 292, 331, 333
Thermal treatment, 251
Thermally induced phase separation, 55
Thermodynamic, 247-248, 250
Thermogravimetry/thermogravimetric analysis, 91, 94, 97, 99-100, 110-110, 180, 197, 200, 210, 232, 234, 236, 247, 331, 333
Thermoplastic(s), 1, 4, 12, 21-22, 25-28, 33-34, 37, 45, 48, 61-62, 65-68, 74-75, 78, 80-82, 87, 105, 117-118, 120, 122, 138, 145-147, 149, 151, 153-165, 172, 177, 179, 181, 184, 193-195, 216, 245, 248, 253, 261-262, 268-269, 285, 323, 328-329, 333
  elastomer, 248
  polymers, 1, 21-22, 118, 147, 323
Thermoset, 21-22, 61-62, 64-65, 72, 80, 120, 180
  Thermosetting, 21, 23, 25, 36-37, 64, 68, 71, 85, 87-89, 91, 93, 95, 97, 99, 101, 103, 106, 147, 149, 179, 261-262, 324
Thickness, 19, 32, 38-40, 64, 67, 79, 128, 130-131, 133, 148, 152, 195, 281, 318
Thin film, 293
Three-dimensional, 21, 49, 55, 87, 171, 285, 292, 294, 311, 334
Time, 11, 18, 21-22, 63-65, 67-68, 70, 79, 123, 146-147, 150-151, 155, 157, 160-161, 178, 183, 186, 194-195, 199-200, 204-210, 214-218, 221-222, 231-232, 248-250, 262, 314-316, 329-330, 333
Tip, 311, 318
Tissue, 51-55, 57, 199, 279, 294-295
  engineering, 51-52, 57, 279, 294-295
Titanium dioxide, 267
Titration, 315
Tool, 63, 76-77, 79-81, 216, 328

Tough, 239-240, 283
Toughness, 61, 128, 147, 176, 183, 246, 248, 291
Toxic, 1, 5, 111, 262, 286
Transfer, 20, 29-32, 34, 54, 64, 69-70, 74, 105, 108, 146, 152-153, 160, 162, 175, 254, 283, 322
  moulding, 64, 69-70, 74
Transformation, 313
Transition, 6, 26, 33, 35, 48, 181, 265, 286, 296, 310, 328, 331-332
Transmission, 26, 246, 292, 316, 318, 335
  electron micrographs, 316-317
  electron microscopy, 318
  Transmittance, 286
Transparency, 282, 284, 286-287, 291, 336
  Transparent, 282, 286, 288, 292-293, 324
Transverse, 40, 307, 311, 313
Trend, 31-33, 119, 139, 176, 215, 218, 220-221, 233, 296, 330
True density, 242
Tube, 195, 294
Tyes, 123, 186

U

Ultimate elongation, 15
Ultimate tensile strength, 23
Ultrasonic, 80, 280, 314
Ultraviolet, 55, 132-133, 173, 194-195, 197, 202, 207-209, 212, 218, 220-223, 260, 268, 270
Uncured, 73
Uniformity, 66, 70, 326
Unmodified, 94-96, 100, 178, 324, 330, 332, 335
Unsaturated, 20, 36-37, 47, 64-65, 75, 89, 97, 135, 290
  polyester resin, 36, 64-65, 89, 97
Untreated, 88-92, 97-99, 184-185
Urea-formaldehyde, 36, 230, 232, 236

V

Vacuum, 27, 69, 71, 74, 196-197, 199, 206, 324
  oven, 69, 197
Vapour, 13, 99, 246, 287, 335-336
Vertical, 20, 37
Vibration, 26, 150, 155, 202

Vinyl group, 203-204
Virgin high density polyethylene, 127-128, 131, 133
Virgin low density polyethylene, 133
Virgin polypropylene, 126-131, 133
Viscosity, 21, 37, 70, 200, 216, 218-223, 290, 329
  Viscous, 33, 87-88, 217-218, 287
Volatile, 121, 243
Voltage, 199
Volume, 34, 42-44, 64, 66, 68, 78, 122, 125, 176-177, 182-183, 238, 243-244, 252, 279, 334
  fraction, 64, 66, 78, 176-177, 182-183
  Volumetric, 161

W

Wall, 3, 18-19, 48, 53, 86-87, 168-170, 195, 245, 283, 307-308, 316
Warpage, 87
Washing, 122, 265
Water, 2-3, 13, 15, 18-19, 27, 29-30, 35-37, 39-40, 50, 69-71, 86-89, 97, 99, 116-118, 126-127, 130-132, 145, 172, 183-184, 196, 199, 207-209, 230-234, 240, 242, 244-246, 248-252, 257, 264, 266, 270, 279, 284-287, 289, 292, 294, 312-313, 316, 320-323, 329, 331, 333-336
  adsorption, 248, 250
  content, 183
  uptake, 118, 199, 208-209, 246, 248, 287, 333-334
  vapour, 13, 99, 246, 287, 335-336
  vapour permeability, 287, 335-336
  vapour transmission rate, 335
  -dispersible, 320
  -soluble high molecular weight chitosan, 286
Waterborne, 133, 290, 320, 322
  polyurethane, 290, 322
Wavelength, 197, 284
Weathering tests, 197
Weighing, 199, 333
  Weight, 4-5, 18, 21, 28, 30, 32-34, 48, 56, 61-63, 72-73, 75, 91, 99-100, 111, 134, 145, 159-160, 164, 171, 176, 181-182, 186, 199, 202, 210-212, 230-232, 235, 264-266, 286-287, 292, 309, 333, 335
  fraction, 33-34, 182
  loss, 91, 99-100, 211-212, 232, 235
  ratio, 61, 145, 309

Welding, 22, 283
Wetting, 19, 25, 153, 175, 207
White rice husk ash, 239, 242-251
Width, 318
Wood, 4, 11-12, 15, 18-20, 22-23, 25, 29, 38, 40-42, 45, 48-49, 73, 102, 117, 122, 124, 126-127, 130-132, 134-135, 139, 141, 167, 170-172, 177, 180, 193-195, 201, 204, 216, 224-225, 236, 259, 262-267, 270, 274, 287, 306-309, 311, 319, 337, 342, 346
  flour, 125-133, 194-195, 231
  plastic composite(s), 127-129, 133, 136, 194-195, 204, 215

X

X-ray diffraction, 15, 95, 243, 310-311, 332
X-ray fluorescence, 122

Y

Yellowing, 195
Yield, 13, 51, 98, 181, 195, 216, 248, 312, 315
  Yielding, 311, 322
Young's modulus, 49, 53, 117, 174, 178, 181, 183, 198, 214-215, 270, 282, 285-286, 306, 309-312, 326-327

Lightning Source UK Ltd.
Milton Keynes UK
UKOW02n1536241013

219717UK00001B/15/P